Debating the Canon: A Reader from Addison to Nafisi

Edited by

Lee Morrissey

LEARNING RESOURCES CENTRE

Havering College
of Further and Higher education

palgrave
macmillan

DEBATING THE CANON: A READER FROM ADDISON TO NAFISI
© Lee Morrissey, 2005.

First published in 2005 by
PALGRAVE MACMILLAN™
175 Fifth Avenue, New York, N.Y. 10010 and
Houndmills, Basingstoke, Hampshire, England RG21 6XS
Companies and representatives throughout the world.

PALGRAVE MACMILLAN is the global academic imprint of the Palgrave Macmillan division of St. Martin's Press, LLC and of Palgrave Macmillan Ltd. Macmillan® is a registered trademark in the United States, United Kingdom and other countries. Palgrave is a registered trademark in the European Union and other countries.

ISBN 1–4039–6818–7

Library of Congress Cataloging-in-Publication Data is available from the Library of Congress.

A catalogue record for this book is available from the British Library.

Design by Newgen Imaging Systems (P) Ltd., Chennai, India.

First edition: September 2005

10 9 8 7 6 5 4 3 2 1

Printed in the United States of America.

Contents

Introduction: "The Canon Brawl: Arguments over the Canon"

Lee Morrissey

In every era the attempt must be made anew to wrest tradition away from a conformism that is about to overpower it.

Walter Benjamin, *"Theses on the Philosophy of History"*[1]

In *Beyond the Culture Wars* (excerpted in chapter 32), Gerald Graff claims "what first made literature, history, and other intellectual pursuits seem attractive to me was exposure to critical debates."[2] Reflecting on his own experience as a student, Graff argues persuasively that we ought to aim to generate discussion in the classroom so that all involved might realize that we already represent a range of critical positions, and that, by extension, we are already taking part in an extended, critical debate, even in ways we are not fully aware of. Graff calls the process "teaching the controversies," and it can make for an exciting classroom. It is also, though, an excellent description of the dynamic of debate that is embedded within literary traditions, and of how the influential texts of those traditions can shape the terms for subsequent participants. Indeed, traditions are made up of debates, diachronically (as past addresses present, and present the past), anachronically (as something ancient seems to matter for the present, and vice versa), and pluralistically (as an extraordinary range of voices make up a tradition, and the readings of that tradition). Like Graff, I too believe that we ought to "teach the controversies," although to Graff's sense of critical debate in the classroom I would add the importance of heightening awareness that such debate is also central to the texts we read and discuss. In this sense, "we" are no longer teachers and students, but the larger community of readers and writers involved in the discussion.

Over the last two decades, there has probably been no debate within the academy more notable or visible than that over the state of and the stakes involved in the study of the humanities. The controversy has sustained a prominent discussion both inside the academy and outside it. The resulting articles, essays, and books represent an extended meditation on what college students ought to learn. Central to this controversy are the lists of major works of "Western Civilization," sometimes called "The Great Books," and sometimes thought to constitute "The Canon." Books by several participants in this debate, such as Allan Bloom's *The Closing of the American Mind* (1987), E.D. Hirsch's *Cultural Literacy* (1987), Roger Kimball's *Tenured Radicals* (1990, 1998), Harold Bloom's *The Western Canon* (1994), and David Denby's *The Great Books* (1996), became best-sellers, attracting both the general reader and faculty at every level of higher education. Consequently, it is not overstating the case to say that this canon debate represents the central public controversy regarding the cultural aspect of higher education. Indeed, some contributions to the canon debate have achieved "canonical" status themselves. While the debate dovetails with what are sometimes called the Culture Wars, it tends to come back to one recurring question: what should college students read? This anthology, *Debating the Canon*, offers an introduction to the controversy over the nature and content of the cultural component of a college education. Because the debate over the Great Books has a long, complicated history, likely unknown to the general reader, and often overlooked by participants in the recent controversies, this anthology does more than provide selections from the recent debates—it also provides primary-source material for understanding the debate across the history of English-language literary criticism.

It was with Graff's idea of teaching the controversies in mind that I set out to compile this collection of major contributions to the debate over the canon. Over the years, I have addressed quite a few groups of students, or I should say emerging readers, who were unaware of the controversies Graff would have us engage—and I am not referring simply to their not knowing about a debate over the canon. While most of my students do not know about any canon controversy, they also and more importantly do not know about the various critical positions that produced the canon debate. That is, like so many of us when we first started to read—and debate—they did not know that their critical positions already had a history. This anthology began as supplementary readings for those students. That course packet then, and this anthology now, are intended to do the work of introducing the controversies, so that in class we can actually teach—by which I mean *engage*—them. To read these essays is to be introduced to literary theory put into practice, as the most abstract points of literary-critical theory became involved in the most practical of practical criticisms. Moreover, as there has probably not been another issue over the last two decades that has involved so many different figures and literary critical positions, participants, cognizant of the possibility of a wide audience, made their claims with diction as colloquial as we had ever seen them use. For all the concern about the texts of the past, the debate itself represents an intellectual and cultural history of the present. In this way, the debate over the canon mimics an important quality of the canon itself: the sense of an extended debate over recurring questions, with shifting answers, sometimes shifting even with the same terms.

Of course, to read these essays is to be invited to consider why we might (or might not) read and teach what are sometimes called the "Great Books." My own sense is that the titles often included in a list of "the" Great Books remain acutely relevant for us as readers, as students, and as teachers, although reading them often involves deciding for ourselves—repeatedly—whether they ought to be considered "Great." Rather than reading them because they are (or are said to be) Great, one reason for reading these books is to be found in how they raise the question of Greatness so acutely. Some say the Great Books do so in themselves, on account of some quality of how they are written. Some argue the Great Books are great on account of the extraordinary influence they have exerted. Others might combine the two positions and argue that the quality of the writing leads to the influence. Each of these positions, though, has a history within the development of literary criticism. Indeed, the selections included in this anthology suggest that we cannot understand the recent debate over the canon without seeing them as related to the development of literary criticism in English. There is a way in which the list of the so-called Great Books is produced by an emergent modern institution of literary criticism in the eighteenth-, nineteenth- and early twentieth centuries, or long after most of the books then judged to be Great had been written. In this way, the Great Books are to books what literature is to the literary: a codification, and a classification of an almost indescribable quality that both precedes and survives the designation.

To those of us living in twenty-first century America, with the protection afforded by the First Amendment or with the variety of means of publication (broadcast, virtual) now available, the idea that books might be dangerous seems merely imaginary or dystopian. However, until at least the eighteenth century, many people believed there was nothing more destabilizing than print. As evidence, they could point to the upheaval of the Protestant Reformation and the seventeenth-century wars of religion as unfortunate consequences of making books, and especially The Book, so widely available to so many new readers. It is no coincidence that modern English literary criticism—in its periodical, textual editing, and incipient academic varieties—emerges during the late seventeenth and eighteenth centuries, after the Restoration that ends the period of the Civil Wars and Interregnum. One way of reading this post-Restoration emergence of English literary criticism is to see it as an attempt to make reading safer, or maybe to make the world safe for printing. Developing a list of important authors and texts, and providing readings of their importance, is part of that eighteenth-century contribution to safer reading. Although, to be fair, the literary critics involved did not call such lists "Canons," they were laying the groundwork for what would become the Great Books approach to literary history.

In this anthology, therefore, I situate the recent debate over the canon in a larger span of English literary criticism. The sense, today, that there is something called a literary Canon, or, similarly, a set of texts that are Great, the reading of which might improve the reader, can be traced back to these earlier developments in literary criticism. Joseph Addison, for example, argues in *The Tatler* (1709) "It is impossible to read a page in Plato, Tully, and a thousand other ancient moralists, without being a greater and better man for it" (15). Through a literary-critical lineage that includes Samuel Johnson, Matthew Arnold, and F.R. Leavis, this claim is reiterated today by some and refuted by others in the canon debates, nearly 300 years after Addison first proposed it. David Hume's "Of the Standard of Taste" (excerpted in chapter 2), although somewhat neglected in recent histories of literary criticism, nonetheless makes several important contributions to the development of criticism—and to the idea of the canon. First, of course, there is the question of a standard raised by the title. Eventually, we know, subsequent critics will treat the Great Books as the standard. But in this essay Hume's discussion of a standard is both tentative and more complicated than we might assume looking back—a standard is conventional. Hume's essay is neither about natural laws of taste, nor about universal rules of literature. Unfortunately, though, by arguing that even the "coarsest daubing . . . would affect the mind of a peasant or Indian," Hume the skeptic treats Indians and the poor as if each were a rare instance of a natural category, with, some in the debate will later argue, important consequences for the relative accessibility of literary history.

The recent debates over the canon assume this institutional history, although often only by mentioning its most significant figures in passing. Samuel Johnson, the most important English literary critic of the eighteenth century, contributes more than any English critic before him to the development of a Great Books sense of literary history. His *Lives of the English Poets*, for example, articulate a claim for a canon of English literature as well as principles for approaching that canon. His periodical essays, such as *The Rambler*, combine what we would now recognize as the topicality of literary criticism for a wide audience with reflection on method. But it is with the "Preface" to the *Plays of William Shakespeare* (chapter 3) that Johnson makes his most important contribution to the canon debate. In his edition, Johnson addresses the concerns of several influential previous critics who had tried to correct perceived faults in Shakespeare's style and structure. After Johnson, the question of whether Shakespeare is worthy is much less important than it was before. Moreover, in the process of responding to Shakespeare's critics, Johnson also adds to what we might call, after Hume, the "standards" for determining whether texts should be considered valuable: "what has been longest known has been most considered, and what is most considered is best understood" (22). The longer the better, presumably, but Johnson does say that a century is "the term commonly fixed as the test of literary merit" (22).

In his criticism, Matthew Arnold, represented in this collection by selections from "The Function of Criticism at the Present Time" (1865; chapter 5), builds on Addison, Hume, and Johnson. He actually selects a few of Johnson's *Lives*, publishing them with a new Preface of his own, a helpful image of a kind of continuity across the century that separates these two major critics. In "The Function of Criticism at the Present Time," Arnold adds a layer of conviction that Addison, Hume, and Johnson, perhaps closer to the destabilizing effects of print in the seventeenth century, either lacked or downplayed. Like Hume, for example, Arnold believed that "English criticism should clearly discern what rule for its course" (27). For Arnold, that rule can be summarized in one word: "disinterest." The job of the critics is "to know the best that is known and thought in the world," without regard to politics or consequence (27). Elsewhere, Arnold refers to "touchstones," which represent and measure the best that is known and thought in the world. Eventually, what Arnold calls touchstones come to be known as the Great Books.

During the first decades of the twentieth century, an important part of the transition from standards, rules, and touchstones to "The Great Books" takes place within the academy, as Columbia University, and later the University of Chicago, design what Columbia now calls "Literature Humanities" and "Contemporary Civilization" courses. Part of the story of this development has to include Mortimer Adler's studying at Columbia, and his being hired by Chicago in 1930. At Chicago, Adler helped to reorganize the undergraduate education, remodeling it on what he had experienced at Columbia. His 1940 book, *How to Read a Book* (excerpted in chapter 8), remains in print today. By creating and editing *Great Books of the Western World*, published by Encyclopedia Britannica starting in 1952, Adler probably did more than anyone to popularize the idea of The Great Books (and the idea that they were Western). Given how familiar the term, and the sense of continuity it has come to represent, would become, it is all the more important to remember, then, that in 1919 T.S. Eliot could claim that "in English writing we seldom speak of tradition, though we occasionally apply its name in deploring its absence" (29). At the very time that Columbia was piecing together what is now seen as the first Great Books general education curriculum, Eliot is claiming that tradition either is not spoken of or does not exist.

In "Tradition and the Individual Talent," Eliot could be read as speaking for an older generation, one that precedes this creation of a Great Books curriculum. On the other hand, his comment also speaks to his modernist sense of the disturbing fragmentation of life in modernity. In *Mass Civilization and Minority Culture* (1933; see chapter 7), F.R. Leavis is quite clear about this: "it is a breach in continuity that threatens."[3] Concerned about the upheaval caused by the machine in "the modern phase of human history," Leavis points to "how the automobile (to take one instance) has, in a few years, radically affected religion, broken up the family, and revolutionized social custom."[4] The destructive conditions and experience of modernization demand a synthesis, Leavis believes. His claim for continuity echoes the arguments of several others we have seen. Like Hume, Leavis contends, "culture has always been in minority keeping" (35). Like Arnold with the touchstones, Leavis refers to gold, claiming that the relationship between this minority and the great number of texts is similar to the process whereby "accepted valuations are a kind of paper currency based upon a very small proportion of gold."[5] Unlike either of them, though, Leavis believes that he is living at a time when "the landmarks have shifted."[6] For many at the time, it was as if the touchstones—and many landmarks are stones—have been displaced, or now only disorient an age that whizzes by in machines. At the same time, the way in which Leavis distinguishes between what he regrets having to call "highbrow" and culture's other forms in modernity points toward the work of Theodor Adorno (represented in this anthology by his 1962 essay, "Commitment," chapter 12). For they both share a concern about the passivity they see encouraged by mass culture, or what Leavis calls "that deliberate exploitation of the cheap response."[7]

If modernization looked like a threat to continuity in 1933, imagine the situation by the mid-1940s. The systematic, mechanized extermination of peoples, the development of attack by rocket, and the deployment of two atomic bombs further convinced people that modernity had removed the landmarks. In this context, it is difficult to overstate the importance of Erich Auerbach's *Mimesis*. Published in the year after the war ended, the book's copyright page includes one unusual line referring to what would be taken as the book's most poignant and important circumstance: "written in Istanbul between May 1942 and April 1945." As the book begins immediately with its readings, first of Homer's *Odyssey* ("Odysseus' Scar," chapter 9), it was not until the Epilogue that readers would learn that Auerbach had composed the book while in exile, "where the libraries were not well equipped for European studies."[8] It was an important enough book simply for trying to tell the story of the rise of the novel, from Homer's *Odyssey* to Virginia Woolf's *To the Lighthouse*. But the fact that its author claimed the book was more or less written from memory made it all the more remarkable. It suggested that despite the concerns, the landmarks had survived. And the idea that a book about "texts ranging over 3000 years" would be written not only from memory, but more or less in the very place where both the book and the tradition it wound up recording had begun made it all the more significant.[9]

It suggested that the landmarks were still where they had been. But the fact that this remarkable synthesis of "European literature" had been written more or less from memory by some who was forced into exile because he had been born Jewish in what would become Nazi Germany made *Mimesis* all the more important.[10] It meant, among other things, that these books were "universal." It is an interesting question whether Encyclopedia Britannica would have started to publish a series on *The Great Books* had not an exiled Jew testified to the importance of these books by writing about them largely from memory on the eastern edge of Europe.

At the same time, the extraordinary effort that went into *Mimesis* also stands for the passing of an era; in several senses, it is a monumental work. For, the same moment, that is, the end of World War II, thanks to the same forces at play would also represent the beginning of the end for this idea of the universal and for this idea of the Great Book. The end of World War II revealed a Europe that was exhausted and overstretched from two massive international wars in just 30 years. Bankruptcy was both an economic and a moral issue: countries were decimated for reasons that remain a matter of debate. The great works of the European intellectual heritage did not seem to have been able to stave off mass devastation. Some might debate whether that is because, as Leavis believed, there were simply too few who understood that intellectual lineage. Nonetheless, the idea of Europe, even a rebuilding Europe, did not inspire confidence for a new universal system. Indeed, as the overstretched countries of Europe regrouped and rebuilt, they also revealed what was at least one practical legacy of the idea of the universal: colonization. In the aftermath of World War II, there were mass movements around the world against the idea and effects of colonization and claims to universality. The most important of these was in India, of course, which the British returned, partitioned, to self-determination in 1947. This inspired decolonization movements around the world. Indeed, support for territorial self-determination was written into the Charter of the new United Nations.

After World War II, liberation movements swept the globe: postcolonial, feminist, and civil rights. Each of them brought not only a literature, but also a criticism. Together, this literature and criticism tackled directly the assumptions of universality implicit in the idea of the Great Books that had come together over the preceding two centuries. Particularly important in this is the work of Martinique-born writer Frantz Fanon, represented in this anthology by "On National Culture" (chapter 11). For Fanon, national culture is an important rebuke to the universalizing claims of the European colonizing powers, and the educational system that "by a kind of perverted logic . . . turns to the past of the oppressed people, and distorts, disfigures, and destroys it." (66). The imported culture of the colonizing country, for Fanon, threatens to overwhelm the unrecognized culture that had been in place prior to the colonizer's arrival. Thus Fanon describes the "anxiety shared by native intellectuals to shrink away from that Western culture in which they all risk being swamped" (65). In "Colonialist Criticism" (chapter 13), Nigerian author Chinua Achebe builds on Fanon's nationalist vision, focusing more specifically than Fanon on the question of literature and criticism in a decolonizing country. Beginning with a book review that dismisses his first novel, Achebe points out the limitations of this view of literature from the supposed center. Specifically, he argues "I should like to see the word 'universal' banned altogether from discussions of African literature until such time as people cease to use it as a synonym for the narrow, self-serving parochialism of Europe, until their horizon extends to include all the world" (77).

To the north, or what is often called The West, other liberation movements, feminism being among the most prominent, were similarly reshaping curricula and the study of literature. Although this movement, like postcolonialism, is larger than this anthology, it is also central to it. The influence of feminism on literary studies and on the question of the canon is diffuse; just a few representative essays from an extraordinarily extensive bibliography are included here. In "A Map for Rereading" (1980; chapter 15), Annette Kolodny responds to Harold Bloom's 1975 book, *A Map of Misreading*, by pointing out that poems can be " 'about poems' " (93), as Bloom claims, within a tradition whose relative homogeneity would require excluding any work that might constitute an alternative to that tradition. For Kolodny, Bloom's argument overlooks the possibility that "we are calling attention to interpretive strategies that are learned, historically determined, and thereby necessarily gender-inflected" (94). Not only are most Great Books written by men, the very process of reading them is itself, Kolodny argues, appropriated by and structured for men. There is another tradition; books were written by women prior to the seventeenth century. But those titles, and the different ways of reading them, are not included in the literary history on which Bloom focuses. What Kolodny proposes, then, "challenges fundamentally not only the shape of our canon of major American authors but, indeed, that very 'continuity [writes Kolodny, quoting Bloom] that began in the sixth century BC when Homer first became a schoolbook for the Greeks' " (99).

Perhaps it was inevitable that three decades of such critiques might result in a reactive defense of the Canon. Still, there is something uncanny about the fact that the defense coalesced in the mid-1980s. Particularly important in this regard is the 1985 report, "To Reclaim a Legacy," by the then Secretary of Education William J. Bennett. There Bennett claimed, "the humanities, and particularly the study of Western civilization, have lost their central place in the undergraduate curriculum" (112). Bennett's report highlighted the consequences of the extraordinary variety of institutions in higher education. According to Bennett's surveys, for example, "fewer than half of all colleges and universities now require foreign language study for the bachelor's degree" (112). This report at least coincides with—and may even be said to have triggered—the most visible, hotly contested, widespread and public phase of the canon debate, approximately a decade spanning the late 1980s and the early 1990s. In 1987, two years after Bennett's report, E.D. Hirsch's *Cultural Literacy* (chapter 22) and Allan Bloom's *Closing of the American Mind* (chapter 21) were published. Although they are very different books, with Hirsch focused on primary and secondary school, and Bloom on postsecondary education, both authors share Bennett's concern about the fragmentation of the curriculum in American education. In a way, their concern with fragmentation transfers to the curriculum the concern that Modernist critics such as Eliot and Leavis had for modernity.

In *The Death of Literature* (chapter 27), Alvin Kernan argues that the canon debate reveals that "what has passed, or is passing, is the romantic and modernist literature or Wordsworth and Goethe, Valéry and Joyce, that flourished in capitalistic society in the high age of print, between the mid-eighteenth-century and the mid twentieth" (167).

This "paradigm shift," according to Kernan, represents but one shift in a transformation affecting "traditional institutions" (169). However, to the degree that literature can be said to be involved in those changes, Kernan is on surer ground when he emphasizes the end of "the high age of print." The controversy over the canon follows a logic that is internal to the history of English literary criticism. Some might say this idea of print found its most prominent symbol in the Great Books. However, stability is central to the vision of print that accompanies literature in the high age of print. There were literary texts before the institution of literature that Kernan describes as developing between the eighteenth and the twentieth centuries. What has passed, then, is a particular treatment of print. That is, when Kernan argues that "the Anglo-Saxon tendency has been to keep literature and politics as widely separated as possible" (168), his is a contemporary formulation of a principle that goes back to the eighteenth century. England in particular, having had a seventeenth-century Civil War connected to the democratization of print, was concerned that print and access to the press not have such devastating consequences again.

If Kernan announces the death of Literature, it is Harold Bloom who provides its eulogy, in "An Elegy for the Canon" (chapter 37), the first chapter of his 1994 book, *The Western Canon*. Bloom stocks this essay with a series of crisp formulations of his sense of the canon. On one level, he welcomes the death of literature, thinking that the emergence of cultural studies departments will return literature to those few readers who used to enjoy it, and that in the process literature will regain its counter-cultural élan. Among those readers who love literature, the Great Books are part of the art of Memory, and are related to our finitude, our mortality. Had we but world enough and time, we would read all the books; we would not need to make selections. At the same time, though, Bloom coins a phrase, "The School of Resentment," to describe those who have criticized the Canon for the preceding 30 years: "Feminists, Afrocentrists, Marxists, Foucault-inspired New Historicists, or Deconstructors" (244). Consistent with his earlier work on the importance of agonistic relationships in literary history, Bloom believes that those whom he names under these rubrics are bothered by the "originality" that characterizes the great, canonical work (247). For Bloom, The School of Resentment cannot bear what Bloom sees as the overpowering strength of originality. That is, Bloom implies that there is an agonistic relationship between such critics and the "great literature [that] will insist upon its self-sufficiency" (249).

Bloom summarizes what had by then become a political binary: liberals opposed to the Canon and conservatives concerned with upholding immutable truths and value in support of it. Of course, Bloom was neither the inventor of this opposition, nor the first to note it. Different versions of it can be seen in the work of others, such as Roger Kimball (chapter 28), Paul Lauter (chapter 29), and others. But Bloom's formulation ties the positions in the debate to questions of historical exclusion: white men for the Canon versus African American women opposed to it, according to Bloom. Bloom might have tried to valorize debate as part of the process of canon—"aesthetic value emanates from the struggle between text: in the reader, in language, in the classroom, in arguments with a society" (256)—but his book had the unfortunate effect of popularizing a polarized version of the canon debate. It is no exaggeration to say that we still live with the consequences of such polarities. We live with them in the way we live with the results of so many conflicts: unresolved, adjusting to a truce, wondering what had happened, and how it had escalated. Those who might want to teach Bloom's "baker's dozen, Homer, Virgil, Dante, Chaucer, Shakespeare, Cervantes, Montaigne, Milton, Goethe, Tolstoy, Ibsen, Kafka, and Proust," will also need to consider, thanks to Bloom's binary, whether they want to be identified by colleagues and students as unsympathetic to historically marginalized voices. With the quiet offered by the truce and the perspective offered by reflection (and by seeing the Canon Brawl in the context of the development and diffusion of a vision of literary criticism), this anthology might provide us the opportunity to consider how it has come to this.

Moreover, although Bloom does not think so, there are many ways in which the polarization of the debate can make for better readings of the Great Books. After the Canon Brawl, and with knowledge of the canon debate, it is possible to return to the Great Books with a new awareness, a new level of sensitivity to the range of issues involved in reading these books. The fact of the Canon Debate contributes, I believe, to the importance of rereading the Canon. There is yet another level of responsibility involved, in this case a responsibility to the terms of the debate through which we now cannot help but see the Great Books. When read in this way, the Great Books are already crossing some of the most prominent lines in the debates over them: for example, the prominence and importance of Arabic authors in Dante; Augustine, not a white European but rather a North African, who emigrates to Italy; Milton's Eve so overdetermined in her choice as to invite our considering the possibility that Milton, author of a defense of divorce, might think she has done the right thing. What I am interested in, then, is a reading of the canon that takes into account the critiques of it. Unfortunately, too many of those who have defended the canon also defend a vision of it that rejects the critique out of hand. There is also a tendency among those who critique the canon not to be particularly interested in reconsidering it in the light of their own critique. With his idea of "heterodoxy," Arnold Krupat points to this possibility in his essay (chapter 25) (160). After the Canon Brawl, terms such as "conservative," "liberal," "modern," and "postmodern" cannot fit the kind of reading of the Great Books, informed by the debate, that could occur. That is, after the Canon Brawl, it may be possible then to see the debates over the Canon as matching the heterodoxy of the canonical texts themselves.

In "Canon to the Right of Me" (1991; chapter 30) Katha Pollitt carved out an important middle position, effectively pointing out that the so-called conservative and liberal positions shared a sense of the importance of reading. The difference devolved largely to what they would recommend to readers. And, Pollitt points out, such recommendations take on particular urgency for literary critics and academic faculty, as it is quite likely that the reading accomplished in college might be the only chance for a substantial program of reading young people might undertake, ever. With the decline in reading, the syllabus looked like the last best chance to get the next generation to read whatever it was that either side thought should be read. For Pollitt, the concern is not only, as Bloom would have it, that mortality forces readers to search for what Arnold called the best of what has been thought. Rather, a major shift in value was underway, and reading seemed to have been losing its value. In *Cultural Capital* (1993; excerpted in chapter 33), John Guillory—influenced by Pierre Bourdieu (see chapter 16)—offers a theory of how this change was occurring.

For Guillory, literature had lost its status as cultural capital, or, some might argue conversely was clinging to it all too tightly and was losing its relevance even to education as well. Guillory points to the rise of composition and writing programs as evidence of how literature's once pervasive, and instrumentalist, justification had eroded, as students prepare through writing courses for professional "sociolects."

In the decade or more since Pollitt and Guillory articulated their sense that reading was somehow slipping away, if not in general then maybe from its formerly central place in college education, the pressures on such reading have only increased. As Pollitt was writing, Mosaic, the pioneering web-browser was being created. If "tenured radicals" were leading students away from the humanities, as Roger Kimball claimed in 1990 (see chapter 28), it would seem that by the end of the decade the University's adoption of business models was having an equally powerful dampening effect on the humanities. The student consumer entered the entrepreneurial university strapped with student loans and having been told that the principal advantage of college education was the great increase in lifetime earnings when compared with the holder of a high school diploma. In this situation, although literature retains its cultural capital—especially at those institutions that already have the most cultural (and other kinds of) capital—the Great Books simply do not fit with the self-declared objectives of the young person who will need to be trained for a job so that, understandably, she can begin to pay off student loans. Of course, actually, the Great Books do fit, in all the same old ways: they provide historical perspective; reading, discussing, and writing about these works trains the mind to think quickly, flexibly, and critically; and, when read in the light of the canon debate, they can illustrate the complexities of conflicting values in a pluralistic society. These are all skills that young people will need both as citizens and in whatever career they find themselves. Unfortunately, though, it seems that the Canon Brawl left the academy both too polarized and too tired to put forward this vision for education. Although there are exceptions, there has not, yet, been a national discussion on how these new pressures are affecting college education, and the role of the humanities and the Great Books within it.

Perhaps especially in these circumstances, the question is how to recognize and transmit the excitement, excesses, and occasionally even dangers, of that indescribable quality of texts, despite the weight of tradition. In this activity of rediscovery, the selection from Derrida's *The Gift of Death* included in this anthology is particularly helpful (chapter 38). There, while discussing Abraham's decision, Derrida describes an unrepresentable moment beyond knowledge in which, Derrida believes, decisions occur. He calls this moment the secret, and connects it (and decisions) to faith. As a secret, faith cannot be transmitted, except as a secret. For Derrida, "This untransmissibility of the highest passion, the normal condition of a faith which is thus bound to secrecy, nevertheless dictates to us the following: we must always start over" (261). Although Derrida is addressing religion, there is a way in which Derrida's "faith" operates like meaning in literary texts. Meaning, like the secret, is not stated, although we learn how to see it anyway. Thus, there are those in the essays that follow who propose that we read the Great Books with an eye on what they do *not* say. In Derrida's terms, the re-reading of the canon they propose focuses on the secret. Others add that it is only through reading non-canonical works also alongside the canonical that we can get access to the secrets of the Great Books. In either case,

Derrida would remind us, with each rereading we start over. At the same time, Derrida's point refers to more than the individual reader. For Derrida, "each generation must begin again to involve itself in it" (261). For Derrida, the perpetual, repeated renewal characterizes the history of faith, just as, we could say, it characterizes literary tradition as well. For this generation, the opportunity is to reread the Great Books renewed by Canon Debate.

In *Reading Lolita in Tehran* (excerpted in chapter 42), Azar Nafisi recounts teaching *The Great Gatsby* in Iran during the first months of the Iranian revolution. While Nafisi does not set out to contribute to the debate over the Great Books per se, her description of her experience in Tehran participates in aspects of the canon debate and can help us imagine new possibilities for some of the debate's key terms. Confronted by one of her students, a young revolutionary named Nyazi, about the problems of reading and teaching *Gatsby*, Nafisi decides to take the student very much at his word and organize the class around a trial of *The Great Gatsby*, with students adopting different courtroom roles and writing research-paper briefs. In other words, Nafisi teaches the controversies, bringing them directly into the classroom. We need not be teaching during a revolution for the classroom to be, as Nafisi puts it, "as electric and important as the ideological conflicts raging over the country" (300). Although I would not want to suggest that the classroom embody ideological conflicts solely, Nafisi reminds us of something that a focus on "tradition" and "canons" might obscure: reading can be that electric, especially once we see the traditions and canon as debates themselves. Moreover, Nafisi's teaching *Gatsby* in Iran, during the Revolution, represents a wonderful laboratory experiment with many of the central claims in the Canon Debate. During one of the most postcolonial of postcolonial moments, Nafisi elects to teach literature from the country seen as the colonizer. In the process, of course, the student/prosecutor Nyazi rejects the universality some participants in the canon debate attribute to literature generally, and the Great Books in particular. Equally striking, though, is the relationship between gender and argument during the trial. Mr. Nyazi, of course, rejects the book and the American immorality for which he believes it stands. A young woman, Zarrin, is called upon to be the book's defendant. At one point, despite Professor Nafisi calling for a ten-minute break, the debate spills out of the room and into the hall, where Zarrin continues to debate the constraining gender roles on which she believed Nyazi's rejection of the book rested.

At the end of "Colonialist Criticism" (chapter 13) Achebe asks, "did not the black people in America, deprived of their own musical instruments, take the trumpets and the trombone and blow them as they had never been blown before, as indeed they were not designed to be blown" (84)? If we imagine the canonical text as if it were that trombone, we might say that it is not clear we can know how the canonical texts will be read. Just because they have been read a certain way for some time does not mean that texts cannot be read another way. Indeed, in music, the soloist will play against our expectations, taking the melody we thought we know, and turning it into an improvisation we had not anticipated. The same can happen in reading. We cannot predict what the results might be if we, and our students, were to approach the Great Books, again, through the lens provided by the debate. For Rorty, this is the "inspirational value of literature" (chapter 40) that can continue to be rediscovered in the Great Books. Like Rorty, I believe that this inspirational quality of literature attracts readers, and more importantly prompts readers, to stake a claim for a different, and Rorty believes, better world—one that must first be imagined. I would add, though, that such staking imagines the Great Books as a debate, both a debate across the canon and a debate over the canon. By providing a primary-source history of the Canon and its critiques, this anthology, like the Great Books themselves, offers theme and variation, on, I would say, what it means to be educated. But, again like the Great Books, it also calls out for a response. That is, the canon debate cannot be concluded. This suggests, then, that we are not actually debating a canon. Canons are closed, even if the interpretation of them might not be. The literary "canon" is always being added to, and is also always thereby being rearranged, as Eliot reminds us. Of course, it might also be that such additions to the canon are themselves but interpretations of it. In that case, then, the canon may already be closed. But, paradoxically, our continual rediscovery of that seeming closure is also what keeps it open, always subject to addition and revision.

A Note on the Texts

To edit an anthology on the canon debate is to step into some of the most familiar predicaments associated with the canon generally: what to include, what not to include, how to arrange it, and so on. In the selections, I have aimed for a combination of critical range and influence. By range, I mean the breadth of the critical approaches involved in the debate. In some ways, influence might seem more difficult to measure, but as it turns out certain figures recur within the critical works that constitute the canon debate, and I have tried to include essays by those figures. There is a way in which the book is intrinsically cross-referenced, as critics allude to, argue with, and rework the arguments of various other critics. In keeping with at least some critics' sense of the canon itself, editing this collection was also a confrontation with finitude, page constraints in this case. It was not possible to include every article that relates to the debate over the canon. Although classical literary criticism, on the one hand, and several contributions to the Culture Wars of the late 1980s and early 1990s on the other certainly overlap with the canon debate, my focus is on the development of and debates over Canons in literary criticism in particular. That focus guided my selections. Moreover, as it was impossible to simultaneously represent the history of the debate and provide each article or book chapter in its complete form within the book's page limits, it was necessary to edit almost each of the articles for length. Throughout, I tried to focus my editing of each article so as to highlight the essays' claims. I believe that arguments, from thesis to thesis, survived this editing, although I regret that supporting evidence, such as extended readings of passages, for those claims often did not. Edited articles are indicated by "from," while edits within each chapter are indicated by bracketed ellipses. I have arranged the selections chronologically (and alphabetically within years), to convey a sense of the debate as it has unfolded. For each author, I have provided the briefest of biographical notes—usually just a summary of their most famous institutional affiliation—and an influential claim made in the selected essay. Of course there is much more to be said about the authors and the essays than is provided in these preliminary footnotes. I hope what I have provided will trigger the reader's curiosity. What I might say about these authors is less important in this anthology than what both the authors and the readers of those essays have to say.

The College of Architecture, Arts, and Humanities, the Idol South Fund and my colleagues in the English Department at Clemson University made a significant contribution to the Permissions for this book. I thank them, Janice Schach, Dean, and Mark Charney, Chair, for their generosity. Without the extraordinarily careful help of Mason King, this book would not exist. I hope he will now get a chance to read it, with my thanks. And finally, my thanks to Sanna, again, for taking on the risk of *Debating the Canon* with me.

Notes

1. Walter Benjamin, "Theses on the Philosophy of History," in *Illuminations*, ed. Hannah Arendt, trans. Harry Zohn, (New York: Schocken Books, 1969), pp. 253–264, 255.
2. Gerald Graff, *Beyond the Culture Wars: How Teaching the Conflicts Can Revitalize American Education* (New York: Norton, 1992), p. 66.
3. Leavis, F.R., "Mass Civilization and Minority Culture," in *For Continuity*, 1933 reprint, (Freeport: Books for Libraries Press, 1968), p. 17.
4. Ibid., pp. 16–17.
5. Leavis, "Mass Civilization and Minority Culture," p. 14.
6. Leavis, "Mass Civilization and Minority Culture," p. 31.
7. Leavis, "Mass Civilization and Minority Culture," p. 22.
8. Erich Auerbach, *Mimesis: The Representation of Reality in Western Literature*, trans. Willard R. Trask (Princeton: Princeton University Press, 1953), p. 557.
9. Ibid., p. 557.
10. Ibid., p. 557.

Chapter 1

Joseph Addison (1672–1719) from *The Tatler*, No. 108 (Thursday, December 15, to Saturday, December 17, 1709)

I must confess, there is nothing that more pleases me in all that I read in books, or see among mankind, than such passages as represent human nature in its proper dignity. As man is a creature made up of different extremes, he has something in him very great and very mean: a skilful artist may draw an excellent picture of him in either of these views. The finest authors of antiquity have taken him on the more advantageous side. They cultivate the natural grandeur of the soul, raise in her a generous ambition, feed her with hopes of immortality and perfection, and do all they can to widen the partition between the virtuous and the vicious, by making the difference betwixt them as great as between gods and brutes. In short, it is impossible to read a page in Plato, Tully, and a thousand other ancient moralists, without being a greater and a better man for it. On the contrary, I could never read any of our modish French authors, or those of our own country who are the imitators and admirers of that trifling nation, without being for some time out of humour with myself, and at everything about me. Their business is to depreciate human nature, and consider it under its worst appearances. They give mean interpretations and base motives to the worthiest actions: they resolve virtue and vice into constitution. In short, they endeavour to make no distinction between man and man, or between the species of men and that of brutes. As an instance of this kind of authors, among many others, let anyone examine the celebrated Rochefoucault, who is the great philosopher for administering of consolation to the idle, the envious, and worthless part of mankind. [. . .]

I think it is one of Pythagoras's Golden Sayings, that a man should take care above all things to have a due respect for himself; and it is certain, that this licentious sort of authors, who are for depreciating mankind, endeavour to disappoint and undo

Joseph Addison collaborated with Richard Steele on *The Tatler* (1709–1710/11) and on their subsequent journal, *The Spectator* (1711–1712). Often considered among the earliest English literary critics, Addison makes a claim in this essay that recurs in the canon debate: "it is impossible to read a page in Plato, Tully, and a thousand other ancient moralists, without being a greater and a better man for it."

what the most refined spirits have been labouring to advance since the beginning of the world. The very design of dress, good breeding, outward ornaments, and ceremony, were to lift up human nature, and set it off to advantage. Architecture, painting, and statuary were invented with the same design; as indeed every art and science contributes to the embellishment of life, and to the wearing off or throwing into shades the mean or low parts of our nature. Poetry carries on this great end more than all the rest, as may be seen in the following passage, taken out of Sir Francis Bacon's "Advancement of Learning," which gives a truer and better account of this art than all the volumes that were ever written upon it.

'Poetry, especially heroical, seems to be raised altogether from a noble foundation, which makes much for the dignity of man's nature. For seeing this sensible world is in dignity inferior to the soul of man, poesy seems to endow human nature with that which history denies, and to give satisfaction to the mind, with at least the shadow of things, where the substance cannot be had. For if the matter be thoroughly considered, a strong argument may be drawn from poesy, that a more stately greatness of things, a more perfect order, and a more beautiful variety, delights the soul of man, than any way can be found in nature since the Fall. Wherefore seeing the acts and events, which are the subject of true history, are not of that amplitude as to content the mind of man; poesy is ready at hand to feign acts more heroical. Because true history reports the successes of business not proportionable to the merit of virtues and vices, poesy corrects it, and presents events and fortunes according to desert, and according to the law of Providence. Because true history, through the frequent satiety and similitude of things, works a distaste and misprision in the mind of man, poesy cheereth and refresheth the soul, chanting things rare and various, and full of vicissitudes. So as poesy serveth and conferreth to delectation, magnanimity, and morality; and therefore it may seem deservedly to have some participation of divineness, because it doth raise the mind, and exalt the spirit with high raptures, by proportioning the shows of things to the desires of the mind; and not submitting the mind to things, as reason and history do. And by these allurements and congruities, whereby it cherisheth the soul of man, joined also with consort of music, whereby it may more sweetly insinuate itself; it hath won such access, that it hath been in estimation even in rude times and barbarous nations, when other learning stood excluded.'

Chapter 2

David Hume (1711–1776)
from "Of the Standard of Taste,"
Essays (1757)

It is natural for us to seek a *Standard of Taste*; a rule, by which the various sentiments of men may be reconciled; at least, a decision, afforded, confirming one sentiment, and condemning another. [. . .]

It is evident that none of the rules of composition are fixed by reasonings *a priori*, or can be esteemed abstract conclusions of the understanding, from comparing those habitudes and relations of ideas, which are eternal and immutable. Their foundation is the same with that of all the practical sciences, experience; nor are they any thing but general observations, concerning what has been universally found to please in all countries and in all ages. Many of the beauties of poetry and even of eloquence are founded on falsehood and fiction, on hyperboles, metaphors, and an abuse or per-version of terms from their natural meaning. To check the sallies of the imagination, and to reduce every expression to geometrical truth and exactness, would be the most contrary to the laws of criticism; because it would produce a work, which, by univer-sal experience, has been found the most insipid and disagreeable. But though poetry can never submit to exact truth, it must be confined by rules of art, discovered to the author either by genius or observation. If some negligent or irregular writers have pleased, they have not pleased by their transgressions of rule or order, but in spite of these transgressions: They have possessed other beauties, which were conformable to just criticism; and the force of these beauties has been able to overpower censure, and give the mind a satisfaction superior to the disgust arising from the blemishes. Ariosto pleases; but not by his monstrous and improbable fictions, by his bizarre mixture of the serious and comic styles, by the want of coherence in his stories, or by the con-tinual interruptions of his narration. He charms by the force and clearness of his expression, by the readiness and variety of his inventions, and by his natural pictures of the passions, especially those of the gay and amorous kind: And however his faults

---

David Hume, probably the most important philosopher of the Scottish Enlightenment, is not usually con-sidered a literary critic. Nonetheless, his attempt in this essay to develop a standard of taste will be impor-tant for subsequent literary critics.

may diminish our satisfaction, they are not able entirely to destroy it. Did our pleasure really arise from those parts of his poem, which we denominate faults, this would be no objection to criticism in general: It would only be an objection to those particular rules of criticism, which would establish such circumstances to be faults, and would represent them as universally blameable. If they are found to please, they cannot be faults; let the pleasure, which they produce, be ever so unexpected and unaccountable.

But though all the general rules of art are founded only on experience and on the observation of the common sentiments of human nature, we must not imagine, that, on every occasion, the feelings of men will be conformable to these rules. Those finer emotions of the mind are of a very tender and delicate nature, and require the concurrence of many favourable circumstances to make them play with facility and exactness, according to their general and established principles. The least exterior hindrance to such small springs, or the least internal disorder, disturbs their motion, and confounds the operation of the whole machine. When we would make an experiment of this nature, and would try the force of any beauty or deformity, we must choose with care a proper time and place, and bring the fancy to a suitable situation and disposition. A perfect serenity of mind, a recollection of thought, a due attention to the object; if any of these circumstances be wanting, our experiment will be fallacious, and we shall be unable to judge of the catholic and universal beauty. The relation, which nature has placed between the form and the sentiment, will at least be more obscure; and it will require greater accuracy to trace and discern it. We shall be able to ascertain its influence not so much from the operation of each particular beauty, as from the durable admiration, which attends those works, that have survived all the caprices of mode and fashion, all the mistakes of ignorance and envy.

The same Homer, who pleased at Athens and Rome two thousand years ago, is still admired at Paris and at London. All the changes of climate, government, religion, and language, have not been able to obscure his glory. Authority or prejudice may give a temporary vogue to a bad poet or orator; but his reputation will never be durable or general. When his compositions are examined by posterity or by foreigners, the enchantment is dissipated, and his faults appear in their true colors. On the contrary, a real genius, the longer his works endure, and the more wide they are spread, the more sincere is the admiration which he meets with. Envy and jealousy have too much place in a narrow circle; and even familiar acquaintance with his person may diminish the applause due to his performances: But when these obstructions are removed, the beauties, which are naturally fitted to excite agreeable sentiments, immediately display their energy; and while the world endures, they maintain their authority over the minds of men. [. . .]

One obvious cause, why many feel not the proper sentiment of beauty, is the want of that *delicacy* of imagination, which is requisite to convey a sensibility of those finer emotions. This delicacy everyone pretends to: Everyone talks of it; and would reduce every kind of taste or sentiment to its standard. But as our intention in this essay is to mingle some light of the understanding with the feelings of sentiment, it will be proper to give a more accurate definition of delicacy, than has hithert been attempted. [. . .]

Where the organs are so fine, as to allow nothing to escape them; and at the same time so exact as to perceive every ingredient in the composition: This we call delicacy of taste, whether we employ these terms in the literal or metaphorical sense. Here then the general rules of beauty are of use; being drawn from established models, and from the observation of what pleases or displeases, when presented singly and in a high degree: And if the same qualities, in a continued composition and in a smaller degree, affect not the organs with a sensible delight or uneasiness, we exclude the person from all pretensions to this delicacy. [. . .]

But though there be naturally a wide difference in point of delicacy between one person and another, nothing tends further to encrease and improve this talent, than *practice* in a particular art, and the frequent surveyor contemplation of a particular species of beauty. When objects of any kind are first presented to the eye or imagination, the sentiment, which attends them, is obscure and confused; and, the mind is, in a great measure, incapable of pronouncing concerning their merits or defects. The taste cannot perceive the several excellences of the performance; much less distinguish the particular character of each excellency, and ascertain its quality and degree. If it pronounce the whole in general to be beautiful or deformed, it is the utmost that can be expected; and even this judgment, a person, so unpractised, will be apt to deliver with great hesitation and reserve. But allow him to acquire experience in those objects, his feeling becomes more exact and nice: He not only perceives the beauties and defects of each part, but marks the distinguishing species of each quality, and assigns it suitable praise or blame. A clear and distinct sentiment attends him through the whole survey of the objects; and he discerns that very degree and kind of approbation or displeasure, which each part is naturally fitted to produce. The mist dissipates, which seemed formerly to hang over the object: The organ acquires greater perfection in its operations; and can pronounce, without danger of mistake, concerning the merits of every performance. In a word, the same address and dexterity, which practice gives to the execution of any work, is also acquired by the same means, in the judging of it.

So advantageous is practice to the discernment of beauty, that, before we can give judgment on any work of importance, it will even be requisite, that that very individual performance be more than once perused by us, and be surveyed in different lights with attention and deliberation. There is a flutter or hurry of thought which attends the first perusal of any piece, and which confounds the genuine sentiment of beauty. The relation of the parts is not discerned: The true characters of style are little distinguished: The several perfections and defects seem wrapped up in a species of confusion, and present themselves indistinctly to the imagination. Not to mention, that there is a species of beauty, which, as it is florid and superficial, pleases at first; but being found incompatible with a just expression either of reason or passion, soon palls upon the taste, and is then rejected with disdain, at least rated at a much lower value. [. . .]

It is well known, that in all questions, submitted to the understanding, prejudice is destructive of sound judgment, and perverts all operations of the intellectual faculties: It is no less contrary to good taste; nor has it less influence to corrupt our sentiment of beauty. It belongs to *good sense* to check its influence in both cases; and in this respect, as well as in many others, reason, if not an essential part of taste, is at least requisite to the operations of this latter faculty. In all the nobler productions of genius, there is a mutual relation and correspondence of parts; nor can either the beauties or blemishes be perceived by him, whose thought is not capacious enough to comprehend all those parts, and compare them with each other, in order to perceive the consistence and uniformity of the whole. Every work of art has also a certain end or purpose, for which it is calculated; and is to be deemed more or less perfect, as it is more or less fitted to attain this end. The object of eloquence is to persuade, of history to instruct, of poetry to please by means of the passions and the imagination. These ends we must carry constantly in our view, when we peruse any performance; and we must be able to judge how far the means employed are adapted to their respective purposes. Besides, every kind of composition, even the most poetical, is nothing but a chain of propositions and reasonings; not always, indeed, the justest and most exact, but still plausible and specious, however disguised by the colouring of the imagination. The persons introduced in tragedy and epic poetry, must be represented as reasoning, and thinking, and concluding, and acting, suitably to their character and circumstances; and without judgment, as well as taste and invention, a poet can never hope to succeed in so delicate an undertaking. Not to mention, that the same excellence of faculties which contributes to the improvement of reason, the same clearness of conception, the same exactness of distinction, the same vivacity of apprehension, are essential to the operations of true taste, and are its infallible concomitants. It seldom, or never happens, that a man of sense, who has experience in any art, cannot judge of its beauty; and it is no less rare to meet with a man who has a just taste without a sound understanding. [. . .]

But where are such critics to be found? By what marks are they to be known? How distinguish them from pretenders? These questions are embarrassing; and seem to throw us back into the same uncertainty, from which, during the course of this essay, we have endeavoured to extricate ourselves. [. . .]

Though men of delicate taste be rare, they are easily to be distinguished in society, by the soundness of their understanding and the superiority of their faculties above the rest of mankind. The ascendant, which they acquire, gives a prevalence to that lively approbation, with which they receive any productions of genius, and renders it generally predominant. Many men, when left to themselves, have but a faint and dubious perception of beauty, who yet are capable of relishing any fine stroke, which is pointed out to them. Every convert to the admiration of the real poet or orator is the cause of some new conversion. And though prejudices may prevail for a time, they never unite in celebrating any rival to the true genius, but yield at last to the force of nature and just sentiment. Thus, though a civilized nation may easily be mistaken in the choice of their admired philosopher, they never have been found long to err, in their affection for a favorite epic or tragic author.

Chapter 3

Samuel Johnson (1709–1784) from "Preface to the Plays of William Shakespeare" (1765)

Antiquity, like every other quality that attracts the notice of mankind, has undoubtedly votaries that reverence it, not from reason, but from prejudice. Some seem to admire indiscriminately whatever has been long preserved, without considering that time has sometimes co-operated with chance; all perhaps are more willing to honour past than present excellence; and the mind contemplates genius through the shades of age, as the eye surveys the sun through artificial opacity. The great contention of criticism is to find the faults of the moderns, and the beauties of the ancients. While an author is yet living we estimate his powers by his worst performance, and when he is dead, we rate them by his best.

To works, however, of which the excellence is not absolute and definite, but gradual and comparative; to works not raised upon principles demonstrative and scientifick, but appealing wholly to observation and experience, no other test can be applied than length of duration and continuance of esteem. What mankind have long possessed they have often examined and compared; and if they persist to value the possession, it is because frequent comparisons have confirmed opinion in its favour. As among the works of nature no man can properly call a river deep, or a mountain high, without the knowledge of many mountains, and many rivers; so in the productions of genius, nothing can be stiled excellent till it has been compared with other works of the same kind. Demonstration immediately displays its power, and has nothing to hope or fear from the flux of years; but works tentative and experimental must be estimated by their proportion to the general and collective ability of man, as it is discovered in a long succession of endeavours. Of the first building that was raised, it might be with certainty determined that it was round or square; but whether it was spacious or lofty must have been referred to time. The Pythagorean scale of numbers was at once discovered to be perfect; but the poems of *Homer* we yet know not to transcend the common limits of human intelligence, but by remarking, that nation

---

Dr. Johnson, as he is known, is one of the most influential English literary critics. With his "Preface to Shakespeare," Johnson invokes a sense of a general nature to recuperate Shakespeare from contemporary criticism, which usually disliked his work for violating the Classical rules.

after nation, and century after century, has been able to do little more than transpose his incidents, newname his characters, and paraphrase his sentiments.

The reverence due to writings that have long subsisted arises therefore not from any credulous confidence in the superior wisdom of past ages, or gloomy persuasion of the degeneracy of mankind, but is the consequence of acknowledged and indubitable positions, that what has been longest known has been most considered, and what is most considered is best understood.

The Poet, of whose works I have undertaken the revision, may now begin to assume the dignity of an ancient, and claim the privilege of established fame and prescriptive veneration. He has long outlived his century, the term commonly fixed as the test of literary merit. Whatever advantages he might once derive from personal allusions, local customs, or temporary opinions, have for many years been lost; and every topick of merriment, or motive of sorrow, which the modes of artificial life afforded him, now only obscure the scenes which they once illuminated. The effects of favour and competition are at an end; the tradition of his friendships and his enmities has perished; his works support no opinion with arguments, nor supply any faction with invectives; they can neither indulge vanity nor gratify malignity; but are read without any other reason than the desire of pleasure, and are therefore praised only as pleasure is obtained; yet, thus unassisted by interests or passion, they have past through variations of taste and changes of manners, and, as they devolved from one generation to another, have received new honours at every transmission.

But because human judgment, though it be gradually gaining upon certainty, never becomes infallible; and approbation, though long continued, may yet be only the approbation of prejudice or fashion; it is proper to inquire, by what peculiarities of excellence *Shakespeare* has gained and kept the favour of his countrymen.

Nothing can please many, and please long, but just representations of general nature. Particular manners can be known to few, and therefore few only can judge how nearly they are copied. The irregular combinations of fanciful invention may delight a while, by that novelty of which the common satiety of life sends us all in quest; but the pleasures of sudden wonder are soon exhausted, and the mind can only repose on the stability of truth.

Shakespeare is above all writers, at least above all modern writers, the poet of nature; the poet that holds up to his readers a faithful mirror of manners and of life. His characters are not modified by the customs of particular places, unpractised by the rest of the world; by the peculiarities of studies or professions, which can operate but small numbers; or by the accidents of transient fashions or temporary opinions; they are the genuine progeny of common humanity, such as the world will always supply, and observation will always find. His persons act and speak by the influence of those general passions and principles by which all minds are agitated, and the whole system of life is continued in motion. In the writings of other poets a character is too often an individual; in those of Shakespeare it is commonly a species.

Chapter 4

Red Jacket (c.1750–1830)
"Why not all agree, as you can all read the book?" from a speech to the Boston Missionary Society (1828)

You say that you are sent to instruct us how to worship the Great Spirit agreeably to his mind; and if we do not take hold of the religion which you white people teach, we shall be unhappy hereafter. You say that you are right and we are lost. How do we know this to be true? We understand that your religion is written in a book. If it was intended for us as well as for you, why has not the Great Spirit given it to us; and not only to us, but why did he not give to our forefathers the knowledge of that book, with the means of understanding it rightly? We only know what you tell us about it. How shall we know when to believe, being so often deceived by the white people?

Brother!—You say there is but one way to worship and serve the Great Spirit. If there is but one religion, why do you white people differ so much about it? Why not all agree, as you can all read the book?

Brother!—We do not understand these things. We are told that your religion was given to your forefathers, and has been handed down from father to son. We also have a religion which was given to our forefathers, and has been handed down to us their children. We worship that way. It teaches us to be thankful for all the favors we receive, to love each other, and to be united. We never quarrel about religion.

An Iroquois leader, Red Jacket addresses the Boston Missionary Society in this transcript of an 1828 address. His critique of the white people—their failure to agree despite sharing the book—is also an important insight into the dynamics of the literary canon.

Chapter 5

Matthew Arnold (1822–1888) from "The Function of Criticism at the Present Time," *Essays in Criticism* (1865)

It is undeniable that the exercise of a creative power, that a free creative activity, is the highest function of man; it is proved to be so by man's finding in it his true happiness. But it is undeniable, also, that men may have the sense of exercising this free creative activity in other ways than in producing great works of literature or art; if it were not so, all but a very few men would be shut out from the true happiness of all men. They may have it in well-doing, they may have it in learning, they may have it even in criticizing. This is one thing to be kept in mind. Another is, that the exercise of the creative power in the production of great works of literature or art, however high this exercise of it may rank, is not at all epochs and under all conditions possible; and that therefore labor may be vainly spent in attempting it, which might with more fruit be used in preparing for it, in rendering it possible. This creative power works with elements, with materials; what if it has not those materials, those elements, ready for its use? In that case it must surely wait till they are ready. Now, in literature,—I will limit myself to literature, for it is about literature that the question arises,—the elements with which the creative power works are ideas; the best ideas on every matter which literature touches, current at the time. At any rate we may lay it down as certain that in modern literature no manifestation of the creative power, not working with these can be very important or fruitful. And I say *current* at the time, not merely accessible at the time; for creative literary genius does not principally show itself in discovering new ideas, that is rather the business of the philosopher. The grand work of literary genius is a work of synthesis and exposition, not of analysis and discovery; its gift lies in the faculty of being happily inspired by a certain intellectual and spiritual atmosphere; by a certain order of ideas, when it finds itself in them; of dealing divinely with these ideas, presenting them in the most effective and attractive combinations,—making beautiful works with them, in short. But it must have the

---

Poet, critic, and educational administrator, Matthew Arnold was Professor of Poetry at Oxford. In "The Function of Criticism," he offers an extraordinarily influential defense of studying the arts and culture as "the best that is known and thought in the world, irrespectively of practice, politics, and everything of the kind."

atmosphere, it must find itself amidst the order of ideas, in order to work freely; and these it is not so easy to command. This is why great creative epochs in literature are so rare, this is why there is so much that is unsatisfactory in the productions of many men of real genius; because, for the creation of a master-work of literature two powers must concur, the power of the man and the power of the moment, and the man is not enough without the moment; the creative power has, for its happy exercise, appointed elements, and those elements are not in its own control. [. . .]

The notion of the free play of the mind upon all subjects being a pleasure in itself, being an object of desire, being an essential provider of elements without which a nation's spirit, whatever compensations it may have for them, must, in the long run, die of inanition, hardly enters into an Englishman's thoughts. It is noticeable that the word *curiosity*, which in other languages is used in a good sense, to mean, as a high and fine quality of man's nature, just this disinterested love of a free play of the mind on all subjects, for its own sake,—it is noticeable, I say, that this word has in our language no sense of the kind, no sense but a rather bad and disparaging one. But criticism, real criticism is essentially the exercise of this very quality. It obeys an instinct prompting it to try to know the best that is known and thought in the world, irrespectively of practice, politics, and everything of the kind; and to value knowledge and thought as they approach this best, without the intrusion of any other considerations whatever. This is an instinct for which there is, I think, little original sympathy in the practical English nature, and what there was of it has undergone a long benumbing period of blight and suppression in the epoch of concentration which followed the French Revolution.

But epochs of concentration cannot well endure forever; epochs of expansion, in the due course of things, follow them. Such an epoch of expansion seems to be opening in this country. In the first place all danger of a hostile forcible pressure of foreign ideas upon our practice has long disappeared; like the traveler in the fable, therefore, we begin to wear our cloak a little more loosely. Then, with a long peace, the ideas of Europe steal gradually and amicably in, and mingle, though in infinitesimally small quantities at a time, with our own notions. Then, too, in spite of all that is said about the absorbing and brutalizing influence of our passionate material progress, it seems to me indisputable that this progress is likely, though not certain, to lead in the end to an apparition of intellectual life: and that man, after he has made himself perfectly comfortable and has now to determine what to do with himself next, may begin to remember that he has a mind, and that the mind may be made the source of great pleasure. I grant it is mainly the privilege of faith, at present, to discern this end to our railways, our business, and our fortune-making; but we shall see if, here as elsewhere, faith is not in the end the true prophet. Our ease, our traveling, and our unbounded liberty to hold just as hard and securely as we please to the practice to which our notions have given birth, all tend to beget an inclination to deal a little more freely with these notions themselves, to canvass them a little, to penetrate a little into their real nature. Flutterings of curiosity, in the foreign sense of the word, appear amongst us, and it is in these that criticism must look to find its account. Criticism first; a time of true creative activity, perhaps,—which, as I have said, must inevitably be preceded amongst us by a time of criticism,—hereafter, when criticism has done its work.

It is of the last importance that English criticism should clearly discern what rule for its course, in order to avail itself of the field now opening to it, and to produce fruit for the future, it ought to take. The rule may be summed up in one word— *disinterestedness*. And how is criticism to show disinterestedness? By keeping aloof from what is called "the practical view of things"; by resolutely following the law of its own nature, which is to be a free play of the mind on all subjects which it touches. By steadily refusing to lend itself to any of those ulterior, political, practical considerations about ideas, which plenty of people will be sure to attach to them, which perhaps ought often to be attached to them, which in this country at any rate are certain to be attached to them quite sufficiently, but which criticism has really nothing to do with. Its business is, as I have said, simply to know the best that is known and thought in the world, and by in its turn making this known, to create a current of true and fresh ideas. Its business is to do this with inflexible honesty, with due ability; but its business is to do no more, and to leave alone all questions of practical consequences and applications, questions which will never fail to have due prominence given to them. Else criticism, besides being really false to its own nature, merely continues in the old rut which it has hitherto followed in this country, and will certainly miss the chance now given to it. For what is at present the bane of criticism in this country? It is that practical considerations cling to it and stifle it. It subserves interests not its own. Our organs of criticism are organs of men and parties having practical ends to serve, and with them those practical ends are the first thing and the play of mind the second; so much play of mind as is compatible with the prosecution of those practical ends is all that is wanted.

Chapter 6

T.S. Eliot (1888–1965)
"Tradition and the Individual Talent,"
The Sacred Wood (1919)

I

In English writing we seldom speak of tradition, though we occasionally apply its name in deploring its absence. We cannot refer to "the tradition" or "a tradition"; at most, we employ the adjective in saying that the poetry of So-and-so is "traditional" or even "too traditional." Seldom, perhaps, does the word appear except in a phrase of censure. If otherwise, it is vaguely approbative, with the implication, as to the work approved, of some pleasing archaeological reconstruction. You can hardly make the word agreeable to English ears without this comfortable reference to the reassuring science of archaeology.

Certainly the word is not likely to appear in our appreciations of living or dead writers. Every nation, every race, has not only its own creative, but its own critical turn of mind; and is even more oblivious of the shortcomings and limitations of its critical habits than those of its creative genius. We know, or think we know, from the enormous mass of critical writing that has appeared in the French language the critical method or habit of the French; we only conclude (we are such unconscious people) that the French are "more critical" than we, and sometimes even plume ourselves a little with the fact, as if the French were the less spontaneous. Perhaps they are; but we might remind ourselves that criticism is as inevitable as breathing, and that we should be none the worse for articulating what passes in our minds when we read a book and feel an emotion about it, for criticizing our own minds in their work of criticism. One of the facts that might come to light in this process is our tendency to insist, when we praise a poet, upon those aspects of his work in which he least resembles anyone else. In these aspects or parts of his work we pretend to find what is individual, what is the peculiar essence of the man. We dwell with satisfaction upon

In this essay, poet and critic T.S. Eliot counters a popular sense of tradition as unchanging, and offers a vision of tradition as a dynamic series of relationships. For Eliot, literary work becomes significant through its awareness of a literary past: the individual talent has the chance to recast literary tradition.

the poet's difference from his predecessors, especially his immediate predecessors; we endeavour to find something that can be isolated in order to be enjoyed. Whereas if we approach a poet without his prejudice we shall often find that not only the best, but the most individual parts of his work may be those in which the dead poets, his ancestors, assert their immortality most vigorously. And I do not mean the impressionable period of adolescence, but the period of full maturity.

Yet if the only form of tradition, of handing down, consisted in following the ways of the immediate generation before us in a blind or timid adherence to its successes, "tradition" should positively be discouraged. We have seen many such simple currents soon lost in the sand; and novelty is better than repetition. Tradition is a matter of much wider significance. It cannot be inherited, and if you want it you must obtain it by great labour. It involves, in the first place, the historical sense, which we may call nearly indispensable to anyone who would continue to be a poet beyond his twenty-fifth year; and the historical sense involves a perception, not only of the pastness of the past, but of its presence; the historical sense compels a man to write not merely with his own generation in his bones, but with a feeling that the whole of the literature of Europe from Homer and within it the whole of the literature of his own country has a simultaneous existence and composes a simultaneous order. This historical sense, which is a sense of the timeless as well as of the temporal and of the timeless and of the temporal together, is what makes a writer traditional. And it is at the same time what makes a writer most acutely conscious of his place in time, of his contemporaneity.

No poet, no artist of any art, has his complete meaning alone. His significance, his appreciation is the appreciation of his relation to the dead poets and artists. You cannot value him alone; you must set him, for contrast and comparison, among the dead. I mean this as a principle of aesthetic, not merely historical, criticism. The necessity that he shall conform, that he shall cohere, is not one-sided; what happens when a new work of art is created is something that happens simultaneously to all the works of art which preceded it. The existing monuments form an ideal order among themselves, which is modified by the introduction of the new (the really new) work of art among them. The existing order is complete before the new work arrives; for order to persist after the supervention of novelty, the *whole* existing order must be, if ever so slightly, altered; and so the relations, proportions, values of each work of art toward the whole are readjusted; and this is conformity between the old and the new. Whoever has approved this idea of order, of the form of European, of English literature, will not find it preposterous that the past should be altered by the present as much as the present is directed by the past. And the poet who is aware of this will be aware of great difficulties and responsibilities.

In a peculiar sense he will be aware also that he must inevitably be judged by the standards of the past. I say judged, not amputated, by them; not judged to be as good as, or worse or better than, the dead; and certainly not judged by the canons of dead critics. It is a judgment, a comparison, in which two things are measured by each other. To conform merely would be for the new work not really to conform at all; it would not be new, and would therefore not be a work of art. And we do not quite say that the new is more valuable because it fits in; but its fitting in is a test of its value— a test, it is true, which can only be slowly and cautiously applied, for we are none of us infallible judges of conformity. We say: it appears to conform, and is perhaps individual, or it appears individual, and may conform; but we are hardly likely to find that it is one and not the other.

To proceed to a more intelligible exposition of the relation of the poet to the past: he can neither take the past as a lump, an indiscriminate bolus, nor can he form himself wholly on one or two private admirations, nor can he form himself wholly upon one preferred period. The first course is inadmissible, the second is an important experience of youth, and the third is a pleasant and highly desirable supplement. The poet must be very conscious of the main current, which does not at all flow invariably through the most distinguished reputations. He must be quite aware of the obvious fact that art never improves, but that the material of art is never quite the same. He must be aware that the mind of Europe—the mind of his own country—a mind which he learns in time to be much more important than his own private mind—is a mind which changes, and that this change is a development which abandons nothing *en route*, which does not superannuate either Shakespeare, or Homer, or the rock drawing of the Magdalenian draughtsmen. That this development, refinement perhaps, complication certainly, is not, from the point of view of the artist, any improvement. Perhaps not even an improvement from the point of view of the psychologist or not to the extent which we imagine; perhaps only in the end based upon a complication in economics and machinery. But the difference between the present and the past is that the conscious present is an awareness of the past in a way and to an extent which the past's awareness of itself cannot show.

Some one said: "The dead writers are remote from us because we *know* so much more than they did." Precisely, and they are that which we know.

I am alive to a usual objection to what is clearly part of my programme for the *métier* of poetry. The objection is that the doctrine requires a ridiculous amount of erudition (pedantry), a claim which can be rejected by appeal to the lives of poets in any pantheon. It will even be affirmed that much learning deadens or perverts poetic sensibility. While, however, we persist in believing that a poet ought to know as much as will not encroach upon his necessary receptivity and necessary laziness, it is not desirable to confine knowledge to whatever can be put into a useful shape for examinations, drawing-rooms, or the still more pretentious modes of publicity. Some can absorb knowledge, the more tardy must sweat for it. Shakespeare acquired more essential history from Plutarch than most men could from the whole British Museum. What is to be insisted upon is that the poet must develop or procure the consciousness of the past and that he should continue to develop this consciousness throughout his career.

What happens is a continual surrender of himself as he is at the moment to something which is more valuable. The progress of an artist is a continual self-sacrifice, a continual extinction of personality.

There remains to define this process of personalization and its relation to the sense of tradition. It is in this depersonalization that art may be said to approach the condition of science. I shall, therefore, invite you to consider, as a suggestive analogy, the action which takes place when a bit of finely filiated platinum is introduced into a chamber containing oxygen and sulphur dioxide.

II

Honest criticism and sensitive appreciation is directed not upon the poet but upon the poetry. If we attend to the confused cries of the newspaper critics and the susurrus of popular repetition that follows, we shall hear the names of poets in great numbers; if we seek not Blue-book knowledge but the enjoyment of poetry, and ask for a poem, we shall seldom find it. In the last article I tried to point out the importance of the relation of the poem to other poems by other authors, and suggested the conception of poetry as a living whole of all the poetry that has ever been written. The other aspect of this Impersonal theory of poetry is the relation of the poem to its author. And I hinted, by an analogy, that the mind of the mature poet differs from that of the immature one not precisely in any valuation of "personality," not being necessarily more interesting, or having "more to say," but rather by being a more finely perfected medium in which special, or very varied, feelings are at liberty to enter into new combinations.

The analogy was that of the catalyst. When the two gases previously mentioned are mixed in the presence of a filament of platinum, they form sulphurous acid. This combination takes place only if the platinum is present; nevertheless the newly formed acid contains no trace of platinum, and the platinum itself is apparently unaffected; has remained inert, neutral, and unchanged. The mind of the poet is the shred of platinum. It may partly or exclusively operate upon the experience of the man himself; but, the more perfect the artist, the more completely separate in him will be the man who suffers and the mind which creates; the more perfectly will the mind digest and transmute the passions which are its material.

The experience, you will notice, the elements which enter the presence of the transforming catalyst, are of two kinds: emotions and feelings. The effect of a work of art upon the person who enjoys it is an experience different in kind from any experience not of art. It may be formed out of one emotion, or may be a combination of several; and various feelings, inhering for the writer in particular words or phrases or images, may be added to compose the final result. Or great poetry may be made without the direct use of any emotion whatever: composed out of feelings solely. Canto XV of the *Inferno* (Brunetto Latini) is a working up of the emotion evident in the situation; but the effect, though single as that of any work of art, is obtained by considerable complexity of detail. The last quatrain gives an image, a feeling attaching to an image, which "came," which did not develop simply out of what precedes, but which was probably in suspension in the poet's mind until the proper combination arrived for it to add itself to. The poet's mind is in fact a receptacle for seizing and storing up numberless feelings, phrases, images, which remain there until all the particles which can unite to form a new compound are present together.

If you compare several representative passages of the greatest poetry you see how great is the variety of types of combination, and also how completely any semi-ethical criterion of "sublimity" misses the mark. For it is not the "greatness," the intensity, of the emotions, the components, but the intensity of the artistic process, the pressure, so to speak, under which the fusion takes place, that counts. The episode of Paolo and Francesca employs a definite emotion, but the intensity of the poetry is something quite different from whatever intensity in the supposed experience it may give the impression of. It is no more intense, furthermore, than Canto XXVI, the voyage of Ulysses, which has not the direct dependence upon an emotion. Great variety is possible in the process of transmution of emotion: the murder of Agamemnon, or the agony of Othello, gives an artistic effect apparently closer to a possible original than the scenes from Dante. In the *Agamemnon*, the artistic emotion approximates to the emotion of an actual spectator; in *Othello* to the emotion of the protagonist himself. But the difference between art and the event is always absolute; the combination which is the murder of Agamemnon is probably as complex as that which is the voyage of Ulysses. In either case there has been a fusion of elements. The ode of Keats contains a number of feelings which have nothing particular to do with the nightingale, but which the nightingale, partly, perhaps, because of its attractive name, and partly because of its reputation, served to bring together.

The point of view which I am struggling to attack is perhaps related to the metaphysical theory of the substantial unity of the soul: for my meaning is, that the poet has, not a "personality" to express, but a particular medium, which is only a medium and not a personality, in which impressions and experiences combine in peculiar and unexpected ways. Impressions and experiences which are important for the man may take no place in the poetry, and those which become important in the poetry may play quite a negligible part in the man, the personality.

I will quote a passage which is unfamiliar enough to be regarded with fresh attention in the light-or darkness-of these observations:

> And now methinks I could e'en chide myself
> For doating on her beauty, though her death
> Shall be revenged after no common action.
> Does the silkworm expend her yellow labours
> For thee? For thee does she undo herself?
> Are lordships sold to maintain ladyships
> For the poor benefit of a bewildering minute?
> Why does yon fellow falsify highways,
> And put his life between the judge's lips,
> To refine such a thing—keeps horse and men
> To beat their valours for her? . . .

In this passage (as is evident if it is taken in its context) there is a combination of positive and negative emotions: an intensely strong attraction toward beauty and an equally intense fascination by the ugliness which is contrasted with it and which destroys it. This balance of contrasted emotion is in the dramatic situation to which the speech is pertinent, but that situation alone is inadequate to it. This is, so to speak, the structural emotion, provided by the drama. But the whole effect, the dominant tone, is due to the fact that a number of floating feelings, having an affinity to this emotion by no means superficially evident, have combined with it to give us a new art emotion.

It is not in his personal emotions, the emotions provoked by particular events in his life, that the poet is in any way remarkable or interesting. His particular emotions may be simple, or crude, or flat. The emotion in his poetry will be a very complex thing, but not with the complexity of the emotions of people who have very complex or unusual emotions in life. One error, in fact, of eccentricity in poetry is to seek for new human emotions to express; and in this search for novelty in the wrong place it discovers the perverse. The business of the poet is not to find new emotions, but to use the ordinary ones and, in working them up into poetry, to express feelings which are not in actual emotions at all. And emotions which he has never experienced will serve his turn as well as those familiar to him. Consequently, we must believe that "emotion recollected in tranquility" is an inexact formula. For it is neither emotion, nor recollection, nor, without distorting of meaning, tranquility. It is a concentration, and a new thing resulting from the concentration, of a very great number of experiences which to the practical and active person would not seem to be experiences at all; it is a concentration which does not happen consciously or of deliberation. These experiences are not "recollected," and they finally unite in an atmosphere which is "tranquil" only in that it is a passive attending upon the event. Of course this is not quite the whole story. There is a great deal, in the writing of poetry which must be conscious and deliberate. In fact, the bad poet is usually unconscious where he ought to be conscious, and conscious where he ought to be unconscious. Both errors tend to make him "personal." Poetry is not a turning loose of emotion, but an escape from emotion; it is not the expression of personality, but an escape from personality. But, of course, only those who have personality and emotions know what it means to want to escape from these things.

III

*ὁ δὲ νοῦς ἴσως θειότερόν τι καὶ ἀπαθές ἐστιγ**

This essay proposes to halt at the frontier of metaphysics or mysticism, and confine itself to such practical conclusions as can be applied by the responsible person in poetry. To divert interest from the poet to the poetry is a laudable aim: for it would conduce to a juster estimation of actual poetry, good and bad. There are many people who appreciate the expression of sincere emotion in verse, and there is a smaller number of people who can appreciate technical excellence. But very few know when there is expression of *significant* emotion, emotion which has its life in the poem and not in the history of the poet. The emotion of art is impersonal. And the poet cannot reach this impersonality without surrendering himself wholly to the work to be done. And he is not likely to know what is to be done unless he lives in what is not merely the present, but the present moment of the past, unless he is conscious, not of what is dead, but of what is already living.

---

* *The mind is doubtless something more divine and unaffected.*

Chapter 7

F.R. Leavis (1895–1978)
from *Mass Civilization and Minority Culture* (1933)

"High-brow" is an ominous addition to the English language. I have said earlier that culture has always been in minority keeping. But the minority now is made conscious, not merely of an uncongenial, but of a hostile environment. "Shakespeare," I once heard Mr. Dover Wilson say, "was not a high-brow." True: there were no "high-brows" in Shakespeare's time. It was possible for Shakespeare to write plays that were at once popular drama and poetry that could be appreciated only by an educated minority. *Hamlet* appealed at a number of levels of response, from the highest downwards. The same is true of *Paradise Lost, Clarissa, Tom Jones, Don Juan, The Return of the Native.* The same is not true, Mr. George A. Birmingham might point out, of *The Waste Land, Hugh Selwyn Mauberley, Ulysses* or *To the Lighthouse.* These works are read only by a very small specialised public and are beyond the reach of the vast majority of those who consider themselves educated. The age in which the finest creative talent tends to be employed in works of this kind is the age that has given currency to the term "high-brow." But it would be as true to say that the attitude implicit in "high-brow" causes this use of talent as the converse. The minority is being cut off as never before from the powers that rule the world; and as Mr. George A. Birmingham and his friends succeed in refining and standardising and conferring authority upon "the taste of the bathos implanted by nature in the literary judgments of man" (to use Matthew Arnold's phrase), they will make it more and more inevitable that work expressing the finest consciousness of the age should be so specialised as to be accessible only to the minority.

Leavis, who studied and taught at Cambridge University, was the author of influential works of literary criticism such as *New Bearings on English Poetry* (1932) and *The Great Tradition* (1948), which addressed and reshaped the English literary canon. In this brief excerpt from *Mass Civilization and Minority Culture,* Leavis argues that culture has not always been something for a minority of the population.

Chapter 8

Mortimer J. Adler (1902–2001)
"Reading and the Growth of the Mind,"
How to Read a Book (1940)

We have defined active reading as the asking of questions, and we have indicated what questions must be asked of any book, and how those questions must be answered in different ways for different kinds of books.

We have identified and discussed the four levels of reading, and shown how these are cumulative, earlier or lower levels being contained in later or higher ones. Consequent upon our stated intention, we have laid more stress upon the later and higher levels of reading than upon the earlier and lower ones, and we have therefore emphasized analytical and syntopical reading. Since analytical reading is probably the most unfamiliar kind for most readers, we have discussed it at greater length than any of the other levels, giving its rules and explaining them in the order in which they must be applied. But almost everything that was said of analytical reading also applies, with certain adaptations that were mentioned in the last chapter, to syntopical reading as well.

We have completed our task, but you may not have completed yours. We do not need to remind you that this is a practical book, nor that the reader of a practical book has a special obligation with respect to it. If, we said, the reader of a practical book accepts the ends it proposes and agrees that the means recommended are appropriate and effective, then he must act in the way proposed. You may not accept the primary aim we have endorsed—namely, that you should be able to read as well as possible— nor the means we have proposed to reach it—namely, the rules of inspectional, analytical, and syntopical reading. (In that case, however, you are not likely to be reading this page.) But if you do accept that aim and agree that the means are appropriate, then you must make the effort to read as you probably have never read before.

At Columbia University, Mortimer Adler taught with John Erskine, who created the first Great Books Honors Seminar, laying the foundation for what would become Columbia College's Core Curriculum. Hired by the University of Chicago in 1930, Mortimer Adler worked with Robert Hutchins to organize undergraduate education there around a Great Books Core. The co-founder of the Great Books Foundation and of the Aspen Institute, Adler would team with *Encyclopedia Britannica* in 1952 to produce "The Great Books of the Western World" series.

That is your task and your obligation. Can we help you in it in any way?

We think we can. The task falls mainly on you—it is you who, henceforth, must do all the work (and obtain all the benefits). But there are several things that remain to be said, about the end and the means. Let us discuss the latter first.

What Good Books Can Do for Us

"Means" can be interpreted in two ways. In the previous paragraph, we interpreted the term as referring to the rules of reading, that is, the *method* by which you become a better reader. But "means" can also be interpreted as referring to *the things you read*. Having a method without materials to which it can be applied is as useless as having the materials with no method to apply to them.

In the latter sense of the term, the means that will serve you in the further improvement of your reading are the books you will read. We have said that the method applies to anything you read, and that is true, if you understand by the statement any *kind* of book—whether fiction or nonfiction, imaginative or expository, practical or theoretical. But in fact, the method, at least as it is exemplified in our discussion of analytical and syntopical reading, *does not apply to every book*. The reason is that some books do not require it.

We have made this point before, but we want to make it now again because of its relevance to the task that lies before you. *If you are reading in order to become a better reader, you cannot read just any book or article.* You will not improve as a reader if all you read are books that are well within your capacity. You must tackle books that are beyond you, or, as we have said, books that are over your head. Only books of that sort will make you stretch your mind. And unless you stretch, you will not learn.

Thus, it becomes of crucial importance for you not only to be able to read well but also to be able to identify those books that make the kinds of demands on you that improvement in reading ability requires. A book that can do no more than amuse or entertain you may be a pleasant diversion for an idle hour, but you must not expect to get anything but amusement from it. We are not against amusement in its own right, but we do want to stress that *improvement in reading skill does not accompany it*. The same goes for a book that merely informs you of facts that you did not know without adding to your understanding of those facts. Reading for information does not stretch your mind any more than reading for amusement. It may seem as though it does, but that is merely because your mind is fuller of facts than it was before you read the book. However, your mind is essentially in the same condition that it was before. There has been a quantitative change, but no improvement in your skill.

We have said many times that the good reader makes demands on himself when he reads. He reads actively, effortfully. Now we are saying something else. The books that you will want to practice your reading on, particularly your analytical reading, *must also make demands on you*. They must seem to you to be beyond your capacity. You need not fear that they really are, because there is no book that is completely out of your grasp if you apply the rules of reading to it that we have described. This does not mean, of course, that these rules will accomplish immediate miracles for you.

There are certainly some books that will continue to extend you no matter how good a reader you are. Actually, those are the very books that you must seek out, because they are the ones that can best help you to become an ever more skillful reader.

Some readers make the mistake of supposing that such books—the ones that provide a constant and never-ending challenge to their skill—are always ones in relatively unfamiliar fields. In practice, this comes down to believing, in the case of most readers, that only scientific books, and perhaps philosophical ones, satisfy the criterion. But that is far from the case. We have already remarked that the great scientific books are in many ways easier to read than nonscientific ones, because of the care with which scientific authors help you to come to terms, identify the key propositions, and state the main arguments. These helps are absent from poetical works, and so in the long run they are quite likely to be the hardest, the most demanding, books that you can read. Homer, for example, is in many ways harder to read than Newton, despite the fact that you may get more out of Homer the first time through. The reason is that Homer deals with subjects that are harder to write well about.

The difficulties that we are talking about here are very different from the difficulties that are presented by a bad book. It is hard to read a bad book, too, for it defies your efforts to analyze it, slipping through your fingers whenever you think you have it pinned down. In fact, in the case of a bad book, there is really nothing *to* pin down. It is not worth the effort of trying. You receive no reward for your struggle.

A good book does reward you for trying to read it. The best books reward you most of all. The reward, of course, is of two kinds. First, there is the improvement in your reading skill that occurs when you successfully tackle a good, difficult work. Second—and this in the long run is much more important—a good book can teach you about the world and about yourself. You learn more than how to read better; you also learn more about life. You become wiser. Not just more knowledgeable—books that provide nothing but information can produce that result. But wiser, in the sense that you are more deeply aware of the great and enduring truths of human life.

There are some human problems, after all, that have no solution. There are some relationships, both among human beings and between human beings and the non-human world, about which no one can have the last word. This is true not only in such fields as science and philosophy, where it is obvious that final understanding about nature and its laws, and about being and becoming, has not been achieved by anyone and never will be; it is also true of such familiar and everyday matters as the relation between men and women, or parents and children, or man and God. These are matters about which you cannot think too much, or too well. The greatest books can help you to think better about them, because they were written by men and women who thought better than other people about them.

The Pyramid of Books

The great majority of the several million books that have been written in the Western tradition alone—more than 99 per cent of them—will not make sufficient demands on you for you to improve your skill in reading. This may seem like a distressing fact, and the percentages may seem an overestimate. But obviously, considering the numbers involved, it is true. These are the books that can be read only for amusement or information. The amusement may be of many kinds, and the information may be interesting in all sorts of ways. But you should not expect to learn anything of importance from them. In fact, you do not have to read them—analytically—at all. Skimming will do.

There is a second class of books from which you can learn—both how to read and how to live. Less than one out of every hundred books belongs in this class—probably it is more like one in a thousand, or even one in ten thousand. These are the good books, the ones that were carefully wrought by their authors, the ones that convey to the reader significant insights about subjects of enduring interest to human beings. There are in all probably no more than a few thousand such books. They make severe demands on the reader. They are worth reading analytically—once. If you are skillful, you will be able to get everything out of them that they can give in the course of one good reading. They are books that you read once and then put away on your shelf. You know that you will never have to read them again, although you may return to them to check certain points or to refresh your memory of certain ideas or episodes. (It is in the case of such books that the notes you make in the margin or elsewhere in the volume are particularly valuable.)

How do you know that you do not ever have to read such books again? You know it by your own mental reaction to the experience of reading them. Such a book stretches your mind and increases your understanding. But as your mind stretches and your understanding increases, you realize, by a process that is more or less mysterious, that you are not going to be changed any more in the future by this book. You realize that you have grasped the book in its entirety. You have milked it dry. You are grateful to it for what it has given you, but you know it has no more to give.

Of the few thousand such books there is a much smaller number—here the number is probably less than a hundred—that cannot be exhausted by even the very best reading you can manage. How do you recognize this? Again it is rather mysterious, but when you have closed the book after reading it analytically to the best of your ability, and place it back on the shelf, you have a sneaking suspicion that there is more there than you got. We say "suspicion" because that may be all it is at this stage. If you knew what it was that you had missed, your obligation as an analytical reader would take you back to the book immediately to seek it out. In fact, you cannot put your finger on it, but you know it is there. You find that you cannot forget the book, that you keep thinking about it and your reaction to it. Finally, you return to it. And then a very remarkable thing happens.

If the book belongs to the second class of books to which we referred before, you find, on returning to it, that there was *less there than you remembered*. The reason, of course, is that you yourself have grown in the meantime. Your mind is fuller, your understanding greater. The book has not changed, but you have. Such a return is inevitably disappointing.

But if the book belongs to the highest class—the very small number of inexhaustible books—you discover on returning that *the book seems to have grown with you*. You see new things in it—whole sets of new things—that you did not see before. Your previous understanding of the book is not invalidated (assuming that you read it well the first time); it is just as true as it ever was, and in the same ways that it was true before. But now it is true in still other ways, too.

How can a book grow as you grow? It is impossible, of course; a book, once it is written and published, does not change. But what you only now begin to realize is that the book was so far above you to begin with that it has remained above you, and probably always will remain so. Since it is a really good book—a great book, as we might say—it is accessible at different levels. Your impression of increased understanding on your previous reading was not false. The book truly lifted you then. But now, even though you have become wiser and more knowledgeable, it can lift you again. And it will go on doing this until you die.

There are obviously not many books that can do this for any of us. Our estimate was that the number is considerably less than a hundred. But the number is *even less than that for any given reader*. Human beings differ in many ways other than in the power of their minds. They have different tastes; different things appeal more to one person than to another. You may never feel about Newton the way you feel about Shakespeare, either because you may be able to read Newton so well that you do not have to read him again, or because mathematical systems of the world just do not have all that appeal to you. Or, if they do—Charles Darwin is an example of such a person—then Newton may be one of the handful of books that are great for you, and not Shakespeare.

We do not want to state authoritatively that any particular book or group of books must be great for you, in this sense, although in our first Appendix we do list those books that experience has shown are capable of having this kind of value for many readers. Our point, instead, is that *you should seek out the few books that can have this value for you*. They are the books that will teach you the most, both about reading and about life. They are the books to which you will want to return over and over. They are the books that will help you to grow.

The Life and Growth of the Mind

There is an old test—it was quite popular a generation ago—that was designed to tell you which books are the ones that can do this for you. Suppose, the test went, that you know in advance that you will be marooned on a desert island for the rest of your life, or at least for a long period. Suppose, too, that you have time to prepare for the experience. There are certain practical and useful articles that you would be sure to take with you. You will also be allowed ten books. Which ones would you select?

Trying to decide on a list is instructive, and not only because it may help you to identify the books that you would most like to read and reread. That, in fact, is probably of minor importance, compared with what you can learn about yourself when you imagine what life would be like if you were cut off from all the sources of amusement, information, and understanding that ordinarily surround you. Remember, there would be no radio or television on the island, and no lending library. There would be just you and ten books.

This imagined situation seems bizarre and unreal when you begin to think about it. But is it actually so unreal? We do not think so. We are all to some extent persons marooned on a desert island. We all face the same challenge that we would face if we really were there—the challenge of finding the resources within ourselves to live a good human life.

There is a strange fact about the human mind, a fact that differentiates the mind sharply from the body. The body is limited in ways that the mind is not. One sign of this is that the body does not continue indefinitely to grow in strength and develop in skill and grace. By the time most people are thirty years old, their bodies are as good as they will ever be; in fact, many persons' bodies have begun to deteriorate by that time. *But there is no limit to the amount of growth and development that the mind can sustain.* The mind does not stop growing at any particular age; only when the brain itself loses its vigor, in senescence, does the mind lose its power to increase in skill and understanding.

This is one of the most remarkable things about human beings, and it may actually be the major difference between *homo sapiens* and the others [*sic*] animals, which do not seem to grow mentally beyond a certain stage in their development. But this great advantage that man possesses carries with it a great peril. *The mind can atrophy*, like the muscles, *if it is not used*. Atrophy of the mental muscles is the penalty that we pay for not taking mental exercise. And this is a terrible penalty, for there is evidence that atrophy of the mind is a mortal disease. There seems to be no other explanation for the fact that so many busy people die so soon after retirement. They were kept alive by the demands of their work upon their minds; they were propped up artificially, as it were, by external forces. But as soon as those demands cease, having no resources within themselves in the way of mental activity, they cease thinking altogether, and expire.

Television, radio, and all the sources of amusement and information that surround us in our daily lives are also artificial props. They can give us the impression that our minds are active, because we are required to react to stimuli from outside. But the power of those external stimuli to keep us going is limited. They are like drugs. We grow used to them, and we continuously need more and more of them. Eventually, they have little or no effect. Then, if we lack resources within ourselves, we cease to grow intellectually, morally, and spiritually. And when we cease to grow, we begin to die.

Reading well, which means reading actively, is thus not only a good in itself, nor is it merely a means to advancement in our work or career. It also serves to keep our minds alive and growing.

Appendix A: A Recommended Reading List

On the following pages appears a list of books that it would be worth your while to read. We mean the phrase "worth your while" quite seriously. Although not all of the books listed are "great" in any of the commonly accepted meanings of the term, all of them will reward you for the effort you make to read them. All of these books are over most people's heads—sufficiently so, at any rate, to force most readers to stretch their minds to understand and appreciate them. And that, of course, is the kind of book you should seek out if you want to improve your reading skills, and at the same time discover the best that has been thought and said in our literary tradition.

Some of the books are great in the special sense of the term that we employed in the last chapter. On returning to them, you will always find something new, often many things. They are endlessly re-readable. Another way to say this is that some of the books—we will not say exactly how many, nor will we try to identify them, since to some extent this is an individual judgment—are over the heads of all readers, no matter how skillful. As we observed in the last chapter, these are the works that everyone should make a special effort to seek out. They are the truly great books; they are the books that anyone should choose to take with him to his own desert island.

The list is long, and it may seem a little overwhelming. We urge you not to allow yourself to be abashed by it. In the first place, you are likely to recognize the names of most of the authors. There is nothing here that is so recondite as to be esoteric. More important, we want to remind you that it is wise to begin with those books that interest you most, for whatever reason. As we have pointed out several times, the primary aim is to read well, not widely. You should not be disappointed if you read no more than a handful of the books in a year. The list is not something to be gotten through in any amount of time. It is not a challenge that you can meet only by finishing every item on it. Instead, it is an invitation that you can accept graciously by beginning wherever you feel at home.

The authors are listed chronologically, according to the known or supposed date of their birth. When several works of an author are listed, these too are arranged chronologically, where that is possible. Scholars do not always agree about the first publication of a book, but this need not concern you. The point to remember is that the list as a whole moves forward through time. That does not necessarily mean that you should read it chronologically, of course. You might even start with the end of the list and read backward to Homer and the Old Testament.

We have not listed all the works of every author. We have usually cited only the more important titles, selecting them, in the case of expository books, to show the diversity of an author's contribution to different fields of learning. In some instances, we have listed an author's Works and specified, in brackets, those titles that are especially important or useful.

In drawing up a list of this kind, the greatest difficulty always arises with respect to the relatively contemporary items. The closer an author is to our own time, the harder it is to exercise a detached judgment about him. It is all very well to say that time will tell, but we may not want to wait. Thus, with regard to the more recent writers and books, there is much room for differences of opinion, and we would not claim for the later items on our list the degree of authority that we can claim for the earlier ones.

There may be differences of opinion about some of the earlier items too, and we may be charged with being prejudiced against some authors that we have not listed at all. We are willing to admit that this may be true, in some cases. This is our list, and it may differ in some respects from lists drawn up by others. But it will not differ very significantly if everyone concurs seriously in the aim of making up a reading program that is worth spending a lifetime on. Ultimately, of course, you should make up your own list, and then go to work on it. It is wise, however, to read a fair number of the books that have been unanimously acclaimed before you branch off on your own. This list is a place to begin.

We want to mention one omission that may strike some readers as unfortunate. The list contains only Western authors and books; there are no Chinese, Japanese, or Indian works. There are several reasons for this. One is that we are not particularly knowledgeable outside of the Western literary tradition, and our recommendations would carry little weight. Another is that there is in the East no single tradition, as there is in the West, and we would have to be learned in all Eastern traditions in order to do the job well. There are very few scholars who have this kind of acquaintance with all the works of the East. Third, there is something to be said for knowing your own tradition before trying to understand that of other parts of the world. Many persons who today attempt to read such books as the *I Ching* or the *Bhagavad-Gita* are baffled, not only because of the inherent difficulty of such works, but also because they have not learned to read well by practicing on the more accessible works—more accessible to them—of their own culture. And finally, the list is long enough as it is.

One other omission requires comment. The list, being one of books, includes the names of few persons known primarily as lyric poets. Some of the writers on the list wrote lyric poems, of course, but they are best known for other, longer works. This fact is not to be taken as reflecting a prejudice on our part against lyric poetry. But we would recommend starting with a good anthology of poetry rather than with the collected works of a single author. Palgrave's *The Golden Treasury* and *The Oxford Book of English Verse* are excellent places to start. These older anthologies should be supplemented by more modern ones—for example, Selden Rodman's *One Hundred Modern Poems*, a collection widely available in paperback that extends the notion of a lyric poem in interesting ways. Since reading lyric poetry requires special skill, we would also recommend any of several available handbooks on the subject—for example, Mark Van Doren's *Introduction to Poetry*, an anthology that also contains short discussions of how to read many famous lyrics.

We have listed the books by author and title, but we have not attempted to indicate a publisher or a particular edition. Almost every work on the list is available in some form, and many are available in several editions, both paperback and hard cover. However, we have indicated which authors and titles are included in two sets that we ourselves have edited. *Titles* included in *Great Books of the Western World* are identified by a single asterisk; *authors* represented in *Gateway to the Great Books* are identified by a double asterisk.

1. Homer (9th century B.C.?)
   Iliad
   Odyssey
2. The Old Testament
3. Aeschylus (*c.* 525–456 B.C.)
   Tragedies
4. Sophocles (*c.* 495–406 B.C.)
   Tragedies
5. Herodotus (*c.* 484–425 B.C.)
   History (of the Persian Wars)
6. Euripides (*c.* 485–406 B.C.)
   Tragedies
   (esp. *Medea, Hippolytus, The Bacchae*)
7. Thucydides (*c.* 460–400 B.C.)
   History of the Peloponnesian War
8. Hippocrates (*c.* 460–377? B.C.)
   Medical writings
9. Aristophanes (*c.* 448–380 B.C.)
   Comedies
   (esp. *The Clouds, The Birds, The Frogs*)
10. Plato (*c.* 427–347 B.C.)
    Dialogues
    (esp. *The Republic, Symposium, Phaedo, Meno, Apology, Phaedrus, Protagoras, Gorgias, Sophist, Theaetetus*)
11. Aristotle (384–322 B.C.)
    Works
    (esp. *Organon, Physics, Metaphysics, On the Soul, The Nichomachean Ethics, Politics, Rhetoric, Poetics*)
12. **Epicurus (*c.* 341–270 B.C.)
    Letter to Herodotus
    Letter to Menoeceus
13. Euclid (*fl.c.* 300 B.C.)
    Elements (*of Geometry*)
14. Archimedes (*c.* 287–212 B.C.)
    Works
    (esp. *On the Equilibrium of Planes, On Floating Bodies, The Sand-Reckoner*)
15. Apollonius of Perga (*fl.c.* 240 B.C.)
    On Conic Sections
16. **Cicero (106–43 B.C.)
    Works
    (esp. *Orations, On Friendship, On Old Age*)

17. Lucretius (*c.* 95–55 B.C.)
    On the Nature of Things
18. Virgil (70–19 B.C.)
    *Works
19. Horace (65–8 B.C.)
    Works
    (esp. *Odes and Epodes, The Art of
    Poetry*)
20. Livy (59 B.C.–A.D. 17)
    History of Rome
21. Ovid (43 B.C.–A.D. 17)
    Works
    (esp. *Metamorphoses*)
22. **Plutarch (*c.* 45–120)
    *Lives of the Noble Grecians and
    Romans
    Moralia*
23. **Tacitus (*c.* 55–117)
    *Histories
    *Annals
    Agricola
    Germania*
24. Nicomachus of Gerasa (*fl.c.* 100 A.D.)
    Introduction to Arithmetic
25. **Epictetus (*c.* 60–20)
    *Discourses
    Encheiridion (Handbook)*
26. Ptolemy (*c.* 100–178; *fl.* 127–151)
    Almagest
27. **Lucian (*c.* 120–*c.* 190)
    Works
    (esp. *The Way to Write History,
    The True History, The Sale of Creeds*)
28. Marcus Aurelius (121–180)
    Meditations
29. Galen (*c.* 130–200)
    On the Natural Faculties
30. The New Testament
31. Plotinus (205–270)
    The Enneads
32. St.Augustine (354–430)
    Works
    (esp. *On the Teacher, *Confessions,
    *The City of God, *Christian
    Doctrine*)
33. *The Song of Roland* (12th century?)
34. *The Nibelungenlied* (13th century)
    (The *Volsunga Saga* is the
    Scandinavian version
    of the same legend.)
35. *The Saga of Burnt Nial*

36. St. Thomas Aquinas (*c.* 1225–1274)
    Summa Theologica
37. **Dante Alighieri (1265–1321)
    Works
    (esp. *The New Life, On Monarchy,
    The Divine Comedy)
38. Geoffrey Chaucer (*c.* 1340–1400)
    Works
    (esp. *Troilus and Criseyde,
    Canterbury Tales)
39. Leonardo da Vinci (1452–1519)
    Notebooks
40. Niccolo Machiavelli (1469–1527)
    *The Prince
    Discourses on the First Ten Books of
    Livy*
41. Desiderius Erasmus (*c.* 1469–1536)
    The Praise of Folly
42. Nicolaus Copernicus (1473–1543)
    *On the Revolutions of the Heavenly
    Spheres*
43. Sir Thomas More (*c.* 1478–1535)
    Utopia
44. Martin Luther (1483–1546)
    *Three Treatises
    Table-Talk*
45. François Rabelais (*c.* 1495–1553)
    Gargantua and Pantagruel
46. John Calvin (1509–1564)
    Institutes of the Christian Religion
47. Michel de Montaigne (1533–1992)
    Essays
48. William Gilbert (1540–1603)
    *On the Loadstone and Magnetic
    Bodies*
49. Miguel de Cervantes (1547–1616)
    Don Quixote
50. Edmund Spenser (*c.* 1552–1599)
    *Prothalamion
    The Faerie Queene*
51. **Francis Bacon (1561–1626)
    *Essays
    *Advancement of Learning
    *Novum Organum
    New Atlantis
52. William Shakespeare (1564–1616)
    *Works
53. **Galileo Galilei (1564–1642)
    *The Starry Messenger
    *Dialogues Concerning Two New
    Sciences*

54. Johannes Kepler (1571–1630)
    *Epitome of Copernican Astronomy
    *Concerning the Harmonies of the
    World
55. William Harvey (1578–1657)
    *On the Motion of the Heart and
    Blood in Animals
    *On the Circulation of the Blood
    *On the Generation of Animals
56. Thomas Hobbes (1588–1679)
    *The Leviathan
57. René Descartes (1596–1650)
    *Rules for the Direction of the Mind
    *Discourse on Method
    *Geometry
    *Meditations on First Philosophy
58. John Milton (1608–1674) Works
    (esp. °the minor poems,
    *Areopagitica, *Paradise Lost,
    *Samson Agonistes)
59. ** Molière (1622–1673)
    Comedies
    (esp. The Miser, The School for
    Wives, The Misanthrope, The Doctor
    in Spite of Himself, Tartuffe)
60. Blaise Pascal (1623–1662)
    *The Provincial Letters
    *Pensées
    *Scientific treatises
61. Christiaan Huygens (1629–1695)
    *Treatise on Light
62. Benedict de Spinoza (1632–1677)
    *Ethics
63. John Locke (1632–1704)
    *Letter Concerning Toleration
    *"Of Civil Government" (second
    treatise in
    Two Treatises on Government)
    *Essay Concerning Human
    Understanding
    Thoughts Concerning Education
64. Jean Baptiste Racine (1639–1699)
    Tragedies
    (esp. Andromache, Phaedra)
65. Isaac Newton (1642–1727)
    *Mathematical Principles of Natural
    Philosophy
    *Optics
66. Gottfried Wilhelm von Leibniz
    (1646–1716)

Discourse on Metaphysics
New Essays Concerning Human
Understanding
Monadology
67. **Daniel Defoe (1660–1731)
    Robinson Crusoe
68. **Jonathan Swift (1667–1745)
    A Tale of a Tub
    Journal to Stella
    *Gulliver's Travels
    A Modest Proposal
69. William Congreve (1670–1729)
    The Way of the World
70. George Berkeley (1685–1753)
    *Principles of Human Knowledge
71. Alexander Pope (1688–1744)
    Essay on Criticism
    Rape of the Lock
    Essay on Man
72. Charles de Secondat, Baron de
    Montesquieu (1689–1755)
    Persian Letters
    *Spirit of Laws
73. **Voltaire (1694–1778)
    Letters on the English
    Candide
    Philosophical Dictionary
74. Henry Fielding (1707–1754)
    Joseph Andrews
    *Tom Jones
75. **Samuel Johnson (1709–1784)
    The Vanity of Human Wishes
    Dictionary
    Rasselas
    The Lives of the Poets
    (esp. the essays on Milton
    and Pope)
76. **David Hume (1711–1776)
    Treatise of Human Nature
    Essays Moral and Political
    *An Inquiry Concerning Human
    Understanding
77. **Jean Jacques Rousseau
    (1712–1778)
    *On the Origin of Inequality
    *On Political Economy
    Emile
    *The Social Contract
78. Laurence Sterne (1713–1768)
    *Tristram Shandy

*A Sentimental Journey Through
France and Italy*

79. Adam Smith (1723–1790)
    The Theory of the Moral Sentiments
    **Inquiry into the Nature and Causes
    of the Wealth of Nations*

80. **Immanuel Kant (1724–1804)
    *Critique of Pure Reason
    **Fundamental Principles of the
    Metaphysics of Morals*
    *Critique of Practical Reason
    *The Science of Right
    *Critique of Judgment
    Perpetual Peace

81. Edward Gibbon (1737–1794)
    **The Decline and Fall of the Roman
    Empire*
    Autobiography

82. James Boswell (1740–1795)
    Journal
    (esp. *London Journal*
    *Life of Samuel Johnson Ll.D.

83. Antoine Laurent Lavoisier
    (1743–1794)
    *Elements of Chemistry

84. John Jay (1745–1829), James
    Madison (1751–1836), and
    Alexander Hamilton (1757–1804)
    *Federalist *Papers*
    (together with the **Articles of
    Confederation*, the **Constitution of
    the United States*, and the
    *Declaration of Independence)
    *Lectures on the Philosophy of History

85. Jeremy Bentham (1748–1832)
    *Introduction to the Principles of
    Morals and Legislation*
    Theory of Fictions

86. Johann Wolfgang von Goethe
    (1749–1832)
    *Faust
    Poetry and Truth

87. Jean Baptiste Joseph Fourier
    (1768–1830)
    *Analytical Theory of Heat

88. Georg Wilhelm Friedrich Hegel
    (1770–1831)
    Phenomenology of Spirit
    *Philosophy of Right
    *Lectures on the Philosophy of History

89. William Wordsworth (1770–1850)
    Poems
    (esp. *Lyrical Ballads*, Lucy poems,
    sonnets;
    The Prelude)

90. Samuel Taylor Coleridge
    (1772–1834)
    Poems
    (esp. "Kubla Khan,"
    Rime of the Ancient Mariner)
    Biographia Literaria

91. Jane Austen (1775–1817)
    Pride and Prejudice
    Emma

92. **Karl von Clausewitz
    (1780–1831)
    On War

93. Stendhal (1783–1842)
    The Red and the Black
    The Charterhouse of Parma
    On Love

94. George Gordon, Lord Byron
    (1788–1824)
    Don Juan

95. **Arthur Schopenhauer
    (1788–1860)
    Studies in Pessimism

96. **Michael Faraday (1791–1867)
    Chemical History of a Candle
    **Experimental Researches in
    Electricity*

97. **Charles Lyell (1797–1875)
    Principles of Geology

98. Auguste Comte (1798–1857)
    The Positive Philosophy

99. **Honoré de Balzac (1799–1850)
    Pére Goriot
    Eugénie Grandet

100. **Ralph Waldo Emerson
    (1803–1882)
    Representative Men
    Essays
    Journal

101. **Nathaniel Hawthorne
    (1804–1864)
    The Scarlet Letter

102. **Alexis de TocqueviIle
    (1805–1859)
    Democracy in America

103. **John Stuart Mill (1806–1873)

A System of Logic
On Liberty
Representative Government
Utilitarianism
The Subjection of Women
Autobiography

104. **Charles Darwin (1809–1882)
The Origin of Species
The Descent of Man
Autobiography

105. **Charles Dickens (1812–1870)
Works
(esp. *Pickwick Papers, David Copperfield, Hard Times*)

106. **Claude Bernard (1813–1878)
Introduction to the Study of Experimental Medicine

107. **Henry David Thoreau (1817–1862)
Civil Disobedience
Walden

108. Karl Marx (1818–1883)
Capital
(together with the *Communist Manifesto*)

109. George Eliot (1819–1880)
Adam Bede
Middlemarch

110. **Herman Melville (1819–1891)
Moby Dick
Billy Budd

111. **Fyodor Dostoevsky (1821–1881)
Crime and Punishment
The Idiot
The Brothers Karamazov

112. **Gustave Flaubert (1821–1880)
Madame Bovary
Three Stories

113. **Henrik Ibsen (1828–1906)
Plays
(esp. *Hedda Gabler, A Doll's House, The Wild Duck*)

114. **Leo Tolstoy (1828–1910)
War and Peace
Anna Karenina
What Is Art?
Twenty-three Tales

115. **Mark Twain (1835–1910)

The Adventures of Huckleberry Finn
The Mysterious Stranger

116. **William James (1842–1910)
The Principles of Psychology
The Varieties of Religious Experience Pragmatism
Essays in Radical Empiricism

117. **Henry James (1843–1916)
The American
The Ambassadors

118. Friedrich Wilhelm Nietzsche (1844–1900)
Thus Spoke Zarathustra
Beyond Good and Evil
The Genealogy of Morals
The Will to Power

119. Jules Henri Poincaré (1854–1912)
Science and Hypothesis
Science and Method

120. Sigmund Freud (1856–1939)
The Interpretation of Dreams
Introductory Lectures on Psychoanalysis
Civilization and Its Discontents
New Introductory Lectures on Psychoanalysis

121. **George Bernard Shaw (1856–1950)
Plays (and Prefaces)
(esp. *Man and Superman, Major Barbara, Caesar and Cleopatra, Pygmalion, Saint Joan*)

122. **Max Planck (1858–1947)
Origin and Development of the Quantum Theory
Where Is Science Going?
Scientific Autobiography

123. Henri Bergson (1859–1941)
Time and Free Will
Matter and Memory
Creative Evolution
The Two Sources of Morality and Religion

124. **John Dewey (1859–1952)
How We Think
Democracy and Education
Experience and Nature
Logic, the Theory of Inquiry

125. **Alfred North Whitehead (1861–1947)
An Introduction to Mathematics
Science and the Modern World
The Aims of Education and Other Essays
Adventures of Ideas

126. **George Santayana (1863–1952)
The Life of Reason
Skepticism and Animal Faith
Persons and Places

127. Nikolai Lenin (1870–1924)
The State and Revolution

128. Marcel Proust (1871–1922)
Remembrance of Things Past

129. **Bertrand Russell (1872–1970)
The Problems of Philosophy
The Analysis of Mind
An Inquiry into Meaning and Truth
Human Knowledge; Its Scope and Limits

130. **Thomas Mann (1875–1955)
The Magic Mountain
Joseph and His Brothers

131. ** Albert Einstein (1879–1955)
The Meaning of Relativity
On the Method of Theoretical Physics
The Evolution of Physics (with L. Infeld)

132. **James Joyce (1882–1941)
"The Dead" in *Dubliners*
Portrait of the Artist as a Young Man
Ulysses

133. Jacques Maritain (1882–)
Art and Scholasticism
The Degrees of Knowledge
The Rights of Man and Natural Law
True Humanism

134. Franz Kafka (1883–1924)
The Trial
The Castle

135. Arnold Toynbee (1889–)
A Study of History
Civilization on Trial

136. Jean Paul Sartre (1905–)
Nausea
No Exit
Being and Nothingness

137. Aleksandr I. Solzhenitsyn (1918–)
The First Circle
Cancer Ward

Chapter 9

Erich Auerbach (1892–1957)
from "Odysseus' Scar," *Mimesis*
(1946; trans. 1953)

Readers of the *Odyssey* will remember the well-prepared and touching scene in book 19, when Odysseus has at last come home, the scene in which the old house-keeper Euryclea, who had been his nurse, recognizes him by a scar on his thigh. The stranger has won Penelope's good will; at his request she tells the housekeeper to wash his feet, which, in all old stories, is the first duty of hospitality toward a tired traveler. Euryclea busies herself fetching water and mixing cold with hot, meanwhile speaking sadly of her absent master, who is probably of the same age as the guest, and who perhaps, like the guest, is even now wandering somewhere, a stranger; and she remarks how astonishingly like him the guest looks. Meanwhile Odysseus, remembering his scar, moves back out of the light; he knows that, despite his efforts to hide his identity, Euryclea will now recognize him, but he wants at least to keep Penelope in ignorance. No sooner has the old woman touched the scar than, in her joyous surprise, she lets Odysseus' foot drop into the basin; the water spills over, she is about to cry out her joy; Odysseus restrains her with whispered threats and endearments; she recovers herself and conceals her emotion. Penelope, whose attention Athena's foresight had diverted from the incident, has observed nothing.

All this is scrupulously externalized and narrated in leisurely fashion. The two women express their feelings in copious direct discourse. Feelings though they are, with only a slight admixture of the most general considerations upon human destiny, the syntactical connection between part and part is perfectly clear, no contour is blurred. There is also room and time for orderly, perfectly well-articulated, uniformly illuminated descriptions of implements, ministrations, and gestures; even in the dramatic moment of recognition, Homer does not omit to tell the reader that it is with his right hand that Odysseus takes the old woman by the throat to keep her from speaking, at the same time that he draws her closer to him with his left. Clearly outlined,

In 1936, Erich Auerbach left Germany for Istanbul, where he taught until moving to the United States in 1947. While in Istanbul, and without access to extensive library holdings, Auerbach wrote *Mimesis*, a sweeping comparative study of the representation of reality in Western literature. In the first chapter of *Mimesis*, excerpted here, Auerbach contrasts what he sees as Greek and Hebrew conventions of representation.

brightly and uniformly illuminated, men and things stand out in a realm where everything is visible; and not less clear—wholly expressed, orderly even in their ardor—are the feelings and thoughts of the persons involved. [. . .]

The first thought of a modern reader—that this is a device to increase suspense—is, if not wholly wrong, at least not the essential explanation of this Homeric procedure. For the element of suspense is very slight in the Homeric poems; nothing in their entire style is calculated to keep the reader or hearer breathless. The digressions are not meant to keep the reader in suspense, but rather to relax the tension. And this frequently occurs, as in the passage before us. The broadly narrated, charming, and subtly fashioned story of the hunt, with all its elegance and self-sufficiency, its wealth of idyllic pictures, seeks to win the reader over wholly to itself as long as he is hearing it, to make him forget what had just taken place during the foot-washing. But an episode that will increase suspense by retarding the action must be so constructed that it will not fill the present entirely, will not put the crisis, whose resolution is being awaited, entirely out of the reader's mind, and thereby destroy the mood of suspense; the crisis and the suspense must continue, must remain vibrant in the background. But Homer—and to this we shall have to return later—knows no background. What he narrates is for the time being the only present, and fills both the stage and the reader's mind completely. So it is with the passage before us. When the young Euryclea (vv. 401ff.) sets the infant Odysseus on his grandfather Autolycus' lap after the banquet, the aged Euryclea, who a few lines earlier had touched the wanderer's foot, has entirely vanished from the stage and from the reader's mind. [. . .]

The true cause of the impression of "retardation" appears to me to lie elsewhere—namely, in the need of the Homeric style to leave nothing which it mentions half in darkness and unexternalized. [. . .]

And this procession of phenomena takes place in the foreground—that is, in a local and temporal present which is absolute. One might think that the many interpolations, the frequent moving back and forth, would create a sort of perspective in time and place; but the Homeric style never gives any such impression. [. . .] But any such subjectivistic-perspectivistic procedure, creating a foreground and background, resulting in the present lying open to the depths of the past, is entirely foreign to the Homeric style; the Homeric style knows only a foreground, only a uniformly illuminated, uniformly objective present. And so the excursus does not begin until two lines later, when Euryclea has discovered the scar—the possibility for a perspectivistic connection no longer exists, and the story of the wound becomes an independent and exclusive present.

The genius of the Homeric style becomes even more apparent when it is compared with an equally ancient and equally epic style from a different world of forms. I shall attempt this comparison with the account of the sacrifice of Isaac, a homogeneous narrative produced by the so-called Elohist. The King James version translates the opening as follows (Genesis 22:1): "And it came to pass after these things, that God did tempt Abraham, and said to him, Abraham! and he said, Behold, here I am." Even this opening startles us when we come to it from Homer. Where are the two speakers? We are not told. The reader, however, knows that they are not normally to be found together in one place on earth, that one of them, God, in order to speak to Abraham, must come from somewhere, must enter the earthly realm from some unknown heights or depths. Whence does he come, whence does he call to Abraham? We are not told. He does not come, like Zeus or Poseidon, from the Aethiopians, where he has been enjoying a sacrificial feast. Nor are we told anything of his reasons for tempting Abraham so terribly. He has not, like Zeus, discussed them in set speeches with other gods gathered in council; nor have the deliberations in his own heart been presented to us; unexpected and mysterious, he enters the scene from some unknown height or depth and calls: Abraham! It will at once be said that this is to be explained by the particular concept of God which the Jews held and which was wholly different from that of the Greeks. True enough—but this constitutes no objection. For how is the Jewish concept of God to be explained? Even their earlier God of the desert was not fixed in form and content, and was alone; his lack of form, his lack of local habitation, his singleness, was in the end not only maintained but developed even further in competition with the comparatively far more manifest gods of the surrounding Near Eastern world. The concept of God held by the Jews is less a cause than a symptom of their manner of comprehending and representing things. [. . .]

A journey is made, because God has designated the place where the sacrifice is to be performed; but we are told nothing about the journey except that it took three days, and even that we are told in a mysterious way: Abraham and his followers rose "early in the morning" and "went unto" the place of which God had told him; on the third day he lifted up his eyes and saw the place from afar. [. . .]

Thus the journey is like a silent progress through the indeterminate and the contingent, a holding of the breath, a process which has no present, which is inserted, like a blank duration, between what has passed and what lies ahead, and which yet is measured: three days! Three such days positively demand the symbolic interpretation which they later received. [. . .]

We find the same contrast if we compare the two uses of direct discourse. The personages speak in the Bible story too; but their speech does not serve, as does speech in Homer, to manifest, to externalize thoughts—on the contrary, it serves to indicate thoughts which remain unexpressed. God gives his command in direct discourse, but he leaves his motives and his purpose unexpressed; Abraham, receiving the command, says nothing and does what he has been told to do. The conversation between Abraham and Isaac on the way to the place of sacrifice is only an interruption of the heavy silence and makes it all the more burdensome. The two of them, Isaac carrying the wood and Abraham with fire and a knife, "went together." Hesitantly, Isaac ventures to ask about the ram, and Abraham gives the well-known answer. Then the text repeats: "So they went both of them together." Everything remains unexpressed.

It would be difficult, then, to imagine styles more contrasted than those of these two equally ancient and equally epic texts. On the one hand, externalized, uniformly illuminated phenomena, at a definite time and in a definite place, connected together without lacunae in a perpetual foreground; thoughts and feeling completely expressed; events taking place in leisurely fashion and with very little of suspense. On the other hand, the externalization of only so much of the phenomena as is necessary for the purpose of the narrative, all else left in obscurity; the decisive points of the narrative alone are emphasized, what lies between is nonexistent; time and place are undefined and call for interpretation; thoughts and feeling remain unexpressed, are only suggested by the silence and the fragmentary speeches; the whole, permeated with the most unrelieved suspense and directed toward a single goal (and to that extent far more of a unity), remains mysterious and "fraught with background." [. . .]

The Homeric poems, then, though their intellectual, linguistic, and above all syntactical culture appears to be so much more highly developed, are yet comparatively simple in their picture of human beings; and no less so in their relation to the real life which they describe in general. Delight in physical existence is everything to them, and their highest aim is to make that delight perceptible to us. Between battles and passions, adventures and perils, they show us hunts, banquets, palaces and shepherds' cots, athletic contests and washing days—in order that we may see the heroes in their ordinary life, and seeing them so, may take pleasure in their manner of enjoying their savory present, a present which sends strong roots down into social usages, landscape, and daily life. And thus they bewitch us and ingratiate themselves to us until we live with them in the reality of their lives; so long as we are reading or hearing the poems, it does not matter whether we know that all this is only legend, "make-believe." The oft-repeated reproach that Homer is a liar takes nothing from his effectiveness, he does not need to base his story on historical reality, his reality is powerful enough in itself; it ensnares us, weaving its web around us, and that suffices him. And this "real" world into which we are lured, exists for itself, contains nothing but itself; the Homeric poems conceal nothing, they contain no teaching and no secret second meaning. Homer can be analyzed, as we have essayed to do here, but he cannot be interpreted. Later allegorizing trends have tried their arts of interpretation upon him, but to no avail. He resists any such treatment; the interpretations are forced and foreign, they do not crystallize into a unified doctrine. The general considerations which occasionally occur (in our episode, for example, v. 360: that in misfortune men age quickly) reveal a calm acceptance of the basic facts of human existence, but with no compulsion to brood over them, still less any passionate impulse either to rebel against them or to embrace them in an ecstasy of submission.

It is all very different in the Biblical stories. Their aim is not to bewitch the senses, and if nevertheless they produce lively sensory effects, it is only because the moral, religious, and psychological phenomena which are their sole concern are made concrete in the sensible matter of life. But their religious intent involves an absolute claim to historical truth. The story of Abraham and Isaac is not better established than the story of Odysseus, Penelope, and Euryclea; both are legendary. But the Biblical narrator, the Elohist, had to believe in the objective truth of the story of Abraham's sacrifice—the existence of the sacred ordinances of life rested upon the truth of this and similar stories. He had to believe in it passionately; or else (as many rationalistic interpreters believed and perhaps still believe) he had to he a conscious liar—no harmless liar like Homer, who lied to give pleasure, but a political liar with a definite end in view, lying in the interest of a claim to absolute authority. [. . .]

If the text of the Biblical narrative, then, is so greatly in need of interpretation on the basis of its own content, its claim to absolute authority forces it still further in the same direction. Far from seeking, like Homer, merely to make us forget our own reality for a few hours, it seeks to overcome our reality: we are to fit our own life into its world, feel ourselves to be elements in its structure of universal history. This becomes increasingly difficult the further our historical environment is removed from that of the Biblical books; and if these nevertheless maintain their claim to absolute authority, it is inevitable that they themselves be adapted through interpretative transformation. This was for a long time comparatively easy; as late as the European Middle Ages it was possible to represent Biblical events as ordinary phenomena of contemporary life, the methods of interpretation themselves forming the basis for such a treatment. But when, through too great a change in environment and through the awakening of a critical consciousness, this becomes impossible, the Biblical claim to absolute authority is jeopardized; the method of interpretation is scorned and rejected, the Biblical stories become ancient legends, and the doctrine they had contained, now dissevered from them, becomes a disembodied image.

As a result of this claim to absolute authority, the method of interpretation spread to traditions other than the Jewish. The Homeric poems present a definite complex of events whose boundaries in space and time are clearly delimited; before it, beside it, and after it, other complexes of events, which do not depend upon it, can be conceived without conflict and without difficulty. The Old Testament, on the other hand, presents universal history: it begins with the beginning of time, with the creation of the world, and will end with the Last Days, the fulfilling of the Covenant, with which the world will come to an end. Everything else that happens in the world can only be conceived as an element in this sequence; into it everything that is known about the world, or at least everything that touches upon the history of the Jews, must be fitted as an ingredient of the divine plan; and as this too became possible only by interpreting the new material as it poured in, the need for interpretation reaches out beyond the original Jewish–Israelitish realm of reality—for example to Assyrian, Babylonian, Persian, and Roman history; interpretation in a determined direction becomes a general method of comprehending reality; the new and strange world which now comes into view and which, in the form in which it presents itself, proves to be wholly unutilizable within the Jewish religious frame, must be so interpreted that it can find a place there. But this process nearly always also reacts upon the frame, which requires enlarging and modifying. The most striking piece of interpretation of this sort occurred in the first century of the Christian era, in consequence of Paul's mission to the Gentiles: Paul and the Church Fathers reinterpreted the entire Jewish tradition as a succession of figures prognosticating the appearance of Christ, and assigned the Roman Empire its proper place in the divine plan of salvation. Thus while, on the one hand, the reality of the Old Testament presents itself as complete truth with a claim to sole authority, on the other hand that very claim forces it to a constant interpretative change in its own content; for millennia it undergoes an incessant and active development with the life of man in Europe. [. . .]

We have compared these two texts, and, with them, the two kinds of style they embody, in order to reach a starting point for an investigation into the literary representation of reality in European culture. The two styles, in their opposition, represent basic types: on the one hand fully externalized description, uniform illumination, uninterrupted connection, free expression, all events in the foreground, displaying unmistakable meanings, few elements of historical development and of psychological perspective; on the other hand, certain parts brought into high relief, others left obscure, abruptness, suggestive influence of the unexpressed, "background" quality, multiplicity of meanings and the need for interpretation, universal-historical claims, development of the concept of the historically becoming, and preoccupation with the problematic.

Homer's realism is, of course, not to be equated with classical-antique realism in general; for the separation of styles, which did not develop until later, permitted no such leisurely and externalized description of everyday happenings; in tragedy especially there was no room for it; furthermore, Greek culture very soon encountered the phenomena of historical becoming and of the "multilayeredness" of the human problem, and dealt with them in its fashion; in Roman realism, finally, new and native concepts are added. We shall go into these later changes in the antique representation of reality when the occasion arises; on the whole, despite them, the basic tendencies of the Homeric style, which we have attempted to work out, remained effective and determinant down into late antiquity.

Since we are using the two styles, the Homeric and the Old Testament, as starting points, we have taken them as finished products, as they appear in the texts; we have disregarded everything that pertains to their origins, and thus have left untouched the question whether their peculiarities were theirs from the beginning or are to be referred wholly or in part to foreign influences. Within the limits of our purpose, a consideration of this question is not necessary; for it is in their full development, which they reached in early times, that the two styles exercised their determining influence upon the representation of reality in European literature.

Chapter 10

Leo Strauss (1899–1973)
from "Persecution and the Art of Writing," *Persecution and the Art of Writing* (1952)

I

All those whose thinking does not follow the rules of *logica equina*, in other words, all those capable of truly independent thinking, cannot be brought to accept the government-sponsored views. Persecution, then, cannot prevent independent thinking. It cannot prevent even the expression of independent thought. For it is as true today as it was more than two thousand years ago that it is a safe venture to tell the truth one knows to benevolent and trustworthy acquaintances, or more precisely, to reasonable friends.[1] Persecution cannot prevent even public expression of the heterodox truth, for a man of independent thought can utter his views in public and remain unharmed, provided he moves with circumspection. He can even utter them in print without incurring any danger, provided he is capable of writing between the lines.

The expression "writing between the lines" indicates the subject of this article. For the influence of persecution on literature is precisely that it compels all writers who hold heterodox views to develop a peculiar technique of writing, the technique which we have in mind when speaking of writing between the lines. This expression is clearly metaphoric. Any attempt to express its meaning in unmetaphoric language would lead to the discovery of a *terra incognita*, a field whose very dimensions are as yet unexplored and which offers ample scope for highly intriguing and even important investigations. One may say without fear of being presently convicted of grave exaggeration that almost the only preparatory work to guide the explorer in this field is buried in the writings of the rhetoricians of antiquity. [. . .]

Born in Germany, Leo Strauss taught at Cambridge, Columbia, and the New School for Social Research before joining the faculty at the University of Chicago, where he taught from 1947 to 1973. Best known as a political philosopher, Strauss argues in *Persecution and the Art of Writing* that authors write "between the lines" to convey their political views in conditions of persecution.

Persecution, then, gives rise to a peculiar technique of writing, and therewith to a peculiar type of literature, in which the truth about all crucial things is presented exclusively between the lines. That literature is addressed, not to all readers, but to trustworthy and intelligent readers only. It has all the advantages of private communication without having its greatest disadvantage—that it reaches only the writer's acquaintances. It has all the advantages of public communication without having its greatest disadvantage—capital punishment for the author. But how can a man perform the miracle of speaking in a publication to a minority, while being silent to the majority of his readers? The fact which makes this literature possible can be expressed in the axiom that thoughtless men are careless readers, and only thoughtful men are careful readers. Therefore an author who wishes to address only thoughtful men has but to write in such a way that only a very careful reader can detect the meaning of his book. But, it will be objected, there may be clever men, careful readers, who are not trustworthy, and who, after having found the author out, would denounce him to the authorities. As a matter of fact, this literature would be impossible if the Socratic dictum that virtue is knowledge, and therefore that thoughtful men as such are trustworthy and not cruel, were entirely wrong.

Another axiom, but one which is meaningful only so long as persecution remains within the bounds of legal procedure, is that a careful writer of normal intelligence is more intelligent than the most intelligent censor, as such. For the burden of proof rests with the censor. It is he, or the public prosecutor, who must prove that the author holds or has uttered heterodox views. In order to do so he must show that certain literary deficiencies of the work are not due to chance, but that the author used a given ambiguous expression deliberately, or that he constructed a certain sentence badly on purpose. That is to say, the censor must prove not only that the author is intelligent and a good writer in general, for a man who intentionally blunders in writing must possess the art of writing, but above all that he was on the usual level of his abilities when writing the incriminating words. But how can that be proved, if even Homer nods from time to time?

II

Suppression of independent thought has occurred fairly frequently in the past. It is reasonable to assume that earlier ages produced proportionately as many men capable of independent thought as we find today, and that at least some of these men combined understanding with caution. Thus, one may wonder whether some of the greatest writers of the past have not adapted their literary technique to the requirements of persecution, by presenting their views on all the then crucial questions exclusively between the lines.

We are prevented from considering this possibility, and still more from considering the questions connected with it, by some habits produced by, or related to, a comparatively recent progress in historical research. This progress was due, at first glance, to the general acceptance and occasional application of the following principles. Each period of the past, it was demanded, must be understood by itself, and must not be judged by standards alien to it. Each author must, as far as possible, be interpreted by himself; no term of any consequence must be used in the interpretation of an author which cannot be literally translated into his language, and which was not used by him or was not in fairly common use in his time. The only presentations of an author's views which can be accepted as true are those ultimately borne out by his own explicit statements. The last of these principles is decisive: it seems to exclude a priori from the sphere of human knowledge such views of earlier writers as are indicated exclusively between the lines. For if an author does not tire of asserting explicitly on every page of his book that *a* is *b*, but indicates between the lines that *a* is not *b*, the modern historian will still demand explicit evidence showing that the author believed *a* not to be *b*. Such evidence cannot possibly be forthcoming, and the modern historian wins his argument: he can dismiss any reading between the lines as arbitrary guesswork, or, if he is lazy, he will accept it as intuitive knowledge. [. . .]

The typical difference between older views and more recent views is due not entirely to progress in historical exactness, but also to a more basic change in the intellectual climate. During the last few decades the rationalist tradition, which was the common denominator of the older views, and which was still rather influential in nineteenth-century positivism, has been either still further transformed or altogether rejected by an ever-increasing number of people. Whether and to what extent this change is to be considered a progress or a decline is a question which only the philosopher can answer.

A more modest duty is imposed on the historian. He will merely, and rightly, demand that in spite of all changes which have occurred or which will occur in the intellectual climate, the tradition of historical exactness shall be continued. Accordingly, he will not accept an arbitrary standard of exactness which might exclude a priori the most important facts of the past from human knowledge, but will adapt the rules of certainty which guide his research to the nature of his subject. He will then follow such rules as these: Reading between the lines is strictly prohibited in all cases where it would be less exact than not doing so. Only such reading between the lines as starts from an exact consideration of the explicit statements of the author is legitimate. The context in which a statement occurs, and the literary character of the whole work as well as its plan, must be perfectly understood before an interpretation of the statement can reasonably claim to be adequate or even correct. One is not entitled to delete a passage, nor to emend its text, before one has fully considered all reasonable possibilities of understanding the passage as it stands—one of these possibilities being that the passage may be ironic. If a master of the art of writing commits such blunders as would shame an intelligent high school boy, it is reasonable to assume that they are intentional, especially if the author discusses, however incidentally, the possibility of intentional blunders in writing. The views of the author of a drama or dialogue must not, without previous proof, be identified with the views expressed by one or more of his characters, or with those agreed upon by all his characters or by his attractive characters. The real opinion of an author is not necessarily identical with that which he expresses in the largest number of passages. In short, exactness is not to be confused with refusal, or inability, to see the wood for the trees. The truly exact historian will reconcile himself to the fact that there is a difference between winning an argument, or proving to practically everyone that he is right, and understanding the thought of the great writers of the past.

It must, then, be considered possible that reading between the lines will not lead to complete agreement among all scholars. If this is an objection to reading between the lines as such, there is the counter-objection that neither have the methods generally used at present led to universal or even wide agreement in regard to very important points. Scholars of the last century were inclined to solve literary problems by having recourse to the genesis of the author's work, or even of his thought. Contradictions or divergences within one book, or between two books by the same author, were supposed to prove that his thought had changed. If the contradictions exceeded a certain limit it was sometimes decided without any external evidence that one of the works must be spurious. That procedure has lately come into some disrepute, and at present many scholars are inclined to be rather more conservative about the literary tradition, and less impressed by merely internal evidence. The conflict between the traditionalists and the higher critics is, however, far from being settled. The traditionalists could show in important cases that the higher critics have not proved their hypotheses at all; but even if all the answers suggested by the higher critics should ultimately prove to be wrong, the questions which led them away from the tradition and tempted them to try a new approach often show an awareness of difficulties which do not disturb the slumber of the typical traditionalist. An adequate answer to the most serious of these questions requires methodical reflection on the literary technique of the great writers of earlier ages, because of the typical character of the literary problems involved obscurity of the plan, contradictions within one work or between two or more works of the same author, omission of important links of the argument, and so on. Such reflection necessarily transcends the boundaries of modern aesthetics and even of traditional poetics, and will, I believe, compel students sooner or later to take into account the phenomenon of persecution. To mention something which is hardly more than another aspect of the same fact, we sometimes observe a conflict between a traditional, superficial and doxographic interpretation of some great writer of the past, and a more intelligent, deeper and monographic interpretation. They are equally exact, so far as both are borne out by explicit statements of the writer concerned. Only a few people at present, however, consider the possibility that the traditional interpretation may reflect the exoteric teaching of the author, whereas the monographic interpretation stops halfway between the exoteric and esoteric teaching of the author.

Modern historical research, which emerged at a time when persecution was a matter of feeble recollection rather than of forceful experience, has counteracted or even destroyed an earlier tendency to read between the lines of the great writers, or to attach more weight to their fundamental design than to those views which they have repeated most often. Any attempt to restore the earlier approach in this age of historicism is confronted by the problem of criteria for distinguishing between legitimate and illegitimate reading between the lines. If it is true that there is a necessary correlation between persecution and writing between the lines, then there is a necessary negative criterion: that the book in question must have been composed in an era of persecution, that is, at a time when some political or other orthodoxy was enforced by law or custom. One positive criterion is this: if an able writer who has a clear mind and a perfect knowledge of the orthodox view and all its ramifications, contradicts surreptitiously and as it were in passing one of its necessary presuppositions or consequences which he explicitly recognizes and maintains everywhere else, we can reasonably suspect that he was opposed to the orthodox system as such and—we must study his whole book all over again, with much greater care and much less naïveté than ever before. In some cases, we possess even explicit evidence proving that the author has indicated his views on the most important subjects only between the lines. Such statements, however, do not usually occur in the preface or other very conspicuous place. Some of them cannot even be noticed, let alone understood, so long as we confine ourselves to the view of persecution and the attitude toward freedom of speech and candor which have become prevalent during the last three hundred years. [. . .]

III

What attitude people adopt toward freedom of public discussion, depends decisively on what they think about popular education and its limits. Generally speaking, premodern philosophers were more timid in this respect than modern philosophers. After about the middle of the seventeenth century an ever increasing number of heterodox philosophers who had suffered from persecution published their books not only to communicate their thoughts but also because they desired to contribute to the abolition of persecution as such. They believed that suppression of free inquiry, and of publication of the results of free inquiry, was accidental, an outcome of the faulty construction of the body politic, and that the kingdom of general darkness could be replaced by the republic of universal light. They looked forward to a time when, as a result of the progress of popular education, practically complete freedom of speech would be possible, or—to exaggerate for purposes of clarification—to a time when no one would suffer any harm from hearing any truth.[2] They concealed their views only far enough to protect themselves as well as possible from persecution; had they been more subtle than that, they would have defeated their purpose, which was to enlighten an ever-increasing number of people who were not potential philosophers. It is therefore comparatively easy to read between the lines of their books.[3] The attitude of an earlier type of writers was fundamentally different. They believed that the gulf separating "the wise" and "the vulgar" was a basic fact of human nature which could not be influenced by any progress of popular education: philosophy, or science, was essentially a privilege of "the few." They were convinced that philosophy as such was suspect to, and hated by, the majority of men.[4] Even if they had had nothing to fear from any particular political quarter, those who started from that assumption would have been driven to the conclusion that public communication of the philosophic or scientific truth was impossible or undesirable, not only for the time being but for all times. They must conceal their opinions from all but philosophers, either by limiting themselves to oral instruction of a carefully selected group of pupils, or by writing about the most important subject by means of "brief indication."[5]

Writings are naturally accessible to all who can read. Therefore a philosopher who chose the second way could expound only such opinions as were suitable for the nonphilosophic majority: all of his writings would have to be, strictly speaking, exoteric.

These opinions would not be in all respects consonant with truth. Being a philosopher, that is, hating "the lie in the soul" more than anything else, he would not deceive himself about the fact that such opinions are merely "likely tales," or "noble lies," or "probable opinions," and would leave it to his philosophic readers to disentangle the truth from its poetic or dialectic presentation. But he would defeat his purpose if he indicated clearly which of his statements expressed a noble lie, and which the still more noble truth. For philosophic readers he would do almost more than enough by drawing their attention to the fact that he did not object to telling lies which were noble, or tales which were merely similar to truth. From the point of view of the literary historian at least, there is no more noteworthy difference between the typical premodern philosopher (who is hard to distinguish from the premodern poet) and the typical modern philosopher than that of their attitudes toward "noble (or just) lies," "pious frauds," the "ductus obliquus"[6] or "economy of the truth." Every decent modern reader is bound to be shocked by the mere suggestion that a great man might have deliberately deceived the large majority of his readers.[7] And yet, as a liberal theologian once remarked, these imitators of the resourceful Odysseus were perhaps merely more sincere than we when they called "lying nobly" what we would call "considering one's social responsibilities."

An exoteric book contains then two teachings: a popular teaching of an edifying character, which is in the foreground; and a philosophic teaching concerning the most important subject, which is indicated only between the lines. This is not to deny that some great writers might have stated certain important truths quite openly by using as mouthpiece some disreputable character: they would thus show how much they disapproved of pronouncing the truths in question. There would then be good reason for our finding in the greatest literature of the past so many interesting devils, madmen, beggars, sophists, drunkards, epicureans and buffoons. Those to whom such books are truly addressed are, however, neither the unphilosophic majority nor the perfect philosopher as such, but the young men who might become philosophers: the potential philosophers are to be led step by step from the popular views which are indispensable for all practical and political purposes to the truth which is merely and purely theoretical, guided by certain obtrusively enigmatic features in the presentation of the popular teaching—obscurity of the plan, contradictions, pseudonyms, inexact repetitions of earlier statements, strange expressions, etc. Such features do not disturb the slumber of those who cannot see the wood for the trees, but act as awakening stumbling blocks for those who can. All books of that kind owe their existence to the love of the mature philosopher for the puppies[8] of his race, by whom he wants to be loved in turn: all exoteric books are "written speeches caused by love."

Exoteric literature presupposes that there are basic truths which would not be pronounced in public by any decent man, because they would do harm to many people who, having been hurt, would naturally be inclined to hunt in turn him who pronounces the unpleasant truths. It presupposes, in other words, that freedom of inquiry, and of publication of all results of inquiry, is not guaranteed as a basic right. This literature is then essentially related to a society which is not liberal. Thus one may very well raise the question of what use it could be in a truly liberal society.

The answer is simple. In Plato's *Banquet*, Alcibiades—that outspoken son of outspoken Athens—compares Socrates and his speeches to certain sculptures which are very ugly from the outside, but within have most beautiful images of things divine. The works of the great writers of the past are very beautiful even from without. And yet their visible beauty is sheer ugliness, compared with the beauty of those hidden treasures which disclose themselves only after very long, never easy, but always pleasant work. This always difficult but always pleasant work is, I believe, what the philosophers had in mind when they recommended education. Education, they felt, is the only answer to the always pressing question, to the political question par excellence, of how to reconcile order which is not oppression with freedom which is not license.

Notes

1. Plato, *Republic*, 450 d3–c1.
2. The question whether that extreme goal is attainable in any but the most halcyon conditions has been raised in our time by Archibald MacLeish, "Post-War Writers and Pre-War Readers," *Journal of Adult Education*, 12 (June 1940) in the following terms: "Perhaps the luxury of the complete confession, the uttermost despair, the farthest doubt should be denied themselves by writers living in any but the most orderly and settled times. I do not know."
3. I am thinking of Hobbes in particular, whose significance for the development outlined above can hardly be overestimated. This was clearly recognized by Tönnies, who emphasized especially these two sayings of his hero: "Paulatim eruditur vulgus" and "Philosophia ut crescat libera esse debet nec metu nec pudore coercenda" (Tönnies, Thomas Hobbes, 3rd ed. (Stuttgart, 1925), pp. iv, p. 195). Hobbes also says: "Suppression of doctrines does but unite and exasperate, that is, increase both the malice and power of them that have already believed them" (*English Works*, Molesworth edition, VI, p. 242). In his *Of Liberty and Necessity*, (1654, 35 fl. London) he writes to the Marquess of Newcastle: "I must confess, if we consider the greatest part of Mankinde, not as they should be, but as they are . . . I must, I say, confess that the dispute of this question will rather hurt than help their piety, and therefore if his Lordship [Bishop Bramhall] had not desired this answer, I should not have written it, nor do I write it but in hopes your Lordship and his, will keep it private."
4. Cicero, *Tusculanae Disputationes*, II, p. 4; Plato, *Phaedo*, p. 64 b; *Republic*, p. 520 b2–3 and p. 494 a4–10.
5. Plato, *Timaeus*, p. 28 c3–5, and *Seventh Letter*, p. 332 d6–7, p. 341 c3–4, and p. 344 d4–e2. That the view mentioned above is reconcilable with the democratic creed is shown most clearly by Spinoza, who was a champion not only of liberalism but also of democracy (*Tractatus politicus*, XI, 2, Bruder edition). See his *Tractatus de intellectus emendatione*, pp. 14 and 17, as well as *Tractatus theologico-politicus*, V, pp. 35–39, XIV, p. 20 and XV end.
6. Sir Thomas More, *Utopia*, latter part of first book.
7. A rather extensive discussion of the "magna quaestio, latebrosa tractatio, disputatio inter doctos alternans," as Augustinus called it, is to be found in Grotius' *De Jure Belli ac Pacis*, III, chap. I, §7 ff., and in particular §17, 3. See also *inter alia* Pascal's ninth and tenth *Provinciales* and Jeremy Taylor, *Ductor Dubitantium*, Book III, chap. 2, rule 5.
8. Compare Plato, *Republic*, p. 539 a5–d1, with *Apology of Socrates*, p. 23 c2–8.

Chapter 11

Frantz Fanon (1925–1961) from "On National Culture," *Wretched of the Earth* (1961)

Inside the political parties, and most often in offshoots from these parties, cultured individuals of the colonised race make their appearance. For these individuals, the demand for a national culture and the affirmation of the existence of such a culture represent a special battlefield. While the politicians situate their action in actual present-day events, men of culture take their stand in the field of history. Confronted with the native intellectual who decides to make an aggressive response to the colonialist theory of pre-colonial barbarism, colonialism will react only slightly, and still less because the ideas developed by the young colonised intelligentsia are widely professed by specialists in the mother country. It is in fact a commonplace to state that for several decades large numbers of research workers have, in the main, rehabilitated the African, Mexican and Peruvian civilisations. The passion with which native intellectuals defend the existence of their national culture may be a source of amazement; but those who condemn this exaggerated passion are strangely apt to forget that their own psyche and their own selves are conveniently sheltered behind a French or German culture which has given full proof of its existence and which is uncontested.

I am ready to concede that on the plane of factual being the past existence of an Aztec civilisation does not change anything very much in the diet of the Mexican peasant of today. I admit that all the proofs of a wonderful Songhai civilization will not change the fact that today the Songhais are under-fed and illiterate, thrown between sky and water with empty heads and empty eyes. But it has been remarked several times that this passionate search for a national culture which existed before the colonial era finds its legitimate reason in the anxiety shared by native intellectuals to shrink away from that Western culture in which they all risk being swamped. Because they realise they are in danger of losing their lives and thus becoming lost to their people, these men, hot-headed and with anger in their hearts, relentlessly determine

Born in Martinique, trained as a psychiatrist in France, and working as a doctor in North Africa, Frantz Fanon delivered a powerful critique of colonialism. Central to that critique, as can be seen in this essay, is Fanon's insistence on recovering national cultures as part of a resistance to the colonial power. In other words, Fanon here rejects universalizing claims made for culture.

to renew contact once more with the oldest and most pre-colonial springs of life of their people.

Let us go further. Perhaps this passionate research and this anger are kept up or at least directed by the secret hope of discovering beyond the misery of today, beyond self-contempt, resignation and abjuration, some very beautiful and splendid era whose existence rehabilitates us both in regard to ourselves and in regard to others. I have said that I have decided to go further. Perhaps unconsciously, the native intellectuals, since they could not stand wonder-struck before the history of today's barbarity, decided to go back further and to delve deeper down; and, let us make no mistake, it was with the greatest delight that they discovered that there was nothing to be ashamed of in the past, but rather dignity, glory and solemnity. The claim to a national culture in the past does not only rehabilitate that nation and serve as a justification for the hope of a future national culture. In the sphere of psycho-affective equilibrium it is responsible for an important change in the native. Perhaps we have not sufficiently demonstrated that colonialism is not simply content to impose its rule upon the present and the future of a dominated country. Colonialism is not satisfied merely with holding a people in its grip and emptying the native's brain of all form and content. By a kind of perverted logic, it turns to the past of the oppressed people, and distorts, disfigures and destroys it. This work of devaluing pre-colonial history takes on a dialectical significance today.

When we consider the efforts made to carry out the cultural estrangement so characteristic of the colonial epoch, we realise that nothing has been left to chance and that the total result looked for by colonial domination was indeed to convince the natives that colonialism came to lighten their darkness. The effect consciously sought by colonialism was to drive into the natives' heads the idea that if the settlers were to leave, they would at once fall back into barbarism, degradation and bestiality.

On the unconscious plane, colonialism therefore did not seek to be considered by the native as a gentle, loving mother who protects her child from a hostile environment, but rather as a mother who unceasingly restrains her fundamentally perverse offspring from managing to commit suicide and from giving free rein to its evil instincts. The colonial mother protects her child from itself, from its ego, and from its physiology, its biology and its own unhappiness which is its very essence.

In such a situation the claims of the native intellectual are no luxury but a necessity in any coherent programme. The native intellectual who takes up arms to defend his nation's legitimacy and who wants to bring proofs to bear out that legitimacy, who is willing to strip himself naked to study the history of his body, is obliged to dissect the heart of his people.

Such an examination is not specifically national. The native intellectual who decides to give battle to colonial lies fights on the field of the whole continent. The past is given back its value. Culture, extracted from the past to be displayed in all its splendour, is not necessarily that of his own country. Colonialism, which has not bothered to put too fine a point on its efforts has never ceased to maintain that the Negro is a savage; and for the colonist, the Negro was neither an Angolan nor a Nigerian, for he simply spoke of "the Negro." For colonialism, this vast continent was the haunt of savages, a country riddled with superstitions and fanaticism, destined for contempt, weighed down by the curse of God, a country of cannibals—in short, the Negro's country. Colonialism's condemnation is continental in its scope. The contention by colonialism that the darkest night of humanity lay over pre-colonial history concerns the whole of the African continent. The efforts of the native to rehabilitate himself and to escape from the claws of colonialism are logically inscribed from the same point of view as that of colonialism. The native intellectual who has gone far beyond the domains of Western culture and who has got it into his head to proclaim the existence of another culture never does so in the name of Angola or of Dahomey. The culture which is affirmed is African culture. The Negro, never so much a Negro as since he has been dominated by the whites, when he decides to prove that he has a culture and to behave like a cultured person, comes to realise that history points out a well-defined path to him: he must demonstrate that a Negro culture exists.

And it is only too true that those who are most responsible for this racialisation of thought, or at least for the first movement towards that thought, are and remain those Europeans who have never ceased to set up white culture to fill the gap left by the absence of other cultures. Colonialism did not dream of wasting its time in denying the existence of one national culture after another. Therefore the reply of the colonised peoples will be straight away continental in its breadth. In Africa, the native literature of the last twenty years is not a national literature but a Negro literature. The concept of Negro-ism, for example, was the emotional if not the logical antithesis of that insult which the white man flung at humanity. This rush of Negro-ism against the white man's contempt showed itself in certain spheres to be the one idea capable of lifting interdictions and anathemas. Because the New Guinean or Kenyan intellectuals found themselves above all up against a general ostracism and delivered to the combined contempt of their overlords, their reaction was to sing praises in admiration of each other. The unconditional affirmation of African culture has succeeded the unconditional affirmation of European culture. On the whole, the poets of Negro-ism oppose the idea of an old Europe to a young Africa, tiresome reasoning lyricism, oppressive logic to a high-stepping nature, and on one side stiffness, ceremony, etiquette and scepticism, while on the other frankness, liveliness, liberty and—why not?—luxuriance: but also irresponsibility.

The poets of Negro-ism will not stop at the limits of the continent. From America, black voices will take up the hymn with fuller unison. The "black world" see the light and Busia from Ghana, Birago Diop from Senegal, Hampaté Ba from the Soudan and Saint-Clair Drake from Chicago will not hesitate to assert the existence of common ties and a motive power that is identical. [. . .]

This historical necessity in which the men of African culture find themselves to racialise their claims and to speak more of African culture than of national culture will tend to lead them up a blind alley. Let us take for example the case of the African Cultural Society. This society had been created by African intellectuals who wished to get to know each other and to compare their experiences and the results of their respective research work. The aim of this society was therefore to affirm the existence of an African culture, to evaluate this culture on the plane of distinct nations and to reveal the internal motive forces of each of their national cultures. But at the same time this society fulfilled another need: the need to exist side by side with the European Cultural Society, which threatened to transform itself into a Universal Cultural Society. There was therefore at the bottom of this decision the anxiety to be present at the universal trysting place fully armed, with a culture springing from the very heart of the African continent. Now, this Society will very quickly show its inability to shoulder these different tasks, and will limit itself to exhibitionist demonstrations, while the habitual behaviour of the members of this Society will be confined to showing Europeans that such a thing as African culture exists, and opposing their ideas to those of ostentatious and narcissistic Europeans. We have shown that such an attitude is normal and draws its legitimacy from the lies propagated by men of Western culture. But the degradation of the aims of this Society will become more marked with the elaboration of the concept of Negro-ism. The African Society will become the cultural society of the black world and will come to include the Negro dispersion, that is to say the tens of thousands of black people spread over the American continents.

The Negroes who live in the United States and in Central or Latin America in fact experience the need to attach themselves to a cultural matrix. Their problem is not fundamentally different from that of the Africans. The whites of America did not mete out to them any different treatment from that of the whites that ruled over the Africans. We have seen that the whites were used to putting all Negroes in the same bag. During the first congress of the African Cultural Society which was held in Paris in 1956, the American Negroes of their own accord considered their problems from the same standpoint as those of their African brothers. Cultured Africans, speaking of African civilisations, decreed that there should be a reasonable status within the state for those who had formerly been slaves. But little by little the American Negroes realised that the essential problems confronting them were not the same as those that confronted the African Negroes. The Negroes of Chicago only resemble the Nigerians or the Tanganykans in so far as they were all defined in relation to the whites. But once the first comparisons had been made and subjective feelings were assuaged, the American Negroes realised that the objective problems were fundamentally heterogeneous. The test cases of civil liberty whereby both whites and blacks in America try to drive back racial discrimination have very little in common in their principles and objectives with the heroic fight of the Angolan people against the detestable Portuguese colonialism. Thus, during the second congress of the African Cultural Society the American Negroes decided to create an American society for people of black cultures.

Negro-ism therefore finds its first limitation in the phenomena which take account of the formation of the historical character of men. Negro and African-Negro culture broke up into different entities because the men who wished to incarnate these cultures realised that every culture is first and foremost national, and that the problems which kept Richard Wright or Langston Hughes on the alert were fundamentally different from those which might confront Leopold Senghor or Jomo Kenyatta. In the same way certain Arab states, though they had chanted the marvellous hymn of Arab renaissance, had nevertheless to realise that their geographical position and the economic ties of their region were stronger even than the past that they wished to revive. Thus we find today the Arab states organically linked once more with societies which are Mediterranean in their culture. The fact is that these states are submitted to modern pressure and to new channels of trade while the network of trade relations which was dominant during the great period of Arab history has disappeared. But above all there is the fact that the political regimes of certain Arab states are so different, and so far away from each other in their conceptions that even a cultural meeting between these states is meaningless. [. . .]

If the action of the native intellectual is limited historically, there remains nevertheless the fact that it contributes greatly to upholding and justifying the action of politicians. It is true that the attitude of the native intellectual sometimes takes on the aspect of a cult or of a religion. But if we really wish to analyse this attitude correctly we will come to see that it is symptomatic of the intellectual's realisation of the danger that he is running in cutting his last moorings and of breaking adrift from his people. This stated belief in a national culture is in fact an ardent, despairing turning towards anything that will afford him secure anchorage. In order to ensure his salvation and to escape from the supremacy of the white man's culture the native feels the need to turn backwards towards his unknown roots and to lose himself at whatever cost in his own barbarous people. Because he feels he is becoming estranged, that is to say because he feels that he is the living haunt of contradictions which run the risk of becoming insurmountable, the native tears himself away from the swamp that may suck him down and accepts everything, decides to take all for granted and confirms everything even though he may lose body and soul. The native finds that he is expected to answer for everything, and to all comers. He not only turns himself into the defender of his people's past; he is willing to be counted as one of them, and henceforward he is even capable of laughing at his past cowardice.

This tearing away, painful and difficult though it may be, is however necessary. If it is not accomplished there will be serious psycho-affective injuries and the result will be individuals without an anchor, without a horizon, colourless, stateless, rootless— a race of angels. It will be also quite normal to hear certain natives declare "I speak as a Senegalese and as a Frenchman . . ." "I speak as an Algerian and as a Frenchman . . ." The intellectual who is Arab and French, or Nigerian and English, when he comes up against the need to take on two nationalities, chooses, if he wants to remain true to himself, the negation of one of these determinations. But most often, since they cannot or will not make a choice, such intellectuals gather together all the historical determining factors which have conditioned them and take up a fundamentally "universal standpoint."

This is because the native intellectual has thrown himself greedily upon western culture. Like adopted children who only stop investigating the new family framework at the moment when a minimum nucleus of security crystallises in their psyche, the native intellectual will try to make European culture his own. He will not be content to get to know Rabelais and Diderot, Shakespeare and Edgar Allen Poe; he will bind them to his intelligence as closely as possible:

La dame n' etait pas seule
Elle avait un mari
Un mari très comme il faut
Qui citait Racine et Corneille
Et Voltaire et Rousseau
Et le Père Hugo et le jeune Musset
Et Gide et Valéry
*Et tant d'autres encore.**

---

* The lady was not alone; she had a most respectable husband, who knew how to quote Racine and Corneille, Voltaire and Rousseau, Victor Hugo and Musset, Gide, Valéry and as many more again. (Rene Depestre: Face à la nuit.)

But at the moment when the nationalist parties are mobilising the people in the name of national independence, the native intellectual sometimes spurns these acquisitions which he suddenly feels make him a stranger in his own land. It is always easier to proclaim rejection than actually to reject. The intellectual who through the medium of culture has filtered into Western civilisation, who has managed to become part of the body of European culture—in other words who has exchanged his own culture for another—will come to realise that the cultural matrix, which now he wishes to assume since he is anxious to appear original, can hardly supply any figure-heads which will bear comparison with those, so many in number and so great in prestige, of the occupying power's civilisation. History, of course, though nevertheless written by the Westerners and to serve their purposes, will be able to evaluate from time to time certain periods of the African past. But, standing face to face with his country at the present time, and observing dearly and objectively the events of today throughout the continent which he wants to make his own, the intellectual is terrified by the void, the degradation and the savagery he sees there. Now he feels that he must get away from white culture. He must seek his culture elsewhere, anywhere at all; and if he fails to find the substance of culture of the same grandeur and scope as displayed by the ruling power, the native intellectual will very often fall back upon emotional attitudes and will develop a psychology which is dominated by exceptional sensitivity and susceptibility. This withdrawal which is due in the first instance to a begging of the question in his internal behaviour mechanism and his own character brings out, above all, a reflex and contradiction which is muscular.

Chapter 12

Theodor Adorno (1903–1969)
from "Commitment: The Politics of Autonomous Art," *New Left Review* (1962)

The type of literature that, in accordance with the tenets of commitment but also with the demands of philistine moralism, exists for man, betrays him by traducing that which alone could help him, if it did not strike a pose of helping him. But any literature which therefore concludes that it can be a law unto itself, and exist only for itself, degenerates into ideology no less. Art, which even in its opposition to society remains a part of it, must close its eyes and ears against it: it cannot escape the shadow of irrationality. But when it appeals to this unreason, making it a *raison d'être*, it converts its own malediction into a theodicy. Even in the most sublimated work of art there is a hidden "it should be otherwise." When a work is merely itself and no other thing, as in a pure pseudo-scientific construction, it becomes bad art—literally pre-artistic. The moment of true volition, however, is mediated through nothing other than the form of the work itself, whose crystallization becomes an analogy of that other condition which should be. As eminently constructed and produced objects, works of art, even literary ones, point to a practice from which they abstain: the creation of a just life. This mediation is not a compromise between commitment and autonomy, nor a sort of mixture of advanced formal elements with an intellectual content inspired by genuinely or supposedly progressive politics. The content of works of art is never the amount of intellect pumped into them: if anything it is the opposite. Nevertheless, an emphasis on autonomous works is itself socio-political in nature. The feigning of a true politics here and now, the freezing of historical relations which nowhere seem ready to melt, oblige the mind to go where it need not degrade itself. Today every phenomenon of culture, even if a model of integrity, is liable to be suffocated in the cultivation of kitsch. Yet paradoxically in the same epoch it is to works of art that has fallen the burden of wordlessly asserting what is barred to politics. Sartre himself has expressed this truth in a passage which does credit to his honesty.[1]

Born in Germany, Adorno emigrated to England and then to the United States in 1934, where he taught at the Institute for Social Research, Princeton and UCLA. In this selection, Adorno argues that "even in the most sublimated work of art there is a hidden 'it should be otherwise,' " and thus indicates a complicated relationship between art works and political arrangements.

This is not a time for political art, but politics has migrated into autonomous art, and nowhere more so than where it seems to be politically dead. An example is Kafka's allegory of toy-guns, in which an idea of non-violence is fused with a dawning aware-ness of the approaching paralysis of politics. Paul Klee too belongs to any debate about committed and autonomous art: for his work, *écriture par excellence*, had roots in literature and would not have been what it was without them—or if it had not consumed them. During the First World War or shortly after, Klee drew cartoons of Kaiser Wilhelm as an inhuman iron-eater. Later, in 1920, these became—the devel-opment can be shown quite clearly—the *Angelus Novus*, the machine angel, who, though he no longer bears any emblem of caricature or commitment, flies far beyond both. The machine angel's enigmatic eyes force the onlooker to try to decide whether he is announcing the culmination of disaster or salvation hidden within it. But, as Walter Benjamin, who owned the drawing, said, he is the angel who does not give but takes.

Note

1. See Jean-Paul Sartre, *L'Existentialisme est un Humanisme* (Paris: 1946), p. 105.

Chapter 13

Chinua Achebe (1930–)
"Colonialist Criticism," *Hopes and Impediments* (1974, 1988)

The word "colonialist" may be deemed inappropriate for two reasons. First, it has come to be associated in many minds with that brand of cheap, demagogic and out-moded rhetoric which the distinguished Ghanaian public servant Robert Gardiner no doubt has in mind when he speaks of our tendency to "intone the colonial litany," implying that the time has come when we must assume responsibility for our problems and our situation in the world and resist the temptation to blame other people. Secondly, it may be said that whatever colonialism may have done in the past, the very fact of a Commonwealth Conference today is sufficient repudiation of it, is indeed a symbol of a new relationship of equality between peoples who were once masters and servants.

Yet in spite of the strength of these arguments one feels the necessity to deal with some basic issues raised by a certain specious criticism which flourishes in African literature today and which derives from the same basic attitude and assumption as colonialism itself and so merits the name "colonialist." This attitude and assumption was crystallized in Albert Schweitzer's immortal dictum in the heyday of colonialism: "The African is indeed my brother, but my junior brother." The latter-day colonial-ist critic, equally given to big-brother arrogance, sees the African writer as a somewhat unfinished European who with patient guidance will grow up one day and write like every other European, but meanwhile must be humble, must learn all he can and while at it give due credit to his teachers in the form of either direct praise or, even better since praise sometimes goes bad and becomes embarrassing, manifest self-contempt. Because of the tricky nature of this subject, I have chosen to speak not in general terms but wherever possible specifically about my own actual experience. In any case, as anyone who has heard anything at all about me may know already, I do have problems with universality and other concepts of that scope, being very

---

Chinua Achebe was born in Nigeria, and educated there and in England, where he also worked for the BBC. In this essay, Achebe responds to European criticism of African literature, indicating the limits of claims for literature's universality.

much a down-to-earth person. But I will hope by reference to a few other writers and critics to show that my concerns and anxieties are perhaps not entirely personal.

When my first novel was published in 1958, a very unusual review of it was written by a British woman, Honor Tracy, who is perhaps not so much a critic as a literary journalist. But what she said was so intriguing that I have never forgotten it. If I remember rightly, she headlined it "Three cheers for mere Anarchy!" The burden of the review itself was as follows: These bright Negro barristers (how barristers came into it remains a mystery to me to this day, but I have sometimes woven fantasies about an earnest white woman and an unscrupulous black barrister) who talk so glibly about African culture, how would they like to return to wearing raffia skirts? How would novelist Achebe like to go back to the mindless times of his grandfather instead of holding the modern job he has in broadcasting in Lagos?

I should perhaps point out that colonialist criticism is not always as crude as this, but the exaggerated grossness of a particular example may sometimes prove useful in studying the anatomy of the species. There are three principal parts here: Africa's inglorious past (raffia skirts) to which Europe brings the blessing of civilization (Achebe's modern job in Lagos) and for which Africa returns ingratitude (sceptical novels like *Things Fall Apart*).

Before I go on to more advanced varieties I must give one more example of the same kind as Honor Tracy's, which on account of its recentness (1970) actually surprised me:

> The British administration not only safeguarded women from the worst tyrannies of their masters, it also enabled them to make their long journeys to farm or market without armed guard, secure from the menace of hostile neighbours . . . The Nigerian novelists who have written the charming and bucolic accounts of domestic harmony in African rural communities, are the sons whom the labours of these women educated; the peaceful village of their childhood to which they nostalgically look back was one which had been purged of bloodshed and alcoholism by an ague-ridden district officer and a Scottish mission lassie whose years were cut short by every kind of intestinal parasite. It is even true to say that one of the most nostalgically convincing of the rural African novelists used as his sourcebook not the memories of his grandfathers but the records of the despised British anthropologists . . . The modern African myth-maker hands down a vision of colonial rule in which the native powers are chivalrously viewed through the eyes of the hard-won liberal tradition of the late Victorian scholar, while the expatriates are shown as schoolboys' blackboard caricatures.[1]

I have quoted this at such length because first of all I am intrigued by Iris Andreski's literary style, which recalls so faithfully the sedate prose of the district officer government anthropologist of sixty or seventy years ago—a tribute to her remarkable powers of identification as well as to the durability of colonialist rhetoric. "Tyrannies of their masters". . . ."menace of hostile neighbours". . . ."purged of bloodshed and alcoholism." But in addition to this, Iris Andreski advances the position taken by Honor Tracy in one significant and crucial direction—its claim to a deeper knowledge and a more reliable appraisal of Africa than the educated African writer has shown himself capable of.

To the colonialist mind it was always of the utmost importance to be able to say: "I know my natives," a claim which implied two things at once: (a) that the native was really quite simple and (b) that understanding him and controlling him went hand in hand—understanding being a pre-condition for control and control constituting adequate proof of understanding. Thus, in the heyday of colonialism any serious incident of native unrest, carrying as it did disquieting intimations of slipping control, was an occasion not only for pacification by the soldiers but also (afterwards) for a royal commission of inquiry—a grand name for yet another perfunctory study of native psychology and institutions. Meanwhile a new situation was slowly developing as a handful of natives began to acquire European education and then to challenge Europe's presence and position in their native land with the intellectual weapons of Europe itself. To deal with this phenomenal presumption the colonialist devised two contradictory arguments. He created the "man of two worlds" theory to prove that no matter how much the native was exposed to European influences he could never truly absorb them; like Prester John he would always discard the mask of civilization when the crucial hour came and reveal his true face. Now, did this mean that the educated native was no different at all from his brothers in the bush? Oh, no! He *was* different; he was worse. His abortive effort at education and culture though leaving him totally unredeemed and unregenerated had nonetheless done something to him—it had deprived him of his links with his own people whom he no longer even understood and who certainly wanted none of his dissatisfaction or pretensions. "I know my natives; they are delighted with the way things are. It's only these half-educated ruffians who don't even know their own people . . ." How often one heard that and the many variations of it in colonial times! And how almost amusing to find its legacy in the colonialist criticism of our literature today! Iris Andreski's book is more than old wives' tales, at least in intention. It is clearly inspired by the desire to undercut the educated African witness (the modern myth-maker, she calls him) by appealing direct to the unspoilt woman of the bush who has retained a healthy gratitude for Europe's intervention in Africa. This desire accounts for all that reliance one finds in modern European travellers' tales on the evidence of "simple natives"— houseboys, cooks, drivers, schoolchildren—supposedly more trustworthy than the smart alecs. An American critic, Charles Larson, makes good use of this kind of evidence not only to validate his literary opinion of Ghana's Ayi Kwei Armah but, even more important, to demonstrate its superiority over the opinion of Ghanaian intellectuals:

> When I asked a number of students at the University of Ghana about their preferences for contemporary African novelists, Ayi Kwei Armah was the writer mentioned most frequently, in spite of the fact that many of Ghana's older writers and intellectuals regard him as a kind of negativist . . . I have for some time regarded Ayi Kwei Armah as Anglophone Africa's most accomplished prose stylist.[2]

In 1962, I published an essay, "Where Angels Fear to Tread,"[3] in which I suggested that the European critic of African literature must cultivate the habit of humility appropriate to his limited experience of the African world and purged of the superiority and arrogance which history so insidiously makes him heir to. That article, though couched in very moderate terms, won for me quite a few bitter enemies. One of them took my comments so badly—almost as a personal affront—that he launched numerous unprovoked attacks against me. Well, he has recently come to grief by his own hand. He published a long abstruse treatise based on an analysis of a number of Igbo proverbs most of which, it turned out, he had so completely misunderstood as to translate "fruit" in one of them as "penis." Whereupon, a merciless native, less charitable than I, proceeded to make mincemeat of him. If only he had listened to me ten years ago!

After the publication of *A Man of the People* in 1966, I was invited to dinner by a British diplomat in Lagos at which his wife, hitherto a fan of mine, admonished me for what she called "this great disservice to Nigeria." She loved Nigeria so much that my criticisms of the country which ignored all the brave efforts it was making left her totally aghast. I told her something not very nice, and our friendship was brought to an end.

Most African writers write out of an African experience and of commitment to an African destiny. For them, that destiny does not include a future European identity for which the present is but an apprenticeship. And let no one be fooled by the fact that we may write in English, for we intend to do unheard of things with it. Already some people are getting worried. This past summer I met one of Australia's leading poets, A. D. Hope, in Canberra, and he said wistfully that the only happy writers today were those writing in small languages like Danish. Why? Because they and their readers understood one another and knew precisely what a word meant when it was used. I had to admit that I hadn't thought of it that way. I had always assumed that the Commonwealth of Nations was a great bonus for a writer, that the English-Speaking Union was a desirable fraternity. But talking with A. D. Hope that evening, I felt somewhat like an illegitimate child face to face with the true son of the house lamenting the excesses of an adventurous and profligate father who had kept a mistress in every port. I felt momentarily nasty and thought of telling A. D. Hope: You ain't seen nothin' yet! But I know he would not have understood. And in any case, there was an important sense in which he was right—that every literature must seek the things that belong unto its peace, must, in other words, speak of a particular place, evolve out of the necessities of its history, past and current, and the aspirations and destiny of its people.

Australia proved quite enlightening. (I hope I do not sound too ungracious. Certainly, I met very many fine and sensitive people in Australia; and the words which the distinguished historian Professor Manning Clark wrote to me after my visit are among the finest tributes I have ever received: "I hope you come back and speak again here, because we need to lose the blinkers of our past. So come and help the young to grow up without the prejudices of their forefathers. . . ."

On another occasion a student at the National University who had taken a course in African literature asked me if the time had not come for African writers to write about "people in general" instead of just Africans. I asked her if by "people in general" she meant *like Australians*, and gave her the bad news that as far as I was concerned such a time would never come. She was only a brash sophomore. But like all the other women I have referred to, she expressed herself with passionate and disarming effrontery. I don't know how women's lib will take it, but I do believe that by and large women are more honest than men in expressing their feelings. This girl was only making the same point which many "serious" critics have been making more tactfully and therefore more insidiously. They dress it up in fine robes which they call universality.

In his book *The Emergence of African Fiction*, Charles Larson tells us a few revealing things about universality. In a chapter devoted to Lenrie Peters's novel, which he finds particularly impressive, he speaks of "its universality, its very limited concern with Africa itself." Then he goes on to spell it all out:

> That it is set in Africa appears to be accidental, for, except for a few comments at the beginning, Peters's story might just as easily take place in the southern part of the United States or in the southern regions of France or Italy. If a few names of characters and places were changed one would indeed feel that this was an American novel. In short, Peters's story is universal.[4]

But Larson is obviously not as foolish as this passage would make him out to be, for he ends it on a note of self-doubt which I find totally disarming. He says (p. 238):

> Or am I deluding myself in considering the work universal? Maybe what I really mean is that *The Second Round* is to a great degree Western and therefore scarcely African at all.

I find it hard after that to show more harshness than merely agreeing about his delusion. But few people I know are prepared to be so charitable. In a recent review of the book in *Okike*, a Nigerian critic, Omolara Leslie, mocks "the shining faith that we are all Americans under the skin."

Does it ever occur to these universalists to try out their game of changing names of characters and places in an American novel, say, a Philip Roth or an Updike, and slotting in African names just to see how it works? But of course it would not occur to them. It would never occur to them to doubt the universality of their own literature. In the nature of things the work of a Western writer is automatically informed by universality. It is only others who must strain to achieve it. So-and-so's work is universal; he has truly arrived! As though universality were some distant bend in the road which you may take if you travel out far enough in the direction of Europe or America, if you put adequate distance between yourself and your home. I should like to see the word "universal" banned altogether from discussions of African literature until such a time as people cease to use it as a synonym for the narrow, self-serving parochialism of Europe, until their horizon extends to include all the world.

If colonialist criticism were merely irritating one might doubt the justification of devoting a whole essay to it. But strange though it may sound, some of its ideas and precepts do exert an influence on our writers, for it is a fact of our contemporary world that Europe's powers of persuasion can be far in excess of the merit and value of her case. Take for instance the black writer who seizes on the theme that "Africa's past is a sadly inglorious one" as though it were something new that had not already been "proved" adequately for him. Colonialist critics will, of course, fall all over him in ecstatic and salivating admiration—which is neither unexpected nor particularly interesting. What is fascinating, however, is the tortuous logic and sophistry they will sometimes weave around a perfectly straightforward and natural enthusiasm.

A review of Yambo Ouologuem's *Bound to Violence* (Heinemann Educational Books, London, 1971) by a Philip M. Allen in the *Pan-African Journal*[5] was an excellent example of sophisticated, even brilliant colonialist criticism. The opening sentence alone would reward long and careful examination; but I shall content myself here with merely quoting it:

> The achievement of Ouologuem's much discussed, impressive, yet over-praised novel has less to do with whose ideological team he's playing on than with the *forcing of moral universality on African civilization* [my italics].

A little later Mr. Allen expounds on this new moral universality:

> This morality is not only "un-African"—denying the standards set by omnipresent ancestors, the solidarity of communities, the legitimacy of social contract: it is a Hobbesian universe that extends beyond the wilderness, beyond the white man's myths of Africa, into all civilization, theirs and ours.

If you should still be wondering at this point how Ouologuem was able to accomplish that Herculean feat of forcing moral universality on Africa or with what gargantuan tools, Mr. Allen does not leave you too long in suspense. Ouologuem is "an African intellectual who has mastered both a style and a prevailing philosophy of French letters," able to enter "the remoter alcoves of French philosophical discourse."

Mr. Allen is quite abrupt in dismissing all the "various polemical factions" and ideologists who have been claiming Ouologuem for their side. Of course they all miss the point,

> . . . for Ouologuem isn't writing their novel. He gives us an Africa cured of the pathetic obsession with racial and cultural confrontation and freed from invidious tradition-mongering . . . His book knows no easy antithesis between white and black, western and indigenous, modern and traditional. Its conflicts are those of the universe, not accidents of history.

And in final demonstration of Ouologuem's liberation from the constraint of local models Mr. Allen tells us:

> Ouologuem does not accept Fanon's idea of liberation, and he calls African unity a theory for dreamers. His Nakem is no more the Mali of Modibo Keita or the continent of Nkrumah than is the golden peace of Emperor Sundiata or the moral parish of Muntu.

Mr. Allen's rhetoric does not entirely conceal whose ideological team *he* is playing on, his attitude to Africa, in other words. Note, for example, the significant antithesis between the infinite space of "a Hobbesian universe" and "the moral parish of Muntu" with its claustrophobic implications. Who but Western man could contrive such arrogance?

Running through Mr. Allen's review is the overriding thesis that Ouologuem has somehow restored dignity to his people and their history by investing them with responsibility for violence and evil. Mr. Allen returns to this thesis again and again, merely changing the form of words. And we are to understand, by fairly clear implication, that this was something brave and new for Africa, this manly assumption of responsibility.

Of course a good deal of colonialist rhetoric always turned on that very question. The moral inferiority of colonized peoples, of which subjugation was a prime consequence and penalty, was most clearly demonstrated in their unwillingness to assume roles of responsibility. As long ago (or as recently, depending on one's historical perspective) as 1910 the popular English novelist John Buchan wrote a colonialist classic, *Prester John*, in which we find the words: "That is the difference between white and black, the gift of responsibility." And the idea did not originate with Buchan, either. It was a foundation tenet of colonialism and a recurrent element of its ideology and rhetoric. Now, to tell a man that he is incapable of assuming responsibility for himself and his actions is of course the utmost insult, to avoid which some Africans will go to any length, will throw anything into the deal; they will agree, for instance, to ignore the presence and role of racism in African history or pretend that somehow it was all the black man's own fault. Which is complete and utter nonsense. For whatever faults the black man may have or whatever crimes he committed (and they were, and are, legion) he did not bring racism into the world. And no matter how emancipated a man may wish to appear or how anxious to please by his largeness of heart, he cannot make history simply go away. Not even a brilliant writer could hope to do that. And as for those who applaud him for trying, who acclaim his bold originality in "restoring historical initiative to his people" when in reality all he does is pander to their racist and colonialist attitudes, they are no more than unscrupulous interrogators taking advantage of an ingratiating defendant's weakness and trust to egg him on to irretrievable self-incrimination.

That a "critic" playing on the ideological team of colonialism should feel sick and tired of Africa's "pathetic obsession with racial and cultural confrontation" should surprise no one. Neither should his enthusiasm for those African works that show "no easy antithesis between white and black." But an African who falls for such nonsense, not only in spite of Africa's so very recent history but, even more, in the face of continuing atrocities committed against millions of Africans in their own land by racist minority regimes, deserves a lot of pity. Certainly anyone, white or black, who chooses to see violence as the abiding principle of African civilization is free to do so. But let him not pass himself off as a restorer of dignity to Africa, or attempt to make out that he is writing about man and about the state of civilization in general. (You could as well claim that fifty years ago Frank Melland's *In Witchbound Africa* was an account of the universality of witchcraft and a vindication of Africa.) The futility of such service to Africa, leaving aside any question of duplicity in the motive, should be sufficiently underscored by one interesting admission in Mr. Allen's review:

> Thus, there is no reason for western reviewers of this book to exult in a black writer's admission of the savagery, sensuality and amorality of his race: he isn't talking about his race as Senghor or Cleaver do: he's talking about us all.

Well, how obtuse of these "western reviewers" to miss that point and draw such wrong conclusions! But the trouble is that not everyone can be as bright as Mr. Allen. Perhaps for most ordinary people what Africa needs is a far less complicated act of restoration. The Canadian novelist and critic Margaret Laurence saw this happening already in the way many African writers are interpreting their world, making it

> . . . neither idyllic, as the views of some nationalists would have had it, nor barbaric, as the missionaries and European administrators wished and needed to believe.[6]

And in the epilogue to the same book she makes the point even more strongly:

> No writer of any quality has viewed the old Africa in an idealized way, but they have tried to regain what is rightly theirs—a past composed of real and vulnerable people, their ancestors, not the figments of missionary and colonialist imaginations.

Ultimately the question of ideological sides which Mr. Allen threw in only to dismiss it again with contempt may not be as far-fetched as he thinks. For colonialism itself was built also on an ideology (although its adherents may no longer realize it) which, despite many setbacks, survives into our own day, and indeed is ready again at the end of a quiescent phase of self-doubt for a new resurgence of proselytization, even, as in the past, among its prime victims!

Fortunately, it can no longer hope for the role of unchallenged arbiter in other people's affairs that it once took so much for granted. There are clear signs that critics and readers from those areas of the world where continuing incidents and recent memories of racism, colonialism and other forms of victimization exist will more and more demand to know from their writers just on whose ideological side they are playing. And we writers had better be prepared to reckon with this questioning. For no amount of prestige or laurels of metropolitan reputation would seem large enough to silence or overawe it. Consider, for instance, a recent judgement on V. S. Naipaul by a fellow Caribbean, Ivan Van Sertima:

> His brilliancy of wit I do not deny but, in my opinion, he has been overrated by English critics whose sensibilities he insidiously flatters by his stock-in-trade: self-contempt.[7]

A Nigerian, Ime Ikiddeh, was even less ceremonious in his dismissal of Naipaul, who he thought did not deserve the attention paid to him by Ngugi wa Thiong'o in his *Homecoming*.[8] One need not accept these judgements in order to see them as signs of things to come.

Meanwhile the seduction of our writers by the blandishments of colonialist criticism is matched by its misdirection of our critics. Thus, an intelligent man like Dr. Sunday Anozie, the Nigerian scholar and critic, is able to dismiss the high moral and social earnestness sometimes expressed by one of our greatest poets, Christopher Okigbo, as only a mark of underdevelopment. In his book *Christopher Okigbo*, the most extensive biographical and critical study of the poet to date, Dr. Anozie tells us of Okigbo's "passion for truth," which apparently makes him sometimes too outspoken, makes him the talkative weaverbird "incapable of whispered secrets." And he proceeds to offer the following explanation:

> No doubt the thrill of actualized prophecy can sometimes lead poets particularly in the young countries to confuse their role with that of seers, and novelists to see themselves as teachers. Whatever the social, psychological, political and economic basis for it in present-day Africa, this interchangeability of role between the creative writer and the prophet appears to be a specific phenomenon of underdevelopment and therefore, like it also, a passing or ephemeral phase.[9]

And he cites the authority of C. M. Bowra in support of his explanation. The fallacy of the argument lies, of course, in its assumption that when you talk about a people's "level of development" you define their total condition and assign them an indisputable and unambiguous place on mankind's evolutionary ladder; in other words, that you are enabled by the authority of that phrase to account for all their material as well as spiritual circumstance. Show me a people's plumbing, you say, and I can tell you their art.

I should have thought that the very example of the Hebrew poet/prophets which Dr. Anozie takes from Bowra to demonstrate underdevelopment and confusion of roles would have been enough to alert him to the folly of his thesis. Or is he seriously suggesting that the poetry of these men—Isaiah, for example—written, it seems, out of a confusion of roles, in an underdeveloped society, is less good than what is written today by poets who are careful to remain within the proper bounds of poetry within developed societies? Personally, I should be quite content to wallow in Isaiah's error and write "For unto us a child is born."

Incidentally, any reader who is at all familiar with some of the arguments that go on around modern African literature will have noticed that in the passage I have just quoted from Dr. Anozie he is not only talking about Okigbo but also alluding (with some disapproval) to a paper I read at Leeds University ten years ago which I called "The Novelist as Teacher"; Anozie thus kills two weaverbirds dexterously with one stone! In his disapproval of what I had to say he follows, of course, in the footsteps of certain Western literary schoolmasters from whom I had already earned many sharp reprimands for that paper, who told me in clear terms that an artist had no business being so earnest.

It seems to me that this matter is of serious and fundamental importance, and must be looked at carefully. Earnestness and its opposite, levity, may be neither good nor bad in themselves but merely appropriate or inappropriate according to circumstance. I hold, however, and have held from the very moment I began to write, that earnestness is appropriate to my situation. Why? I suppose because I have a deep-seated need to alter things within that situation, to find for myself a little more room than has been allowed me in the world. I realize how pompous or even frightening this must sound to delicate sensibilities, but I can't help it.

The missionary who left the comforts of Europe to wander through my primeval forest was extremely earnest. He had to be; he came to change my world. The builders of empire who turned me into a "British protected person" knew the importance of being earnest; they had that quality of mind which Imperial Rome before them understood so well: *gravitas*. Now, it seems to me pretty obvious that if I desire to change the role and identity fashioned for me by those earnest agents of colonialism I will need to borrow some of their resolve. Certainly, I could not hope to do it through self-indulgent levity.

But of course I do appreciate also that the world is large and that all men cannot be, indeed must not be, of one mind. I appreciate that there are people in the world who have no need or desire to change anything. Perhaps they have already accomplished the right amount of change to ensure their own comfort. Perhaps they see the need for change but feel powerless to attempt it, or perhaps they feel it is someone else's business. For these people, earnestness is a dirty word or is simply tiresome. Even the evangelist, once so earnest and certain, now sits back in contemplation of his church, its foundation well and truly laid, its edifice rising majestically where once was jungle; the colonial governor who once brought his provinces so ruthlessly to heel prefers now to speak of the benefits of peace and orderly government. Certainly, they would much rather have easy-going natives under their jurisdiction than earnest ones-unless of course the earnestness be the perverse kind that turns in against itself.

The first nationalists and freedom fighters in the colonies, hardly concerned to oblige their imperial masters, were offensively *earnest*. They had no choice. They needed to alter the arrangement which kept them and their people out in the rain and the heat of the sun. They fought and won some victories. They changed a few things and seemed to secure certain powers of action over others. But quite quickly the great collusive swindle that was independence showed its true face to us. And we were dismayed; but only momentarily for even in our defeat we had gained something of inestimable value—a baptism of fire.

And so our world stands in just as much need of change today as it ever did in the past. Our writers responding to something in themselves and acting also within the traditional concept of an artist's role in society—using his art to control his environment—have addressed themselves to some of these matters in their art. And their concern seems to upset certain people whom history has dealt with differently and who persist in denying the validity of experiences and destinies other than theirs. And worst of all, some of our own critics who ought to guide these people out of their error seem so anxious to oblige them. Whatever the social, psychological, political and economic basis for this acquiescence, one hopes that it is only a passing and an ephemeral phase!

If this earnestness we speak of were manifested by just one or two writers in Africa there might perhaps be a good case for dismissing it out of hand. But look at the evidence:

Amos Tutuola has often given as a reason for his writing the need to preserve his traditional culture. It is true that a foreign critic, Adrian Roscoe, has chosen to jubilate over what appears to him like Tutuola's lack of "an awareness of cultural, national and racial affinities,"[10] but such an opinion may reflect more accurately his own wishful thinking than Tutuola's mind. Certainly a careful reading of *The Palm-Wine Drinkard* will not bear out the assertion that colonialism is "dead for him." Why would he go out of his way to tell us, for example, that "both white and black deads were living in the Dead's Town"[11] unless he considers the information significant, as indeed anyone who lived in the Lagos of the 1950s would readily appreciate? For although Nigeria experienced only "benign colonialism," it required a monumental demonstration of all nationalist organizations in the territory after a particularly blatant incident of racism in a Lagos hotel to compel the administration into token relaxation of the practice whereby Whites and Blacks lived in trim reservations or squalid townships separated by a regulation two-mile cordon sanitaire. Some day a serious critic interested in such matters will assemble and interpret Tutuola's many scattered allusions to colonialism for the benefit of more casual readers. For there are such intriguing incidents as the Drinkard's deliverance of the Red People from an ancient terror which required them to sacrifice one victim every year, for which blessing he lives among them for a while exploiting their cheap labour to develop and extend his plantations "becoming richer than the rest of the people in that town" until the moment of crisis arrives and he causes "the whole of them" to be wiped out. Such a critic will no doubt pay particular attention to the unperturbed and laconic comment of the deliverer's wife: ". . . all the lives of the natives were lost and the life of the non-natives saved."[12] But until that serious critic comes along, Mr. Roscoe can certainly have the comfort of believing that "if Achebe and Soyinka want to write in order to change the world, Tutuola has other reasons."

And then Camara Laye. As recently as 1972 he was saying in an interview:

> In showing the beauty of this culture, my novel testifies to its greatness. People who had not been aware that Africa had its own culture were able to grasp the significance of our past and our civilization. I believe that this understanding is the most meaningful contribution of African literature.[13]

The distinguished and versatile Sierra Leonian, Davidson Abioseh Nicol—scientist, writer and diplomat—explaining why he wrote, said:

> . . . because I found that most of those who wrote about us seldom gave any nobility to their African characters unless they were savages or servants or facing impending destruction. I knew differently.[14]

One could go on citing example after example of earnestness among African writers. But one final quotation—from Kofi Awoonor, the fine Ghanaian poet, novelist and essayist—should suffice:

> An African writer must be a person who has some kind of conception of the society in which he is living and the way he wants the society to go.[15]

All this juvenile earnestness must give unbearable offence to mature people. Have we not heard, they may ask, what Americans say—that the place for sending messages is the Western Union? Perhaps we have; perhaps we haven't. But the plain fact is that we are *not* Americans. Americans have their vision; we have ours. We do not claim that ours is superior; we only ask to keep it. For, as my forefathers said, the firewood which a people have is adequate for the kind of cooking they do. To levy a charge of underdevelopment against African writers today may prove as misguided and unin-formed as a similar dismissal of African art by visitors of an earlier age before the coming of Picasso. Those worthy men saw little good around them, only child-like and grotesque distortions. Frank Willett, in his excellent book *African Art*, tells us of one such visitor to Benin in 1701, a certain David Nyendael, who on being taken to the royal gallery described the objects as being:

> . . . so wretchedly carved that it is hardly possible to distinguish whether they are most like men or beasts; notwithstanding which my guides were able to distinguish them into merchants, soldiers, wild beast hunters, etc.[16]

Most people today would be inclined to ascribe the wretchedness to Nyendael's own mind and taste rather than to the royal art of Benin. And yet, for me, his comment is almost saved by his acknowledgement, albeit grudging, of the very different perception of his guides, the real owners of culture.

The colonialist critic, unwilling to accept the validity of sensibilities other than his own, has made particular point of dismissing the African novel. He has written lengthy articles to prove its non-existence largely on the grounds that the novel is a peculiarly Western genre, a fact which would interest us if our ambition was to write "Western" novels. But, in any case, did not the black people in America, deprived of their own musical instruments, take the trumpet and the trombone and blow them as they had never been blown before, as indeed they were not designed to be blown? And the result, was it not jazz? Is anyone going to say that this was a loss to the world or that those first Negro slaves who began to play around with the discarded instruments of their masters should have played waltzes and foxtrots? No! Let every people bring their gifts to the great festival of the world's cultural harvest and mankind will be all the richer for the variety and distinctiveness of the offerings.

My people speak disapprovingly of an outsider whose wailing drowned the grief of the owners of the corpse. One last word to the owners. It is because our own critics have been somewhat hesitant in taking control of our literary criticism (sometimes—let's face it—for the good reason that we will not do the hard work that should equip us) that the task has fallen to others, some of whom (again we must admit) have been excellent and sensitive. And yet most of what remains to be done can best be tackled by ourselves, the owners. If we fall back, can we complain that others are rushing forward? A man who does not lick his lips, can he blame the harmattan for drying them?

Let us emulate those men of Benin, ready to guide the curious visitor to the gallery of their art, willing to listen with politeness even to his hasty opinions but careful, most careful, to concede nothing to him that might appear to undermine their own position within their heritage or compromise the integrity of their indigenous perception. For supposing the artists of Benin and of Congo and Angola had agreed with Nyendael in 1701 and abandoned their vision and begun to make their images in the style of "developed" Portugal, would they not have committed a grave disservice to Africa and ultimately to Europe herself and the rest of the world? Because they did not, it so happened that after the passage of two centuries other Europeans, more sensitive by far than Nyendael, looked at their work again and learnt from it a new way to see the world.

Notes

1. Iris Andreski, *Old Wives' Tales* (New York: Schocken, 1971), 26.
2. Charles Larson, *Books Abroad* (Norman, Oklahoma: Winter 1974), p. 26.
3. Chinua Achebe, "Where Angles Fear to Tread," *Nigeria Magazine*, No. 75, Lagos (1962).
4. Charles Larson, *The Emergence of African Fiction* (Bloomington: Indiana University Press, 1971), 230.
5. Philip M. Allen, "*Bound to Violence* by Yambo Ouloguem," *Pan-African Journal* ix, No. 4 518–523.
6. Margaret Lawrence, *Long Drums and Cannons* (London: MacMillan, 1968), p. 9.
7. Ivan Van Sertima, *Caribbean Writers* (London: New Beacon, 1968), foreword.

8. Ibid., p. xiv.
9. Sunday O. Anozie, *Cristopher Okigbo* (London: Evans Bros., 1972), p. 17.
10. Adrian A. Roscoe, *Mother is Gold* (Cambridge: Cambridge University Press, 1971), pp. 98–99.
11. Amos Tutuola, *The Palm-Wine Drinkard* (London: Faber, 1952), p. 100.
12. Ibid., p. 92.
13. Camara Laye, interviewed by J. Steven Rubin, *Africa Report*, Washington, D.C. (May 1972), p. 22.
14. Davidson Abioseh Nicol, *The Truly Married Woman and Other Stories* (London: Fontana, 1965), introduction.
15. Kofi Awoonor, quoted in per Wästberg (ed.), *The Writer in Modern Africa* (Upsala: Almqvist & Wiskell).
16. Frank Willet, *African Art* (New York: Praeger, 1971), p. 102.

Chapter 14

Elaine Showalter (1941–) from "The Female Tradition," *A Literature of Their Own* (1977)

English women writers have never suffered from the lack of a reading audience, nor have they wanted for attention from scholars and critics. Yet we have never been sure what unites them as women, or, indeed, whether they share a common heritage connected to their womanhood at all. Writing about female creativity in *The Subjection of Women* (1869), John Stuart Mill argued that women would have a hard struggle to overcome the influence of male literary tradition, and to create an original, primary, and independent art. "If women lived in a different country from men," Mill thought, "and had never read any of their writings, they would have a literature of their own." Instead, he reasoned, they would always be imitators and never innovators. Paradoxically, Mill would never have raised this point had women not already claimed a very important literary place. To many of his contemporaries (and to many of ours), it seemed that the nineteenth century was the Age of the Female Novelist. With such stellar examples as Jane Austen, Charlotte Brontë, and George Eliot, the question of women's aptitude for fiction, at any rate, had been answered. But a larger question was whether women, excluded by custom and education from achieving distinction in poetry, history, or drama, had, in defining their literary culture in the novel, simply appropriated another masculine genre. Both George Henry Lewes and Mill, spokesmen for women's rights and Victorian liberalism in general, felt that, like the Romans in the shadow of Greece, women were overshadowed by male cultural imperialism: "If women's literature is destined to have a different collective character from that of men," wrote Mill, "much longer time is necessary than has yet elapsed before it can emancipate itself from the influence of accepted models, and guide itself by its own impulses."[1]

There is clearly a difference between books that happen to have been written by women, and a "female literature," as Lewes tried to define it, which purposefully and collectively concerns itself with the articulation of women's experience, and which

In this selection from *A Literature of Their Own*, Elaine Showalter, Professor of English at Princeton University, points to women writers of the nineteenth century as evidence of a female literary tradition.

guides itself "by its own impulses" to autonomous self-expression. As novelists, women have always been self-conscious, but only rarely self-defining. While they have been deeply and perennially aware of their individual identities and experiences, women writers have very infrequently considered whether these experiences might transcend the personal and local, assume a collective form in art, and reveal a history. During the intensely feminist period from 1880 to 1910, both British and American women writers explored the theme of an Amazon utopia, a country entirely populated by women and completely isolated from the male world. Yet even in these fantasies of autonomous female communities, there is no theory of female art. Feminist utopias were not visions of primary womanhood, free to define its own nature and culture, but flights from the male world to a culture defined in opposition to the male tradition. Typically the feminist utopias are pastoral sanctuaries, where a population of prelapsarian Eves cultivate their organic gardens, cure water pollution, and run exemplary child care centers, but do not write books.

In contradiction to Mill, and in the absence, until very recently, of any feminist literary manifestoes, many readers of the novel over the past two centuries have nonetheless had the indistinct but persistent impression of a unifying voice in women's literature. [. . .]

There are many reasons why discussion of women writers has been so inaccurate, fragmented, and partisan. First, women's literary history has suffered from an extreme form of what John Gross calls "residual Great Traditionalism"[2] which has reduced and condensed the extraordinary range and diversity of English women novelists to a tiny band of the "great," and derived all theories from them. In practice, the concept of greatness for women novelists often turns out to mean four or five writers—Jane Austen, the Brontës, George Eliot, and Virginia Woolf—and even theoretical studies of "the woman novelist" turn out to be endless recyclings and recombinations of insights about "indispensable Jane and George."[3] Criticism of women novelists, while focusing on these happy few, has ignored those who are not "great," and left them out of anthologies, histories, textbooks, and theories. Having lost sight of the minor novelists, who were the links in the chain that bound one generation to the next, we have not had a very clear understanding of the continuities in women's writing, nor any reliable information about the relationships between the writers' lives and the changes in the legal, economic, and social status of women.

Second, it has been difficult for critics to consider women novelists and women's literature theoretically because of their tendency to project and expand their own culture-bound stereotypes of femininity, and to see in women's writing an eternal opposition of biological and aesthetic creativity. The Victorians expected women's novels to reflect the feminine values they exalted, although obviously the woman novelist herself had outgrown the constraining feminine role. "Come what will," Charlotte Brontë wrote to Lewes, "I cannot, when I write, think always of myself and what is elegant and charming in femininity; it is not on these terms, or with such ideas, that I ever took pen in hand."[4] Even if we ignore the excesses of what Mary Ellmann calls "phallic criticism" and what Cynthia Ozick calls the "ovarian theory of literature," much contemporary criticism of women writers is still prescriptive and circumscribed.[5] Given the difficulties of steering a precarious course between the Scylla of insufficient information and the Charybdis of abundant prejudice, it is not surprising that formalist-structuralist critics have evaded the issue of sexual identity entirely, or dismissed it as irrelevant and subjective. Finding it difficult to think intelligently about women writers, academic criticism has often overcompensated by desexing them.

Yet since the 1960s, and especially since the reemergence of a Women's Liberation Movement in England and in America around 1968, there has been renewed enthusiasm for the idea that "a special female self-awareness emerges through literature in every period."[6] The interest in establishing a more reliable critical vocabulary and a more accurate and systematic literary history for women writers is part of a larger interdisciplinary effort by psychologists, sociologists, social historians, and art historians to reconstruct the political, social, and cultural experience of women.

Scholarship generated by the contemporary feminist movement has increased our sensitivity to the problems of sexual bias or projection in literary history, and has also begun to provide us with the information we need to understand the evolution of a female literary tradition. One of the most significant contributions has been the unearthing and reinterpretation of "lost" works by women writers, and the documentation of their lives and careers. [. . .]

As scholars have been persuaded that women's experience is important, they have begun to see it for the first time. With a new perceptual framework, material hitherto assumed to be nonexistent has suddenly leaped into focus. Interdisciplinary studies of Victorian women have opened up new areas of investigation in medicine, psychology, economics, political science, labor history, and art.[7] Questions of the "female imagination" have taken on intellectual weight in the contexts of theories of Karen Horney about feminine psychology, Erik Erikson about womanhood and the inner space, and R. D. Laing about the divided self. Investigation of female iconography and imagery has been stimulated by the work of art historians like Linda Nochlin, Lise Vogel, and Helene Roberts.[8]

As the works of dozens of women writers have been rescued from what E. P. Thompson calls "the enormous condescension of posterity,"[9] and considered in relation to each other, the lost continent of the female tradition has risen like Atlantis from the sea of English literature. It is now becoming clear that, contrary to Mill's theory, women have had a literature of their own all along. The woman novelist, according to Vineta Colby, was "really neither single nor anomalous," but she was also more than a "register and a spokesman for her age."[10] She was part of a tradition that had its origins before her age, and has carried on through our own. [. . .]

This book is an effort to describe the female literary tradition in the English novel from the generation of the Brontës to the present day, and to show how the development of this tradition is similar to the development of any literary subculture. Women have generally been regarded as "sociological chameleons," taking on the class, lifestyle, and culture of their male relatives. It can, however, be argued that women themselves have constituted a subculture within the framework of a larger society, and have been unified by values, conventions, experiences, and behaviors impinging on each individual. It is important to see the female literary tradition in these broad terms, in relation to the wider evolution of women's self-awareness and to the ways in which any minority group finds its direction of self-expression relative to a dominant society, because we cannot show a pattern of deliberate progress and accumulation. It is true, as Ellen Moers writes, that "women studied with a special closeness the works written by their own sex";[11] in terms of influences, borrowings, and affinities, the tradition is strongly marked. But it is also full of holes and hiatuses, because of what Germaine Greer calls the "phenomenon of the transcience of female literary fame"; "almost uninterruptedly since the Interregnum, a small group of women have enjoyed dazzling literary prestige during their own lifetimes, only to vanish without trace from the records of posterity."[12] Thus each generation of women writers has found itself, in a sense, without a history, forced to rediscover the past anew, forging again and again the consciousness of their sex. Given this perpetual disruption, and also the self-hatred that has alienated women writers from a sense of collective identity, it does not seem possible to speak of a "movement."

I am also uncomfortable with the notion of a "female imagination." The theory of a female sensibility revealing itself in an imagery and form specific to women always runs dangerously close to reiterating the familiar stereotypes. It also suggests permanence, a deep, basic, and inevitable difference between male and female ways of perceiving the world. I think that, instead, the female literary tradition comes from the still-evolving relationships between women writers and their society. Moreover, the "female imagination" cannot be treated by literary historians as a romantic or Freudian abstraction. It is the product of a delicate network of influences operating in time, and it must be analyzed as it expresses itself, in language and in a fixed arrangement of words on a page, a form that itself is subject to a network of influences and conventions, including the operations of the marketplace. In this investigation of the English novel, I am intentionally looking, not at an innate sexual attitude, but at the ways in which the self-awareness of the woman writer has translated itself into a literary form in a specific place and time-span, how this self-awareness has changed and developed, and where it might lead. [. . .]

More important than the question of direct literary influence, however, is the difference between the social and professional worlds inhabited by the eighteenth- and nineteenth-century women. The early women writers refused to deal with a professional role, or had a negative orientation toward it. "What is my life?" lamented the poet Laetitia Landon. "One day of drudgery after another; difficulties incurred for others, which have ever pressed upon me beyond health, which every year, in one severe illness after another, is taxed beyond its strength; envy, malice, and all uncharitableness—these are the fruits of a successful literary career for a woman."[13] These women may have been less than sincere in their insistence that literary success brought them only suffering, but they were not able to see themselves as involved in a vocation that brought responsibilities as well as conflicts, and opportunities as well as burdens. Moreover, they did not see their writing as an aspect of their female experience, or as an expression of it. [. . .]

Feminine, feminist, or female, the woman's novel has always had to struggle against the cultural and historical forces that relegated women's experience to the second rank. In trying to outline the female tradition, I have looked beyond the famous novelists who have been found worthy, to the lives and works of many women who have long been excluded from literary history. I have tried to discover how they felt about themselves and their books, what choices and sacrifices they made, and how their relationship to their profession and their tradition evolved. "What is commonly called literary history," writes Louise Bernikow, "is actually a record of choices. Which writers have survived their time and which have not depends upon who noticed them and chose to record the notice."[14] If some of the writers I notice seem to us to be Teresas and Antigones, struggling with their overwhelming sense of vocation and repression, many more will seem only Dorotheas, prim, mistaken, irreparably minor. And yet it is only by considering them all—Millicent Grogan as well as Virginia Woolf—that we can begin to record new choices in a new literary history, and to understand why, despite prejudice, despite guilt, despite inhibition, women began to write.

Notes

1. John Stuart Mill and Harriet Taylor Mill, "The Subjection of Women," in *Essays on Sex Equality*, ed. Alice S. Rossi (Chicago, 1970), ch. III, p. 207.
2. John Gross, *The Rise and Fall of the Man of Letters* (London: 1969), p. 304.
3. Cynthia Ozick, "Women and Creativity," in *Woman in Sexist Society*, ed. Vivian Gornick and Barbara K. Moran (New York: 1971), p. 436.
4. "Letters of November 1849," Clement Shorter, in *The Brontës: Life and Letters*, II (London: 1908), p. 80.
5. Mary Ellmann, *Thinking About Women* (New York: Harcourt, Brace & World, 1968), pp. 28–54; and Ozick, "Women and Creativity," p. 436.
6. Patricia Meyer Spacks, *The Female Imagination* (New York: 1975), p. 3.
7. See, e.g., Shelia Rowbotham, *Hidden from History* (London: 1973); *Suffer and Be Still: Women in the Victorian Age*, ed. Martha Vicinus (Bloomington: Indiana, 1972); *Clio's Consciousness Raised: New Perspectives on the History of Women*, ed. Mary S. Hartman and Lois N. Banner (New York: 1974): and Françoise Basch, *Relative Creatures: Victorian Women in Society and the Novel* (New York: 1974).
8. Linda Nochlin, "Why Are There No Great Women Artists?" in *Woman in Sexist Society* ed. Vivian Gornick and Barbara K. Moran, (New York: Basic Books, 1971), pp. 344–346.; Lise Vogel, "Fine Arts and Feminism: The Awakening Consciousness," *Feminist Studies II* (1974), pp. 3–37; Helene Roberts, "The Inside, the Surface, the Mass: Some Recurring Images of Women," *Women's Studies II* (1974), pp. 289–308.
9. E. P. Thompson, *The Making of the English Working Class* (New York, 1973), p. 12.
10. Vineta Colby, *The Singular Anomaly: Women Novelists of the Nineteenth Century* (New York: 1970), p. 11.
11. Ellen Moers, "Women's Lit," 28.
12. Germaine Greer, "Flying Pigs and Double Standards," *Times Literary Supplement*, (July 26, 1974) p. 784.
13. Quoted in S.C. Hall, *A Book of Memories of Great Men and Women of the Age* (London: 1877), p. 266.
14. Louise Bernikow, *The World Split Open: Four Centuries of Women Poets in England and America, 1552–1950* (New York: 1974), p. 3.

Chapter 15

Annette Kolodny (1941–)
from "A Map for Rereading: Or, Gender and the Interpretation of Literary Texts," *New Literary History* (1980)

Appealing particularly to a generation still in the process of divorcing itself from the New Critics' habit of bracketing off any text as an entity in itself, as though "it could be read, understood, and criticized entirely in its own terms,"[1] Harold Bloom has proposed a dialectical theory of influence between poets and poets, as well as between poems and poems which, in essence, does away with the static notion of a fixed or knowable text. As he argued in *A Map of Misreading* in 1975, "a poem is a response to a poem, as a poet is a response to a poet, or a person to his parent." Thus, for Bloom, "poems . . . are neither about 'subjects' nor about 'themselves'. They are necessrily about *other poems*."[2]

To read or to know a poem, according to Bloom, engages the reader in an attempt to map the psychodynamic relations by which the poet at hand has willfully misunderstood the work of some precursor (either single or composite) in order to correct, rewrite, or appropriate the prior poetic vision as his own. As first introduced in *The Anxiety of Influence* in 1973, the resultant "wholly different practical criticism . . . give[s] up the failed enterprise of seeking to 'understand' any single poem as an entity in itself" and "pursue[s] instead the quest of learning to read any poem as its poet's deliberate mis-interpretation, *as a poet*, of a precursor poem or of poetry in general."[3] What one deciphers in the process of reading, then, is not any discrete entity but, rather, a complex relational event, "itself a synecdoche for a larger whole including other texts."[4] "Reading a text is necessarily the reading of a whole system of texts," Bloom explains in *Kabbalah and Criticism*, "and meaning is always wandering around between texts" (*KC*, pp. 107–8). [. . .]

What is left out of account, however, is the fact that whether we speak of poets and critics "reading" texts or writers "reading" (and thereby recording for us) the world,

In this essay, Annette Kolodny, Professor of Comparative Cultural and Literary Studies at the University of Arizona, explores the limits of a literary theory, such as Harold Bloom's, that sees literary history as inward-looking series of relationships. At the same time, Kolodny also enlists just such a sense of history to argue for revising literary history and modes of reading.

we are calling attention to interpretive strategies that are learned, historically determined, and thereby necessarily gender-inflected. As others have elsewhere questioned the adequacy of Bloom's paradigm of poetic influence to explain the production of poetry by women,[5] so now I propose to examine analogous limitations in his model for the reading—and hence critical—process (since both, after all, derive from his revisionist rendering of the Freudian family romance). To begin with, to locate that "meaning" which "is always wandering around between texts" (KC, pp. 107–8), Bloom assumes a community of readers (and, thereby, critics) who know the same "whole system of texts" within which the specific poet at hand has enacted his *misprision*. The canonical sense of a shared and coherent literary tradition is thereby essential to the utility of Bloom's paradigm of literary influence as well as to his notions of reading (and misreading). "What happens if one tries to write, or to teach, or to think or even to read without the sense of a tradition?" Bloom asks in *A Map of Misreading*. "Why," as he himself well understands, "nothing at all happens, just nothing. You cannot write or teach or think or even read without imitation, and what you imitate is what another person has done, that person's writing or teaching or thinking or reading. Your relation to what informs that person is tradition, for tradition is influence that extends past one generation, a carrying-over of influence" (MM, p. 32).

So long as the poems and poets he chooses for scrutiny participate in the "continuity that began in the sixth century B.C. when Homer first became a schoolbook for the Greeks" (MM, pp. 33–34), Bloom has a great deal to tell us about the carrying over of literary influence; where he must remain silent is where carrying over takes place among readers and writers who in fact have been, or at least have experienced themselves as, cut off and alien from that dominant tradition. Virginia Woolf made the distinction vividly over a half-century ago, in *A Room of One's Own*, when she described being barred entrance, because of her sex, to a "famous library" in which was housed, among others, a Milton manuscript. Cursing the "Oxbridge" edifice, "venerable and calm, with all its treasures safe locked within its breast," she returns to her room at the inn later that night, still pondering "how unpleasant it is to be locked out; and I thought how it is worse perhaps to be locked in; and, thinking of the safety and prosperity of the one sex and of the poverty and insecurity of the other and of the effect of tradition and of the lack of tradition upon the mind of a writer."[6] And, she might have added, on the mind of a reader as well. For while my main concern here is with reading (albeit largely and perhaps imperfectly defined), I think it worth noting that there exists an intimate interaction between readers and writers in and through which each defines for the other what s/he is about. "The effect . . . of the lack of tradition upon the mind of a writer" will communicate itself, in one way or another, to her readers; and, indeed, may respond to her readers' sense of exclusion from high (or highbrow) culture. [. . .]

From the 1850s on, in America at least, the meanings "wandering around between texts" were wandering around somewhat different groups of texts where male and female readers were concerned.[7] So that with the advent of women "who wished to be regarded as artists rather than careerists,"[8] toward the end of the nineteenth century, there arose the critical problem with which we are still plagued and which Bloom so determinedly ignores: the problem of reading any text as "a synecdoche for a larger whole including other texts" when that necessarily assumed "whole system of texts" in which it is embedded is foreign to one's reading knowledge.

The appearance of Kate Chopin's novel *The Awakening* in 1899, for example, perplexed readers familiar with her earlier (and intentionally "regional") short stories not so much because it turned away from themes or subject matter implicit in her earlier work, nor even less because it dealt with female sensuality and extramarital sexuality, but because her elaboration of those materials deviated radically from the accepted norms of women's fiction out of which her audience so largely derived its expectations. The nuances and consequences of passion and individual temperament, after all, fairly define the focus of most of her preceding fictions. "That the book is strong and that Miss Chopin has a keen knowledge of certain phases of feminine character will not be denied," wrote the anonymous reviewer for the Chicago *Times-Herald*. What marked an unacceptable "new departure" for this critic, then, was the impropriety of Chopin's focus on material previously edited out of the popular genteel novels by and about women which, somewhat inarticulately, s/he translated into the accusation that Chopin had "enter[ed] the overworked field of sex fiction."[9]

Charlotte Perkins Gilman's initial difficulty in seeing "The Yellow Wallpaper" into print repeated the problem, albeit in a somewhat different context: for her story located itself not as any deviation from a previous tradition of women's fiction but, instead, as a continuation of a genre popularized by Poe. And insofar as Americans had earlier learned to follow the fictive processes of aberrant perception and mental breakdown in *his* work, they should have provided Gilman, one would imagine, with a ready-made audience for *her* protagonist's progressively debilitating fantasies of entrapment and liberation. As they had entered popular fiction by the end of the nineteenth century, however, the linguistic markers for those processes were at once heavily male-gendered and highly idiosyncratic, having more to do with individual temperament than with social or cultural situations per se. As a result, it would appear that the reading strategies by which cracks in ancestral walls and suggestions of unchecked masculine willfulness were immediately noted as both symbolically and semantically relevant did not, for some reason, necessarily carry over to "the nursery at the top of the house" with its windows barred, nor even less to the forced submission of the woman who must "take great pains to control myself before" her physician husband.[10]

A reader today seeking meaning in the way Harold Bloom outlines that process might note, of course, a fleeting resemblance between the upstairs chamber in Gilman—with its bed nailed to the floor, its windows barred, and metal rings fixed to the walls—and Poe's evocation of the dungeon chambers of Toledo; in fact, a credible argument might be made for reading "The Yellow Wallpaper" as Gilman's willful and purposeful misprision of "The Pit and the Pendulum." Both stories, after all, involve a sane mind entrapped in an insanity-inducing situation. Gilman's "message" might then be that the equivalent revolution by which the speaking voice of the Poe tale is released to both sanity and freedom is unavailable to her heroine. No *deus ex machina*, no General Lasalle triumphantly entering the city, no "outstretched arm" to prevent Gilman's protagonist from falling into her own internal "abyss" is conceivable, given the rules of the social context in which Gilman's narrative is embedded. When gender is taken into account, then, so this interpretation would run, Gilman is saying that the nature of the trap envisioned must be understood as qualitatively different, and so too the possible escape routes.

Contemporary readers of "The Yellow Wallpaper," however, were apparently unprepared to make such connections. Those fond of Poe could not easily transfer their sense of mental derangement to the mind of a comfortable middle-class wife and mother; and those for whom the woman in the home was a familiar literary character were hard-pressed to comprehend so extreme an anatomy of the psychic price she paid. Horace Scudder, the editor of *The Atlantic Monthly* who first rejected the story, wrote only that "I could not forgive myself if I made others as miserable as I have made myself!" (Hedges, p. 40). And even William Dean Howells, who found the story "chilling" and admired it sufficiently to reprint it in 1920, some twenty-eight years after its first publication (in *The New England Magazine* of May 1892), like most readers, either failed to notice or neglected to report "the connection between the insanity and the sex, or sexual role, of the victim" (Hedges, p. 41). For readers at the turn of the century, then, that "meaning" which "is always wandering around between texts" had as yet failed to find connective pathways linking the fanciers of Poe to the devotees of popular women's fiction, or the shortcut between Gilman's short story and the myriad published feminist analyses of the ills of society (some of them written by Gilman herself). Without such connective contexts, Poe continued as a well-traveled road, while Gilman's story, lacking the possibility of further influence, became a literary dead-end.

In one sense, by hinting at an audience of male readers as ill-equipped to follow the symbolic significance of the narrator's progressive breakdown as was her doctor-husband to diagnose properly the significance of his wife's fascination with the wall-paper's patternings; and by predicating a female readership as yet unprepared for texts which mirrored back, with symbolic exemplariness, certain patterns underlying their empirical reality, "The Yellow Wallpaper" anticipated its own reception. For insofar as writing and reading represent linguistically-based interpretive strategies—the first for the recording of a reality (that has, obviously, in a sense, already been "read") and the second for the deciphering of that recording (and thus also the further decoding of a prior imputed reality)—the wife's progressive descent into madness provides a kind of commentary upon, indeed is revealed in terms of, the sexual politics inherent in the manipulation of those strategies. We are presented at the outset with a protagonist who, ostensibly for her own good, is denied both activities and who, in the course of accommodating herself to that deprivation, comes more and more to experience her self as a text which can neither get read nor recorded. [. . .]

Successively isolated from conversational exchanges, prohibited free access to pen and paper, and thus increasingly denied what Jean Ricardou has called "the local exercise of syntax and vocabulary,"[11] the protagonist of "The Yellow Wallpaper" experiences the extreme extrapolation of those linguistic tools to the processes of perception and response. In fact, it follows directly upon a sequence in which: (1) she acknowledges that John's opposition to her writing has begun to make "the effort . . . greater than the relief"; (2) John refuses to let her "go and make a visit to Cousin Henry and Julia"; and (3) as a kind of punctuation mark to that denial, John carries her "upstairs and laid me on the bed, and sat by me and read to me till it tired my head." It is after these events, I repeat, that the narrator first makes out the dim shape lurking "behind the outside pattern" in the wallpaper: "it is like a woman stooping down and creeping" (pp. 21–22).

From that point on, the narrator progressively gives up the attempt to *record* her reality and instead begins to *read* it—as symbolically adumbrated in her compulsion to discover a consistent and coherent pattern amid "the sprawling outlines" of the wallpaper's apparently "pointless pattern" (pp. 19, 20). Selectively emphasizing one section of the pattern while repressing others, reorganizing and regrouping past impressions into newer, more fully realized configurations—as one might with any complex formal text—the speaking voice becomes obsessed with her quest for meaning, jealous even of her husband's or his sister's momentary interest in the paper. Having caught her sister-in-law "with her hand on it once," the narrator declares, "I know she was studying that pattern, and I am determined that nobody shall find it out but myself!" (p. 27). As the pattern changes with the changing light in the room, so too do her interpretations of it. And what is not quite so apparent by daylight becomes glaringly so at night: "At night in any kind of light, in twilight, candle light, lamplight, and worst of all by moonlight, it becomes bars! The outside pattern I mean, and the woman behind it is as plain as can be." "By daylight," in contrast (like the protagonist herself), "she is subdued, quiet" (p. 26). [. . .]

With the last paragraphs of the story, John faints away—presumably in shock at his wife's now totally delusional state. He has repeatedly misdiagnosed, or misread, the heavily edited behavior with which his wife has presented herself to him; and never once has he divined what his wife sees in the wallpaper. But given his freedom to read (or, in this case, misread) books, people, and the world as he chooses, he is hardly forced to discover for himself so extreme a text. To exploit Bloom's often useful terminology once again, then, Gilman's story represents not so much an object for the recurrent misreadings, or misprisions, of readers and critics (though this, of course, continues to occur) as an exploration, within itself, of the gender-inflected interpretive strategies responsible for our mutual misreadings, and even horrific misprisions, across sex lines. If neither male nor female reading audiences were prepared to decode properly "The Yellow Wallpaper," even less, Gilman understood, were they prepared to comprehend one another.

It is unfortunate that Gilman's story was so quickly relegated to the backwaters of our literary landscape because, coming as it did at the end of the nineteenth century, it spoke to a growing concern among American women who would be serious writers: it spoke, that is, to their strong sense of writing out of nondominant or subcultural traditions (both literary and otherwise), coupled with an acute sensitivity to the fact that since women and men learn to read different worlds, different groups of texts are available to their reading and writing strategies. Had "The Yellow Wallpaper" been able to stand as a potential precursor for the generation of subsequent corrections and revisions, then, as in Bloom's paradigm, it might have made possible a form of fiction by women capable not only of commenting upon but even of overcoming that impasse. That it did not—nor did any other woman's fiction become canonical in the United States[12]—meant that, again and again, each woman who took up the pen had to confront anew her bleak premonition that, both as writers and as readers, women too easily became isolated islands of symbolic significance, available only to, and decipherable only by, one another.[13] If any Bloomian "meaning" wanders around between women's texts, therefore, it must be precisely this shared apprehension. [. . .]

While neither the Gilman nor the Glaspell story necessarily excludes the male as reader—indeed, both in a way are directed specifically at educating him to become a better reader—they do, nonetheless, insist that, however inadvertently, he is a *different kind* of reader and that, where women are concerned, he is often an inadequate reader. In the first instance, because the husband cannot properly diagnose his wife or attend to her reality, the result is horrific: the wife descends into madness. In the second, because the men cannot even recognize as such the very clues for which they search, the ending is a happy one: Minnie Foster is to be set free, no motive having been discovered by which to prosecute her. In both, however, the same point is being made: lacking familiarity with the women's imaginative universe, that universe within which their acts are signs,[14] the men in these stories can neither read nor comprehend the meanings of the women closest to them—and this in spite of the apparent sharing of a common language. It is, in short, a fictive rendering of the dilemma of the woman writer. For while we may all agree that in our daily conversational exchanges men and women speak more or less meaningfully and effectively with one another, thus fostering the illusion of a wholly shared common language, it is also the case that where figurative usage is invoked—that usage which often enough marks the highly specialized language of literature—it "can be inaccessible to all but those who share information about one another's knowledge, beliefs, itentions, and attitudes."[15] Symbolic representations, in other words, depend on a fund of shared recognitions and potential inference. For their intended impact to *take hold* in the reader's imagination, the author simply must, like Minnie Foster, be able to call upon a shared context with her audience; where she cannot, or dare not, she may revert to silence, to the imitation of male forms, or, like the narrator in "The Yellow Wallpaper," to total withdrawal and isolation into madness.

It may be objected, of course, that I have somewhat stretched my argument so as to connate (or perhaps confuse?) *all* interpretive strategies with language processes, specifically *reading*. But in each instance, it is the survival of the *woman as text*— Gilman's narrator and Glaspell's Minnie Foster—that is at stake; and the competence of her reading audience alone determines the outcome. Thus, in my view, both stories intentionally function as highly specialized language acts (called "literature") which examine the difficulty inherent in deciphering other highly specialized realms of meaning—in this case, women's conceptual and symbolic worlds. And further, the intended emphasis in each is the inaccessibility of female meaning to male interpretation.[16] The fact that in recent years each story has increasingly found its way into easily available textbooks, and hence into the Women's Studies and American Literature classroom, to be read and enjoyed by teachers and students of both sexes happily suggests that their fictive premises are attributable not so much to necessity as to contingency.[17] Men can, after all, learn to apprehend the meanings encoded in texts by and about women—just as women have learned to become sensitive readers of Shakespeare and Milton, Hemingway and Mailer.[18] Both stories function, in effect, as a prod to that very process by alerting the reader to the fundamental problem of "reading" correctly within cohabiting but differently structured conceptual worlds.

To take seriously the implications of such relearned reading strategies is to acknowledge that we are embarking upon a revisionist rereading of our entire literary inheritance and, in that process, demonstrating the full applicability of Bloom's second formula for canon-formation, "You are or become what you read" (*KC*, p. 96). To set ourselves the task of learning to read a wholly different set of texts will make of us different kinds of readers (and perhaps different kinds of people as well). But to set ourselves the task of doing this in a public way, on behalf of women's texts specifically, engages us—as the feminists among us have learned—in a challenge to the inevitable issue of "*authority* . . . in all questions of canon-formation" (*KC*, p. 100). It places us, in a sense, in a position analogous to that of the narrator of "The Yellow Wallpaper," bound, if we are to survive, to challenge the (accepted and generally male) authority who has traditionally wielded the power to determine what may be written and how it shall be read. It challenges fundamentally not only the shape of our canon of major American authors but, indeed, that very "continuity that began in the sixth century B.C. when Homer first became a schoolbook for the Greeks" (*MM*, pp. 33–34).

It is no mere coincidence, therefore, that readers as diverse as Adrienne Rich and Harold Bloom have arrived by various routes at the conclusion that *re-vision* constitutes the key to an ongoing literary history. Whether functioning as ephebe/poet or would-be critic, Bloom's reader, as "revisionist," "strives to *see* again, so as to esteem and *estimate* differently, so as then to *aim* 'correctly' " (*MM*, p. 4). For Rich, "re-vision" entails "the act of looking back, of seeing with fresh eyes, of entering an old text from a new critical direction."[19] And each, as a result—though from different motives—strives to make the "literary tradition . . . the captive of the revisionary impulse" (*MM*, p. 36). What Rich and other feminist critics intended by that "revisionism" has been the subject of this essay: not only would such revisionary rereading open new avenues for comprehending male texts but, as I have argued here, it would, as well, allow us to appreciate the variety of women's literary expression, enabling us to take it into serious account for perhaps the first time rather than, as we do now, writing it off as caprice or exception, the irregularity in an otherwise regular design. Looked at this way, feminist appeals to revisionary rereading, as opposed to Bloom's, offer us all a potential enhancing of our capacity to read the world, our literary texts, and even one another, anew.

To end where I began, then, Bloom's paradigm of poetic history, when applied to women, proves useful only in a negative sense: for by omitting the possibility of poet/mothers from his psychodynamic of literary influence (allowing the feminine only the role of Muse—as composite whore and mother), Bloom effectively masks the fact of an *other* tradition entirely—that in which women taught one another how to read and write about and out of their own unique (and sometimes isolated) contexts. In so doing, however, he points up not only the ignorance informing our literary history as it is currently taught in the schools, but, as well, he pinpoints (however unwittingly) what must be done to change our skewed perceptions: all readers, male and female alike, must be taught first to recognize the existence of a significant body of writing by women in America and, second, they must be encouraged to learn how to read it within its own unique and informing contexts of meaning and symbol. *Re-visionary rereading*, if you will. No more must we impose on future generations of readers the inevitability of Norman Mailer's "terrible confession . . .—I have nothing to say about any of the talented women who write today. . . . I do not seem able to read them."[20] Nor should Bloom himself continue to suffer an inability to express useful "judgment upon . . . the 'literature of Women's Liberation.' "[21]

Notes

1. Albert William Levi, "*De Interpretatione*: Cognition and Context in the History of Ideas," *Critical Inquiry*, 3, No. 1 (Autumn 1976) p. 164.

2. Harold Bloom, *A Map of Misreading* (New York: 1975), p. 18 (hereafter cited as *MM*).

3. Harold Bloom, *The Anxiety of Influence: A Theory of Poetry* (New York: 1973), p. 43 (hereafter cited as *AI*).

4. Harold Bloom, *Kabbalah and Criticism* (New York: 1975), p. 106 (hereafter cited as *KC*). This concept is further refined in his *Poetry and Repression: Revisionism from Blake to Stevens* (New Haven: 1976), p. 26, where Bloom describes poems as "defensive processes in constant change, which is to say that poems themselves are acts of reading. A poem is . . . a fierce, proleptic debate *with itself*, as well as with precursor poems."

5. See, for example, Joanne Feit Diehl's attempt to adapt the Bloomian model to the psychodynamics of women's poetic production in " 'Come Slowly—Eden': An Exploration of Women Poets and Their Muse," *Signs*, 3, No. 3 (Spring 1978), pp. 572–587; and the objections to that adaptation raised by Lillian Faderman and Louise Bernikow in their Comments, *Signs*, 4, No. 1 (Autumn 1978), pp. 188–191 and 191–95, respectively. More recently, Sandra M. Gilbert and Susan Gubar have tried to correct the omission of women writers from Bloom's male-centered literary history in *The Madwoman in the Attic: The Woman Writer and the Nineteenth-Century Literary Imagination* (New Haven: 1979).

6. Virginia Woolf, *A Room of One's Own* (1928; rpt. Harmondsworth: 1972), pp. 9–10, 25–26.

7. The problem of audience is complicated by the fact that in nineteenth-century America distinct classes of so-called highbrow and lowbrow readers were emerging, cutting across sex and class lines; and, for each sex, distinctly separate "serious" and "popular" reading materials were also being marketed. Full discussion, however, is beyond the scope of this essay. In its stead, I direct the reader to Henry Nash Smith's clear and concise summation in the introductory chapter to his *Democracy and the Novel: Popular Resistance to Classic American Writers* (New York: 1978), pp. 1–15.

8. Nina Baym, *Woman's Fiction: A Guide to Novels by and About Women in America, 1820–1870* (Ithaca, 1978), p. 178.

9. From "Books of the Day," *Chicago Times-Herald* (June 1, 1899), p. 9; excepted in Kate Chopin, *The Awakening*, ed. Margaret Culley (New York: 1976), p. 149.

10. Charlotte Perkins Gilman, *The Yellow Wallpaper*, with Afterword by Elaine R. Hedges (New York: 1973), pp. 12, 11. Page references to this edition will henceforth be cited parenthetically in the text, with references to Hedges's excellent Afterword preceded by her name.

11. Jean Ricardou, "Composition Discomposed," trans. Erica Freiberg, *Critical Inquiry*, 3, No. 1 (Autumn 1976), p. 90.

12. The possible exception here is Harriet Beecher Stowe's *Uncle Tom's Cabin; or, Life Among the Lowly* (1852).

13. If, to some of the separatist advocates in our current wave of New Feminism, this sounds like a wholly acceptable, even happy circumstance, we must nonetheless understand that, for earlier generations of women artists, acceptance within male precincts conferred the mutually understood marks of success and, in some quarters, vitally needed access to publishing houses, serious critical attention, and even financial independence. That this was *not* the case for the writers of domestic fictions around the middle of the nineteenth century was a fortunate but anomalous circumstance. Insofar as our artist-mothers were separatist, therefore, it was the result of impinging cultural contexts and not (often) of their own choosing.

14. I here paraphrase Clifford Geertz, *The Interpretation of Cultures* (New York: 1973), p. 13, and specifically direct the reader to the parable from Wittgenstein quoted on that same page.

15. Ted Cohen, "Metaphor and the Cultivation of Intimacy," *Critical Inquiry*, 5, No. 1 (Autumn 1978) p. 78.

16. It is significant, I think, that the stories do not suggest any difficulty for the women in apprehending the men's meanings. On the one hand this simply is not relevant to either plot; and on the other, since in each narrative the men clearly control the public realms of discourse, it would, of course, have been incumbent upon the women to learn to understand them. Though masters need not learn the language of their slaves, the reverse is never the case: for survival's sake, oppressed or subdominant groups always study the nuances of meaning and gesture in those who control them.

17. For example, Gilman's "The Yellow Wallpaper" may be found, in addition to the Feminist Press reprinting previously cited, in *The Oven Birds: American Women on Womanhood, 1820–1920*, ed. Gail Parker (Garden City, 1972), pp. 317–334; and Glaspell's "A Jury of Her Peers" is reprinted in *American Voices, American Women*, ed. Lee R. Edwards and Arlyn Diamond (New York, 1973), pp. 359–381.

18. That women may have paid a high psychological and emotional price for their ability to read men's texts is beyond the scope of this essay, but I enthusiastically direct the reader to Judith Fetterley's provocative study of the problem in her *The Resisting Reader: A Feminist Approach to American Fiction* (Bloomington: 1978).

19. Adrienne Rich, "When We Dead Awaken: Writing as Re-Vision," *College English*, 34, No. 1 (October 1972) p. 18; rpt. in *Adrienne Rich's Poetry*, ed. Barbara Charlesworth Gelpi and Albert Gelpi (New York: 1975), p. 90.

20. Norman Mailer, "Evaluations—Quick and Expensive Comments on the Talent in the Room," collected in his *Advertisements for Myself* (New York: 1966), pp. 434–435.

21. *MM*, p. 36. What precisely Bloom intends by the phrase is nowhere made clear; for the purposes of this essay, I have assumed that he is referring to the recently increased publication of new titles by women writers.

Chapter 16

Pierre Bourdieu (1930–2002) from "The Field of Cultural Production, Or: The Economic World Reversed," *The Field of Cultural Production* (1983)

O Poésie, ô ma mère mourante,
Comme tes fils t'aimaient d'un grand amour
Dans ce Paris, en l'an mil huit cent trente:
Pour eux les docks, l'Autrichien, la rente,
Les mots de bourse étaient du pur hébreu.
Th. de Banville, "Ballade de ses regrets pour l'an 1830"

Preliminaries

Few areas more clearly demonstrate the heuristic efficacy of relational thinking than that of art and literature. Constructing an object such as the literary field[1] requires and enables us to make a radical break with the substantialist mode of thought (as Ernst Cassirer calls it) which tends to foreground the individual, or the visible inter-actions between individuals, at the expense of the structural relations—invisible, or visible only through their effects—between social positions that are both occupied and manipulated by social agents, which may be isolated individuals, groups or insti-tutions.[2] There are in fact very few other areas in which the glorification of "great men," unique creators irreducible to any condition or conditioning, is more common or uncontroversial—as one can see, for example, in the fact that most analysts uncrit-ically accept the division of the corpus that is imposed on them by the names of authors ("the work of Racine") or the titles of works (*Phèdre or Bérénice*).

French sociologist Pierre Bourdieu is known for his idea of the *habitus*, a field of expectations and possibilities that set the conditions for personhood. In "The Field of Cultural Production," Bourdieu brings the *habitus* to literary history and production, arguing that literature ought to be seen as part of a symbolic field of cultural and political forces.

To take as one's object of study the literary or artistic field of a given period and society (the field of Florentine painting in the Quattrocento or the field of French literature in the Second Empire) is to set the history of art and literature a task which it never completely performs, because it fails to take it on explicitly, even when it does break out of the routine of monographs which, however interminable, are necessarily inadequate (since the essential explanation of each work lies outside each of them, in the objective relations which constitute this field). The task is that of constructing the space of positions and the space of the position-takings [*prises de position*] in which they are expressed. The science of the literary field is a form of *analysis situs* which establishes that each position—e.g. the one which corresponds to a genre such as the novel or, within this, to a sub-category such as the "society novel" [*roman mondain*] or the "popular" novel—is objectively defined by the system of distinctive properties by which it can be situated relative to other positions; that every position, even the dominant one, depends for its very existence, and for the determinations it imposes on its occupants, on the other positions constituting the field; and that the structure of the field, i.e. of the space of positions, is nothing other than the structure of the distribution of the capital of specific properties which governs success in the field and the winning of the external or specific profits (such as literary prestige) which are at stake in the field.

The *space of literary or artistic position-takings*, i.e. the structured set of the manifestations of the social agents involved in the field—literary or artistic works, of course, but also political acts or pronouncements, manifestoes or polemics, etc.—is inseparable from the *space of literary or artistic positions* defined by possession of a determinate quantity of specific capital (recognition) and, at the same time, by occupation of a determinate position in the structure of the distribution of this specific capital. The literary or artistic field is a field of forces, but it is also a field of struggles tending to transform or conserve this field of forces. The network of objective relations between positions subtends and orients the strategies which the occupants of the different positions implement in their struggles to defend or improve their positions (i.e. their position-takings), strategies which depend for their force and form on the position each agent occupies in the power relations [*rapports de force*].

Every position-taking is defined in relation to the *space of possibles* which is objectively realized as a *problematic* in the form of the actual or potential position-taking corresponding to the different positions; and it receives its distinctive *value* from its negative relationship with the coexistent position-takings to which it is objectively related and which determine it by delimiting it. It follows from this, for example, that a *prise de position* changes, even when it remains identical, whenever there is change in the universe of options that are simultaneously offered for producers and consumers to choose from. The meaning of a work (artistic, literary, philosophical, etc.) changes automatically with each change in the field within which it is situated for the spectator or reader.

This effect is most immediate in the case of so-called classic works, which change constantly as the universe of coexistent works changes. This is seen clearly when the simple *repetition* of a work from the past in a radically transformed field of compossibles produces an entirely automatic *effect of parody* (in the theatre, for example, this effect requires the performers to signal a slight distance from a text impossible to defend as it stands; it can also arise in the presentation of a work corresponding to one extremity of the field before an audience corresponding structurally to the other extremity—e.g.

when an avant-garde play is performed to a bourgeois audience, or the contrary, as more often happens). It is significant that breaks with the most orthodox works of the past, i.e. with the *belief* they impose on the newcomers, often takes the form of *parody* (intentional, this time), which presupposes and confirms *emancipation*. In this case, the newcomers "get beyond" ["dépassent"] the dominant mode of thought and expression not by explicitly denouncing it but by repeating and reproducing it in a sociologically non-congruent context, which has the effect of rendering it incongruous or even absurd, simply by making it perceptible as the arbitrary convention which it is. This form of heretical break is particularly favoured by ex-believers, who use pastiche or parody as the indispensable means of objectifying, and thereby appropriating, the form of thought and expression by which they were formerly possessed.

This explains why writers' efforts to control the reception of their own works are always partially doomed to failure (one thinks of Marx's "I am not a Marxist"); if only because the very effect of their work may transform the condition of its reception and because they would not have had to write many things they did write and write them as they did—e.g. resorting to rhetorical strategies intended to 'twist the stick in the other direction'—if they had been granted from the outset what they are granted retrospectively.

One of the major difficulties of the social history of philosophy, art or literature, is that it has to reconstruct these spaces of original possibles which, because they were part of the self-evident *données* of the situation, remained unremarked and are therefore unlikely to be mentioned in contemporary accounts, chronicles or memoirs. It is difficult to conceive the vast amount of information which is linked to membership of a field and which all contemporaries immediately invest in their reading of works: information about institutions—e.g. academies, journals, magazines, galleries, publishers, etc.—and about persons, their relationships, liaisons and quarrels, information about the ideas and problems which are "in the air" and circulate orally in gossip and rumour. (Some intellectual occupations presuppose a particular mastery of this information.) Ignorance of everything which goes to make up the "mood of the age" produces a derealization of works: stripped of everything which attached them to the most concrete debates of their time (I am thinking in particular of the connotations of words), they are impoverished and transformed in the direction of intellectualism or an empty humanism. This is particularly true in the history of ideas, and especially of philosophy. Here the ordinary effects of derealization and intellectualization are intensified by the representation of philosophical activity as a summit conference between "great philosophers"; in fact, what circulates between contemporary philosophers, or those of different epochs, is not only canonical texts, but a whole philosophical doxa carried along by intellectual rumour—labels of schools, truncated quotations, functioning as slogans in celebration or polemics—by academic routine and perhaps above all by school manuals (an unmentionable reference), which perhaps do more than anything else to constitute the "common sense" of an intellectual generation. Reading, and a fortiori the reading of books, is only one means among others even among professional readers, of acquiring the knowledge that is mobilized in reading.

It goes without saying that, in both cases, change in the space of literary or artistic possibles is the result of change in the power relation which constitutes the space of positions. When a new literary or artistic group makes its presence felt in the field of literary or artistic production, the whole problem is transformed, since its coming into being, i.e. into difference, modifies and displaces the universe of possible options; the previously dominant productions may, for example, be pushed into the status of outmoded (*déclassé*) or classic works.

This theory differs fundamentally from all "systemic" analyses of works of art based on transposition of the phonological model, since it refuses to consider the field of *prises the position* in itself and for itself, i.e. independently of the field of positions which it manifests. This is understandable when it is seen that it applies relational thinking not only to symbolic systems, whether language (like Saussure) or myth (like Lévi-Strauss), or any set of symbolic objects, e.g. clothing, literary works, etc. (like all so-called "structuralist" analyses), but also to the *social relations* of which these symbolic systems are a more or less transformed expression. Pursuing a logic that is entirely characteristic of symbolic structuralism, but realizing that no cultural product exists by itself, i.e. outside the relations of interdependence which link it to other products, Michel Foucault gives the name "field of strategic possibilities" to the regulated system of differences and dispersions within which each individual work defines itself (1968: 40). But—and in this respect he is very close to semiologists such as Trier, and the use they have made of the idea of the "semantic field"—he refuses to look outside the "field of discourse" for the principle which would cast light on each of the discourses within it: "If the Physiocrats' analysis belongs to the same discourses as that of the Utilitarians, this is not because they lived in the same period, not because they confronted one another within the same society, not because their interests interlocked within the same economy, but because their two options sprang from one and the same distribution of the points of choice, one and the same strategic field" (1968: 29). In short, Foucault shifts onto the plane of possible *prises the position* the strategies which are generated and implemented on the sociological plane of positions; he thus refuses to relate works in any way to their social conditions of production, i.e. to positions occupied within the field of cultural production. More precisely, he explicitly rejects as a "doxological illusion" the endeavour to find in the "field of polemics" and in "divergences of interests and mental habits" between individuals the principle of what occurs in the "field of strategic possibilities," which he sees as determined solely by the "strategic possibilities of the conceptual games" (1968: 37). Although there is no question of denying the specific determination exercised by the possibilities inscribed in a given state of the space of *prises de position*—since one of the functions of the notion of the relatively autonomous field with its own history is precisely to account for this—it is not possible, even in the case of the scientific field and the most advanced sciences, to make the cultural order (the "*episteme*") a sort of autonomous, transcendent sphere, capable or developing in accordance with its own laws.

The same criticism applies to the Russian formalists, even in the interpretation put forward by Itamar Even-Zohar in his theory of the "literary polysystem," which seems closer to the reality of the texts if not to the logic of things, than the interpretation which structuralist readings (especially by Todorov) have imposed in France (cf. in particular Tynianov and Jakobson 1965: 138–139; Even-Zohar 1979: 65–74; Erlich 1965). Refusing to consider anything other than the system of works, i.e. the "network of relationships between texts," or "intertextuality," and the—very abstractly defined—relationships between this network and the other systems functioning in the "system-of-systems" which constitutes the society (we are close to Talcott Parsons), these theoreticians of cultural semiology or culturology are forced to seek in the literary system itself the principle of its dynamics. When they make the process of "automatization" and "de-automatization" the fundamental law of poetic change and, more generally, of all cultural change, arguing that a "de-automatization" must necessarily result from the "automatization" induced by repetitive use of the literary means of expression, they forget that the dialectic of orthodoxy which, in Weber's terms, favours a process of "routinization," and of heresy, which "deroutinizes," does not take place in the ethereal realm of ideas, and in the confrontation between "canonized" and "non-canonized" texts; and, more concretely, that the existence, form and direction of change depend not only on the "state of the system," i.e. the "repertoire" of possibilities which it offers, but also on the balance of forces between social agents who have entirely real interests in the different possibilities available to them as stakes and who deploy every sort of strategy to make one set or the other prevail. When we speak of a *field* of *prises de position*, we are insisting that what can be constituted as a system for the sake of analysis is not the product of a coherence-seeking intention or an objective consensus (even if it presupposes unconscious agreement on common principles) but the product and prize of a permanent conflict; or, to put it another way, that the generative, unifying principle of this "system" is the struggle, with all the contradictions it engenders (so that participation in the struggle—which may be indicated objectively by, for example, the attacks that are suffered—can be used as the criterion establishing that a work belongs to the field of *prises the position* and its author to the field of positions).[3]

In defining the literary and artistic field as, inseparably, a field of positions and a field of *prises de position*, we also escape from the usual dilemma of internal ("tautegorical") reading of the work (taken in isolation or within the system of works to which it belongs) and external (or "allegorical") analysis, i.e. analysis of the social conditions of production of the producers and consumers which is based on the—generally tacit—hypothesis of the spontaneous correspondence or deliberate matching of production to demand or commissions. And by the same token we escape from the correlative dilemma of the charismatic image of artistic activity as pure, disinterested creation by an isolated artist, and the reductionist vision which claims to explain the act of production and its product in terms of their conscious or unconscious external functions, by referring them, for example, to the interests of the dominant class or, more subtly, to the ethical or aesthetic values of one or another of its fractions, from which the patrons or audience are drawn.

Here one might usefully point to the contribution of Becker (1974, 1976) who, to his credit, constructs artistic production as a collective action, breaking with the naïve vision of the individual creator. For Becker, "works of art can be understood by viewing them as the result of the co-ordinated activities of all the *people* whose co-operation is necessary in order that the work should occur as it does" (1976: 703). Consequently the inquiry must extend to all those who contribute to this result, i.e. "the people who conceive the idea of the work (e.g. composers or playwrights); people who execute it (musicians or actors); people who provide the necessary equipment and material (e.g. musical instrument makers); and people who make up the audience for the work (playgoers, critics, and so on)" (1976: 703–704). Without elaborating all the differences between this vision of the "art world" and the theory of the literary and artistic field, suffice it to point out that the artistic field is not reducible to a *population*, i.e. a sum of individual agents, linked by simple relations of *interaction*—although the agents and the *volume* of the *population* of producers must obviously be taken into account (e.g. an increase in the number of agents engaged in the field has specific effects).

But when we have to re-emphasize that the principle of *prises de position* lies in the structure and functioning of the field of positions, this is not done so as to return to any form of economism. There is a specific economy of the literary and artistic field, based on a particular form of belief. And the major difficulty lies in the need to make a radical break with this belief and with the deceptive certainties of the language of celebration, without thereby forgetting that they are part of the very reality we are seeking to understand, and that, as such, they must have a place in the model intended to explain it. Like the science of religion, the science of art and literature is threatened by two opposite errors, which, being complementary, are particularly likely to occur since, in reacting diametrically against one of them, one necessarily falls into the other. The work of art is an object which exists as such only by virtue of the (collective) belief which knows and acknowledges it as a work of art. Consequently, in order to escape from the usual choice between celebratory effusions and the reductive analysis which failing to take account of the fact of belief in the work of art and of the social conditions which produce that belief, destroys the work of art as such, a rigorous science of art must, *pace* both the unbelievers and iconoclasts and also the believers, assert the possibility and necessity of understanding the work in its reality as a fetish; it has to take into account everything which helps to constitute the work as such, not least the discourses of direct or disguised celebration which are among the social conditions of production of the work of art *qua* object of belief.

The production of discourse (critical, historical, etc.) about the work of art is one of the conditions of production of the work. Every critical affirmation contains, on the one hand, a recognition of the value of the work which occasions it, which is thus designated as worthy object of legitimate discourse (a recognition sometimes extorted by the logic of the field, as when, for example, the polemic of the dominant confers participant status on the challengers), and on the other hand an affirmation of its own legitimacy. Every critic declares not only his judgement of the work but also his claim to the right to talk about it and judge it. In short, he takes part in a struggle for the monopoly of legitimate discourse about the work of art, and consequently in the production of the value of the work of art. (And one's only hope of producing scientific knowledge—rather than weapons to advance a particular class of specific interests—is to make explicit to oneself one's position in the sub-field of the producers of discourse about art and the contribution of this field to the very existence of the object of study.)

The science of the social representation of art and of the appropriate relation to works of art (in particular, through the social history of the process of autonomization of the intellectual and artistic field) is one of the prerequisites for the constitution of a rigorous science of art, because belief in the value of the work, which is one of the major obstacles to the constitution of a science of artistic production, is part of the full reality of the work of art. There is in fact every reason to suppose that the constitution of the aesthetic gaze as a "pure" gaze, capable of considering the work of art in and for itself, i.e. as a "finality without an end," is linked to the *institution* of the work of art as an object of contemplation, with the creation of private and then public galleries and museums, and the parallel development of a corps of professionals appointed to conserve the work of art, both materially and symbolically. Similarly, the representation of artistic production as a "creation" devoid of any determination or any social function, though asserted from a very early date, achieves its fullest expression in the theories of "art for art's sake"; and correlatively, in the representation of the legitimate relation to the work of art as an act of "re-creation" claiming to replicate the original creation and to focus solely on the work in and for itself, without any reference to anything outside it.

The actual state of the science of works of art cannot be understood unless it is borne in mind that whereas external analyses are always liable to appear crudely reductive, an internal reading, which establishes the charismatic, creator-to-creator relationship with the work that is demanded by the social norms of reception, is guaranteed social approval and reward. One of the effects of this charismatic conception of the relation to the work of art can be seen in the cult of the virtuoso which appeared in the late 19th century and which leads audiences to expect works to be performed and conducted from memory—which has the effect of limiting the repertoire and excluding avant-garde works, which are liable to be played only once (cf. Hanson 1967: 104–105).

The educational system plays a decisive role in the generalized imposition of the legitimate mode of consumption. One reason for this is that the ideology of "re-creation" and "creative reading" supplies teachers—*lectores* assigned to commentary on the canonical texts—with a legitimate substitute for the ambition to act as *auctores*. This is seen most clearly in the case of philosophy, where the emergence of a body of professional teachers was accompanied by the development of a would-be autonomous science of the history of philosophy, and the propensity to read works in and for themselves (philosophy teachers thus tend to identify philosophy with the history of philosophy, i.e. with a pure *commentary* on past works, which are thus invested with a role exactly opposite to that of suppliers of problems and instruments of thought which they would fulfill for original thinking).

Given that works of art exist as symbolic objects only if they are known and recognized, i.e. socially instituted as works of art and received by spectators capable of knowing and recognizing them as such, the sociology of art and literature has to take as its object not only the material production but also the symbolic production of the work, i.e. the production of the value of the work, or, which amounts to the same thing, of belief in the value of the work. It therefore has to consider as contributing to production not only the direct producers of the work in its materiality (artist, writer, etc.) but also the producers of the meaning and value of the work—critics, publishers, gallery directors, and the whole set of agents whose combined efforts produce consumers capable of knowing and recognizing the work of art as such, in particular teachers (but also families, etc.). So it has to take into account not only, as the social history of art usually does, the social conditions of the production of artists, art critics, dealers, patrons, etc., as revealed by indices such as social origin, education or qualifications, but also the social conditions of the production of a set of objects socially constituted as works of art, i.e. the conditions of production of the field of social agents (e.g. museums, galleries, academies, etc.) which help to define and produce the value of works of art. In short, it is question of understanding works of art as a *manifestation* of the field as a whole, in which all the powers of the field, and all the determinisms inherent in its structure and functioning, are concentrated.

Notes

1. Or any other kind of field; art and literature being one area among others for the application of the method of object-construction designated by the concept of the field.
2. Since it is not possible to develop here all that is implied in the notion of the field, one can only refer the reader to earlier works which set out the conditions of the application in the social sciences of the relational mode of thought which has become indispensable in the natural sciences (Pierre Bourdieu, "Structuralism and Theory of Semiological Knowledge," *Social Research*, 35: 4 (1968), pp. 681–706) and the differences between the field as a *structure of objective relations* and the *interactions* studied by Weber's analysis of religious agents or by interactionism (Pierre Bourdieu, "Une interpretation de la sociologie religieuse de Max Weber," *Archives européenes de sociologie*, 12: 1 (1971), pp. 3–21).
3. In this (and only this) respect, the theory of the field could be regarded as a generalized Marxism, freed from the realist mechanism implied in the theory of "instances."

Chapter 17

William J. Bennett (1943–) from "To Reclaim a Legacy," *American Education* (1985)

Our civilization cannot effectively be maintained where it still flourishes, or be restored where it has been crushed, without the revival of the central, continuous and perennial culture of the Western world.

Walter Lippmann, 1941

One reason I wanted to make the gift (was) to remind young people that the liberal arts are still the traditional highway to great thinking and the organization of a life.

James Michener, appearing on the September 26, 1984,
CBS Morning News on the occasion of his
$2 million gift to Swarthmore college.

Although more than 50 percent of America's high school graduates continue their education at American colleges and universities, few of them can be said to receive there an adequate education in the culture and civilization of which they are members. Most of our college graduates remain shortchanged in the humanities—history, literature, philosophy, and the ideals and practices of the past that have shaped the society they enter. The fault lies principally with those of us whose business it is to educate these students. We have blamed others, but the responsibility is ours. Not by our words but by our actions, by our indifference, and by our intellectual diffidence, we have brought about this condition. It is we the educators—not scientists, business people, or the general public—who too often have given up the great task of transmitting a culture to its rightful heirs. Thus, what we have on many of our campuses is an unclaimed legacy, a course of studies in which the humanities have been siphoned off, diluted, or so adulterated that students graduate knowing little of their heritage.

In retrospect, this essay by William Bennett, former Chairman of the National Endowment for the Humanities and Secretary of Education, can be taken as the beginning of the widespread, public debate in the United States over the canon, a period in the late 1980s through the mid-1990s that coincides with the "Culture Wars."

In particular, the study group was disturbed by a number of trends and developments in higher education:

- Many of our colleges and universities have lost a clear sense of the importance of the humanities and the purpose of education, allowing the thickness of their catalogues to substitute for vision and a philosophy of education.
- The humanities, and particularly the study of Western civilization, have lost their central place in the undergraduate curriculum. At best, they are but one subject among many that students might be exposed to before graduating. At worst, and too often, the humanities are virtually absent.
- A student can obtain a bachelor's degree from 75 percent of all American colleges and universities without having studied European history; from 72 percent without having studied American literature or history; and from 86 percent without having studied the civilizations of classical Greece and Rome.
- Fewer than half of all colleges and universities now require foreign language study for the bachelor's degree, down from nearly 90 percent in 1966.
- The sole acquaintance with the humanities for many undergraduates comes during their first two years of college, often in ways that discourage further study.
- The number of students choosing majors in the humanities has plummeted. Since 1970 the number of majors in English has declined by 57 percent, in philosophy by 41 percent, in history by 62 percent, and in modern languages by 50 percent.
- Too many students are graduating from American colleges and universities lacking even the most rudimentary knowledge about the history, literature, art, and philosophical foundations of their nation and their civilization.
- The decline in learning in the humanities was caused in part by a failure of nerve and faith on the part of many college faculties and administrators, and persists because of a vacuum in educational leadership. A recent study of college presidents found that only 2 percent are active in their institutions' academic affairs.

In order to reverse the decline, the study group recommended:

- The nation's colleges and universities must reshape their undergraduate curricula based on a clear vision of what constitutes an educated person, regardless of major, and on the study of history, philosophy, languages, and literature.
- College and university presidents must take responsibility for the educational needs of all students in their institutions by making plain what the institution stands for and what knowledge it regards as essential to a good education.
- Colleges and universities must reward excellent teaching in hiring, promotion, and tenure decisions.
- Faculties must put aside narrow departmentalism and instead work with administrators to shape a challenging curriculum with a core of common studies.
- Study of the humanities and Western civilization must take its place at the heart of the college curriculum.

Why Study the Humanities?

The federal legislation that established the National Endowment for the Humanities in 1965 defined the humanities as specific disciplines: "language, both modern and classical; linguistics; literature; history; jurisprudence; philosophy; archaeology; comparative religion; ethics; the history, criticism, and theory of the arts"; and "those aspects of the social sciences which have humanistic content and employ humanistic methods." But to define the humanities by itemizing the academic fields they embrace is to overlook the qualities that make them uniquely important and worth studying. Expanding on a phrase from Matthew Arnold, I would describe the humanities as the best that has been said, thought, written, and otherwise expressed about the human experience. The humanities tell us how men and women of our own and other civilizations have grappled with life's enduring, fundamental questions: What is justice? What should be loved? What deserves to be defended? What is courage? What is noble? What is base? Why do civilizations flourish? Why do they decline? [. . .]

Further, the humanities can contribute to an informed sense of community by enabling us to learn about and become participants in a common culture, shareholders in our civilization. But our goal should be more than just a common culture—even television and the comics can give us that. We should, instead, want all students to know a common culture rooted in civilization's lasting vision, its highest shared ideals and aspirations, and its heritage. Professor E.D. Hirsch of the University of Virginia calls the beginning of this achievement "cultural literacy" and reminds us that "no culture exists that is ignorant of its own traditions." As the late philosopher Charles Frankel once said, it is through the humanities that a civilized society talks to itself about things that matter most.

How Should the Humanities
Be Taught and Learned?

Mankind's answers to compelling questions are available to us through the written and spoken word—books, manuscripts, letters, plays, and oral traditions—and also in nonliterary forms, which John Ruskin called the book of art. Within them are expressions of human greatness and of pathos and tragedy. In order to tap the consciousness and memory of civilization, one must confront these texts and works of art.

The members of the study group discussed at length the most effective ways to teach the humanities to undergraduates. Our discussion returned continually to two basic prerequisites for learning in the humanities: good teaching and good curriculum. [. . .]

A Good Curriculum

If the teacher is the guide, the curriculum is the path. A good curriculum marks the points of significance so that the student does not wander aimlessly over the terrain, dependent solely on chance to discover the landmarks of human achievement. [. . .]

The choices a college or university makes for its common curriculum should be rooted firmly in its institutional identity and educational purpose. In successful institutions, an awareness of what the college or university is trying to do acts as a unifying principle, a thread that runs through and ties together the faculty, the curriculum, the students, and the administration. If an institution has no clearly conceived and articulated sense of itself, its efforts to design a curriculum will result in little more than an educational garage sale, possibly satisfying most campus factions but serving no real purpose and adding up to nothing of significance. Developing a common curriculum with the humanities at the core is no easy task. In some institutions it will be difficult to attain. But merely being exposed to a variety of subjects and points of view is not enough. Learning to think critically and skeptically is not enough. Being well rounded is not enough if, after all the sharp edges have been filed down, discernment is blunted and the graduate is left to believe without judgment, to decide without wisdom, or to act without standards. [. . .]

What Should be Read?

A curriculum is rarely much stronger than the syllabi of its courses, the arrays of texts singled out for careful reading and discussion. The syllabi should reflect the college's best judgment concerning specific texts with which an educated person should be familiar and should include texts within the competence and interest of its faculty.

Study group members agreed that an institution's syllabi should not be set in stone; indeed, these syllabi should change from time to time to take into account the expertise of available faculty and the result of continuing scrutiny and refinement. The task, however, is not to take faculty beyond their competence and training, nor to displace students' individual interests and career planning, but to reach and inhabit common ground for a while.

We frequently hear that it is no longer possible to reach a consensus on the most significant thinkers, the most compelling ideas, and the books all students should read. Contemporary American culture, the argument goes, has become too fragmented and too pluralistic to justify a belief in common learning. Although it is easier (and more fashionable) to doubt than to believe, it is a grave error to base a college curriculum on such doubt. Also, I have long suspected that there is more consensus on what the important books are than many people have been willing to admit.

In order to test this proposition and to learn what the American public thinks are the most significant works, I recently invited several hundred educational and cultural leaders to recommend ten books that any high school graduate should have read. The general public was also invited in a newspaper column by George F. Will to send me their lists. I received recommendations from more than five hundred individuals. They listed hundreds of different texts and authors, yet four— Shakespeare's plays, American historical documents (the Constitution, Declaration of Independence, and Federalist Papers), *The Adventures of Huckleberry Finn*, and the Bible—were cited at least 50 percent of the time.

I have not done a comparable survey on what college graduates should read, but the point to be made is clear: Many people do believe that some books are more important than others, and there is broader agreement on what those books are than many have supposed. Each college's list will vary somewhat, reflecting the character of the institution and other factors. But there would be, and should be, significant overlap. [. . .]

The works and authors I have in mind include, but are not limited to, the following: from classical antiquity—Homer, Sophocles, Thucydides, Plato, Aristotle, and Vergil; from medieval, Renaissance, and seventeenth-century Europe—Dante, Chaucer, Machiavelli, Montaigne, Shakespeare, Hobbes, Milton, and Locke; from eighteenth through twentieth-century Europe—Swift, Rousseau, Austen, Wordsworth, Tocqueville, Dickens, George Eliot, Dostoyevsky, Marx, Nietzsche, Tolstoy; Mann, and T.S. Eliot; from American literature and historical documents—the Declaration of Independence, the Federalist Papers, the Constitution, the Lincoln-Douglas Debates, Lincoln's Gettysburg Address and Second Inaugural Address, Martin Luther King, Jr.'s "Letter from the Birmingham Jail" and "I have a dream . . ." speech, and such authors as Hawthorne, Melville, Twain, and Faulkner. Finally, I must mention the Bible, which is the basis for so much subsequent history, literature and philosophy. At a college or university, what weight is given to which authors must of course depend on faculty competence and interest. But should not every humanities faculty possess some members qualified to teach at least something of these authors?

Why these particular books and these particular authors? Because an important part of education is learning to read, and the highest purpose of reading is to be in the company of great souls. There are, to be sure, many fine books and important authors not included in the list, and they too deserve the student's time and attention. But to pass up the opportunity to spend time with this company is to miss a fundamental experience of higher education.

Great souls do not express themselves by the written word only; they also paint, sculpt, build, and compose. An educated person should be able not only to recognize some of their works, but also to understand why they embody the best in our culture. Should we be satisfied if the graduates of our colleges and universities know nothing of the Parthenon's timeless classical proportions, of the textbook in medieval faith and philosophy that is Chartres cathedral, of Michelangelo's Sistine ceiling, or of the music of Bach and Mozart? [. . .]

The humanities are important, not to just a few scholars, gifted students, or armchair dilettantes, but to any person who would be educated. They are important precisely because they embody mankind's age-old effort to ask the questions that are central to human existence. As Robertson Davies told a college graduating class, "a university education is meant to enlarge and illuminate your life." A college education worthy of the name must be constructed upon a foundation of the humanities. Unfortunately, our colleges and universities do not always give the humanities their due. All too often teaching is lifeless, arid, and without commitment. On too many campuses the curriculum has become a self-service cafeteria through which students pass without being nourished. Many academic leaders lack the confidence to assert that the curriculum should stand for something more than salesmanship, compromise, or special interest politics. Too many colleges and universities have no clear sense of their educational mission and no conception of what a graduate of their institution ought to know or be.

The solution is not a return to an earlier time when the classical curriculum was the only curriculum and college was available to only a privileged few. American higher education today serves far more people and many more purposes than it did a century ago. Its increased accessibility to women, racial and ethnic minorities, recent immigrants, and students of limited means is a positive accomplishment of which our nation is rightly proud. As higher education broadened, the curriculum became more sensitive to the long-overlooked cultural achievements of many groups, what Janice Harris of the University of Wyoming referred to as "a respect for diversity." This too is a good thing. But our eagerness to assert the virtues of pluralism should not allow us to sacrifice the principle that formerly lent substance and continuity to the curriculum, namely that each college and university should recognize and accept its vital role as conveyor of the accumulated wisdom of our civilization.

We are a part and a product of Western civilization. That our society was founded upon such principles as justice, liberty, government with the consent of the governed, and equality under the law is the result of ideas descended directly from great epochs of Western civilization—Enlightenment England and France, Renaissance Florence, and Periclean Athens. These ideas, so revolutionary in their times yet so taken for granted now, are the glue that binds together our pluralistic nation. The fact that we as Americans—whether black or white, Asian or Hispanic, rich or poor—share these beliefs aligns us with other cultures of the Western tradition. It is not ethnocentric or chauvinistic to acknowledge this. No student citizen of our civilization should be denied access to the best that tradition has to offer.

Ours is not, of course, the only great cultural tradition the world has seen. There are others, and we should expect an educated person to be familiar with them because they have produced art, literature, and thought that are compelling monuments to the human spirit and because they have made significant contributions to our history. Those who know nothing of these other traditions can neither appreciate the uniqueness of their own nor understand how their own fits with the larger world. They are less able to understand the world in which they live. The college curriculum must take the non-Western world into account, not out of political expediency or to appease interest groups, but out of respect for its importance in human history. But the core of the American college curriculum—its heart and soul—should be the civilization of the West, source of the most powerful and pervasive influences on America and all of its people. It is simply not possible for students to understand their society without studying its intellectual legacy. If their past is hidden from them, they will become aliens in their own culture, strangers in their own land.

Chapter 18

Elizabeth Meese (1943–)
from "Sexual Politics and Critical Judgment,"
in Gregory S. Jay and David L. Miller, Eds.,
After Strange Texts (1985)

How the critical community establishes literary reputation is at the heart of the problem for women writers and feminist critics. The complexity of the problem reveals itself easily in the questions it encompasses: What is great literature? How do we know when a book is a "classic"? What works comprise the literary canon and what principles inform the selection of texts? Who decides and by what means? The answers to these questions are, in theory, kept somewhat fluid. Obviously, certain writers like Chaucer, Shakespeare, and Milton enjoy permanence, but then there are numerous others whose reputations remain in a state of flux, waxing and waning in accord with the prevailing interests of the critical moment. In "Literature as an Institution: The View from 1980," Leslie Fiedler cynically observes: "We all know in our hearts that literature is effectively what we teach in departments of English; or conversely, what we teach in departments of English is literature. Within that closed definitional circle, we perform the rituals by which we cast out unworthy pretenders from our ranks and induct true initiates, guardians of the standards by which all song and story ought presumably to be judged."[1] The effects of this kind of exclusion are transparent: it places literature almost entirely in the service of white, male elite culture. The significance of works by writers outside of the mainstream is effectively diminished; as Tillie Olsen explains, "The rule is simple: whenever anyone of that sex, and/or class, and/or color, generally denied enabling circumstances, comes to recognized individual achievement, it is not by virtue of special capacity, courage, determination, will (common qualities), but because of chancy luck, combining with those qualities."[2] As most contemporary writers admit, albeit reluctantly, after the slings and arrows of the marketplace, it is the critics and teachers who create literary reputations, and critical neglect,

---

In this essay, Elizabeth Meese, Professor of English and adjunct professor of Women's Studies at the University of Alabama, argues that the conventions of male critical communities marginalize female writers.

whether occasioned by overt hostility or benign disinterest, produces the same result: women's works are not read, taught, studied, or discussed. [. . .]

If we are to expand our consideration of literature beyond the traditional literary canon, we need to understand the critical dynamics underlying the perpetuation of conventions. In his collection of essays, *Is There a Text in This Class? The Authority of Interpretive Communities*, Stanley Fish presents a view of critical judgments as issuing from an interpretive community, which, when examined from a feminist perspective, provides a useful means of describing the nature of critical bias. Perhaps inadvertently, Fish helps us to see clearly what we have always intuited. A strong insider–outsider dynamic, taking the form of a gender-based literary tribalism, comes into play as a means of control. Critics who permit the possibility of variations in critical interpretation, as opposed to those seeking the *Ur*-reading, immediately face the problem of closing ranks against the extremes of relativism in interpretation. Otherwise, the authority of the mainstream literary tradition could be seriously threatened. Fish guards against this by invoking the concept of "community":

> What will, at any time, be recognized as literature is a function of a communal decision as to what will count as literature. All texts have the potential of so counting, in that it is possible to regard any stretch of language in such a way that it will display those properties presently understood to be literary. In other words, it is not that literature exhibits certain formal properties that compel a certain kind of attention; rather, paying a certain kind of attention (as defined by what literature is understood to be) results in the emergence into noticeability of the properties we know in advance to be literary. The conclusion is that while literature is still a category, it is an open category, not definable by fictionality, or by a disregard of propositional truth, or by a predominance of tropes and figures, but simply by what we decide to put into it.[3]

[. . .] Once the illegitimate children perceive the exclusivity masked in the illusion of objectivity that is perpetuated by this interpretive community, considerable bitterness results. Still resonant today is the outsider's view expressed by Virginia Woolf in *Three Guineas* (Olsen calls this work a "savage" essay emerging from "genius brooding on . . . exclusion"),[4] and epitomized in her fictitious "Outsiders' Society," an anonymous and secret society for the daughters of educated men. In the passage that follows, Woolf's persona is ironically the "insider" (occupant of domestic space) looking "sidelong from an upper window" at the "solemn sight" of the male community in all its awesome symbolic and real power, enrobed and ascendant. They move freely, related to each other in the procession of generations:

> There they go, our brothers who have been educated at public schools and universities, mounting those steps, passing in and out of those doors, ascending those pulpits, preaching, teaching, administering justice, practising medicine, transacting business, making money. It is a solemn sight always—a procession, like a caravanserai crossing the desert. Great-grandfathers, grandfathers, fathers, uncles—they all went that way, wearing their gowns, wearing their wigs, some with ribbons across their breasts, others without. One was a bishop. One was a professor. Another a doctor.[5]

They are self-perpetuating in their authority, these generations of powerful men. The ones who drop out of the procession are excluded in a manner similar to women.

They are cloaked in silence, distant; or, divested, of robes, wigs and ribbons, they wear shabby clothes and hold only menial jobs.

Critics like Fish, Bloom, and Abrams genuinely believe in their community of critics; they march in the procession, speaking the truth from their own positions of privilege but suggesting other truths to feminists, Marxists, and critics of the non majority culture. We see that the "interpretive community" is really the "authoritative community." Even though Fish regards criticism as an "open category," we are forced to see it, like his version of community, as a closed system which excludes us from the arena of its authority. In her time, Woolf perceived a similar circularity in the closed system of the great English universities: "With what other purpose were the universities of Oxford and Cambridge founded, save to protect culture and intellectual liberty? For what other object did your sisters go without teaching or travel or luxuries themselves except that with the money so saved their brothers should go to schools and universities and there learn to protect culture and intellectual liberty?"[6] The fact that literature is simply another cultural institution requiring protection dictates a process of circumscription.

Interpretive communities, like tribal communities, possess the power to ostracize or to embrace, to restrict or to extend membership and participation, and to impose norms—hence their authority. In her article "Dancing through the Minefield: Some Observations on the Theory, Practice and Politics of a Feminist Criticism," Annette Kolodny notes that "the power relations inscribed in the form of conventions within our literary inheritance . . . reify the encodings of those same power relations in the culture at large."[7] The system is mutually reinforcing—designed and chosen to mirror a system of power relationships. Thus, Fish states explicitly that credible interpretations issue not from just any critic but from members of the club: "The reader is identified not as a free agent, making literature in any old way, but as a member of a community whose assumptions about literature determine the kind of attention he pays and thus the kind of literature he 'makes'. . . . The act of recognizing literature . . . proceeds from a collective decision as to what will count as literature, a decision that will be in force only so long as a community of readers and believers continues to abide by it" (p. 11). His remarks contain the answer to the question, Why the failure of so many feminist commentaries aimed at demonstrating the stature of neglected works by women? [. . . .]

Control by such a relatively homogeneous group of critics has resulted in extremely narrow views of what great literature is and what criticism does, not so much because critics enjoy seeing reflections of themselves and their values in what they praise (though this is partially true), but because they pretend to equality, objectivity, and universality. [. . .] The fundamental assumptions underlying judgments are disguised, perhaps even from those who adhere to them, and produce a distortion in the act of reading itself. Margaret Atwood offers a characteristic description of the woman writer's experience with phallic criticism: "A man who reviewed my Procedures for Underground . . . talked about the 'domestic' imagery of the poems, entirely ignoring the fact that seven-eighths of the poems take place outdoors . . . In this case, the theories of what women ought to be writing about, had intruded very solidly between the reader and poems, rendering the poems themselves invisible to him."[8] The result of criticism like this is that we need to consider everything anew, in a complete re-vision of women's work from text to theory. [. . .] By virtue of their pretense to critical objectivity, literary critics have created the need for a criticism of advocacy, espousing special values based on gender, ethnicity, race, and class. Woolf detected this intrusion of gender in critical assumptions at work; as the values of men and women differ in life, so these differences are reflected in literary judgment: "This is an important book, the critic assumes, because it deals with war. This is an insignificant book because it deals with the feelings of women in a drawing room. A scene in a battlefield is more important than a scene in a shop everywhere and much more subtly the difference of value persists."[9] This axiological discrepancy creates the need for feminist criticism to base its work at times on different texts from those designated as the literary canon by representatives of the current regime of truth. [. . .]

The most fortunate circumstance of women's writing in the Western world is that its production, though ignored, devalued, and misrepresented, could never be completely extinguished. Women could be denied education, employment, publication, and honest critical appraisal, but as Woolf keenly observed, ink and paper were the cheapest, most readily available tools for the practice of one's trade. She immediately detected this essential feature of the "profession of literature": "There is no head of the profession; no Lord Chancellor . . . no official body with the power to lay down rules and enforce them. We cannot debar women from the use of libraries; or forbid them to buy ink and paper; or rule that metaphors shall only be used by one sex, as the male only in Academies of music was allowed to play in orchestras."[10] Women could not be programmed or policed thoroughly enough to keep them out of the literary profession. They were saved by the very nature of the craft itself: the act of writing is solitary, accessible to anyone who is literate (still a fact of its elitism), and as such the institution of literature admits its own subversion.

Feminist criticism is a monumental undertaking which involves changing the very structure/sex of knowledge, thereby attempting to liberate us from what Diana Hume George calls an "operational model that artificially . . . dualizes intellectual activity and sexuality."[11] The problem confronting us is both epistemological and political—each equally significant and inseparable from the other. For years feminist critics have hedged on both counts, wanting to believe on the one hand in that "theoretical equality," and fearing on the other the fragmentation that could result from definition and the articulation of methodology (an inchoate ideological map). Consciously or not, we have obscured the terms of the dispute, and with them the need to differentiate between criticism written by women and feminist criticism. Within feminist criticism, we have avoided both the political and the epistemological, as though there were no purpose in recapitulating the politics of gender (which threaten to separate women from men and from the institutions of culture). [. . .]

Some proponents of poststructuralism, engaged in their own attack on the ideological character of discourse, believe that criticism has finally freed itself of its orientation toward objectivity and universality. It is tempting to regard the poststructuralist position as pervasive, characteristic of criticism of the past decade. And yet, far from epitomizing critical activity today, poststructuralists have made only a beginning in their attack on the historically rooted traditions of criticism. [. . .]

It has never been the obligation of literary critics, masked by the pretense of objectivity, to explicate the political origins and implications of their judgments. As a result, feminist critics need to question vigorously the methods and techniques of the inherited critical tradition. For example, Fish, in his notion of the interpretive (authoritative) community, proffers equality: literature is an open system, admitting any text (within reason); variations in interpretation are permitted (within reason); and persuasion is the means by which (reasonable) critics establish consensus. When he reinvests the authority for determining the limits of the reasonable in the profession as it is now constituted, Fish reinscribes the politics of exclusion he might have undone by defining literature as an open category and defending interpretive pluralism. But inherent in Fish's approach is the fact that the right to reason and the power of determination are located where power and reason have always rested in Western civilization—within the community of elite white men. He thereby preserves theoretical access at the expense of actual change. It makes sense to suspect, as Marxist critics have always noted from their vantage point, that our conceptual frameworks mirror ideology. The principle task of feminist criticism, in providing a necessary re-vision of the politics of "truth," is to make its own ideology explicit. If we seek to transform the structures of authority, we must first name them, and in doing so, unmask and expose them for all to see. As we forge a new criticism, our theories and assumptions must stay clear of a hegemonic role reversal that results from unending deconstructions of oppositions like male/female and insider/outsider, where the second term simply replaces the first in an infinite regression within an economy of oppression. The future of feminist criticism rests on defying the oppositional logic currently fostering the very concept of privilege.

Notes

1. Leslie Fiedler, "Literature as an Institution: The View from 1980," in *English Institute: Opening up the Canon*, Selected Papers from the English Institute, 1979, ed. Leslie Fiedler and Houston A. Baker, Jr. (Baltimore: Johns Hopkins University Press, 1981), pp. 73–74.
2. Tillie Olsen, "One Out of Twelve: Writers Who Are Women in Our Century, *Silences* (New York: Seymour Lawrence, 1978), pp. 22–46, 223. Marxist critics, like Fredric Jameson in *The Political Unconscious* (Ithaca, NY: Cornell University Press, 1981), often exclude sex and race as features that also mark a text's exclusion from the literary canon. Most often Marxists consider class the fundamental term in the nexus of sex, race, and class. See, for example, Angela Davis's *Women, Race and Class* (New York: Random House, 1981), in which women's complicity in race and class oppression is considered at the complete expense of knowledge concerning women as an oppressed group.

   In *Sex, Class and Culture* (Bloomington: Indiana University Press, 1978), Lillian S. Robinson makes a useful point concerning the politics of exclusion: "Within the limits of literature, at least, women's exclusion is clearly shared by all non-white and workingclass men" (p. 4). As a Marxist feminist, Robinson is careful not to set exclusionary terms in a hierarchy. As a result, she focuses more attention on women, inclusive of race and class concerns, and calls for a literature reflective of the whole culture.

3. Stanley Fish, *Is There a Text in This Class? The Authority of Interpretive Communities* (Cambridge: Harvard University Press, 1980), pp. 10–11. Further references to this work are in the text.
4. Olsen, "One Out of Twelve," p. 244.
5. Virginia Woolf, *Three Guineas* (New York: Harcourt, Brace and World, 1938), p. 61.
6. Ibid., p. 86.
7. Annette Kolodny, "Dancing through the Minefield: Some Observations on the Theory, Practice and Politics of a Feminist Criticism," *Feminist Studies*, 6, No. 1 (Spring 1980) p. 4.
8. Quoted by Olsen, "One Out of Twelve," p. 229.
9. Virginia Woolf, *A Room of One's Own* (New York: Harcourt, Brace and World, 1929), p. 77.
10. Woolf, *Three Guineas*, p. 90.
11. Diana Hume George, "Stumbling on Melons: Sexual Dialectics and Discrimination in English Departments," *English Literature: Opening up the Canon*, p. 109.

Chapter 19

Jane Tompkins (1940–)
from " 'But Is It Any Good?': The
Institutionalization of Literary Value,"
Sensational Designs (1985)

The objection, as I have phrased it, is never put in exactly this way, but usually takes the form of a question like: but are these works really any *good*? or, what about the *literary* value of *Uncle Tom's Cabin*? or, do you really want to defend Warner's *language*? These questions imply that the standards of judgment to which they refer are not themselves challengeable, but are taken for granted among qualified readers. "You and I know what a good novel is," the objection implies, "and we both know that these novels fall outside that category." But the notion of good literature that the question invokes is precisely what we are arguing about. That tacit sense of what is "good" cannot be used to determine the value of these novels because literary value *is* the point at issue. At this juncture, people will frequently attempt to settle the question empirically by pointing to one or another indisputably "great" work, such as *Moby-Dick* or *The Scarlet Letter*, and asking whether *The Wide, Wide World* is as good as *that*.

But the issue cannot be settled by invoking apparently unquestionable examples of literary excellence such as these as a basis of comparison, because these texts already represent one position in the debate they are being called upon to decide. That is, their value, their identity, and their constituent features have been made available for description by the very modes of perception and evaluation that I am challenging. It is not from any neutral space that we have learned to see the epistemological subtleties of Melville or Hawthorne's psychological acuity. Those characteristics have been made available by critical strategies that have not always been respectable, but had to be explained, illustrated, and argued for (as I am arguing now) against other critical assumptions embodied in other masterpieces that seemed just as invincible, just as unquestionably excellent as these now do. Such strategies do not remain stable and do not emerge in isolation, but are forged in the context of revolutions, revivals,

Jane Tompkins is Professor of English at the University of Illinois, Chicago. In this essay, she focuses on literary anthologies as institutional forces shaping cultural value.

periods of consolidation or reform—in short, in the context of all those historical circumstances by which literary values are supposed to remain unaffected. Even in the last sixty years, the literary canon has undergone more than one major shift as the circumstances within which critics evolved their standards of judgment changed.

The evidence for this assertion becomes dramatically available when one examines the history of literary anthologies.[1] Between the time that Fred Pattee made selections for *Century Readings for a Course in American Literature* (1919) and the time that Perry Miller and his coeditors decided whom to include in *Major Writers of America* (1962), the notion of who counted as a major writer and even the concept of the "major writer" had altered dramatically. Whereas Pattee's single volume, compiled at the close of World War I, contained hundreds of writers, Miller's much larger two-volume work, published at the close of the Cold War, contained only twenty-eight.[2] Three years earlier, Gordon Ray's *Masters of American Literature* had reduced the number to eighteen; the Macmillan anthology, published in the same year as Miller's, had pared the number to twelve; and in 1963, a Norton anthology edited by Norman Foerster and Robert Falk had reduced it to only eight.[3] "In choosing Emerson, Thoreau, Hawthorne, Poe, Melville, Whitman, Mark Twain, James, Emily Dickinson, Frost, Eliot, and Faulkner," the Macmillan editors write, "we can imagine little dispute."[4] But if they had looked back at the literary anthologies published since Pattee's, they might have been less sure about the absence of debate. Howard Mumford Jones and Ernest Leisy, in the preface to *Major American Writers* (1935), state categorically that "there can be no question that Franklin, Cooper, Irving, Bryant, Emerson, Hawthorne, Longfellow, Whittier, Lincoln, Poe, Thoreau, Lowell, Melville, Whitman, and Mark Twain constitute the heart of any course in American literary history."[5] [. . .]

Indeed, if we take Pattee and Miller as representative, we can see that in addition to a sharp narrowing in the range and number of authors, there has been a virtual rewriting of literary history, as entire periods, genres, and modes of classification disappear. Between 1919 and 1962 more than a dozen authors have dropped away in the Colonial period alone, while in the Revolutionary period, only one out of seven makes it through; the Revolutionary songs and ballads are missing entirely. The Federalist period disappears altogether and so does most of the first half of the nineteenth century. Gone are the fin-de-siècle poets—John Trumbull, Timothy Dwight, Joel Barlow—and with them the lyricists of the early century—Richard Henry Dana, Edward Coate Pinckney, Richard Henry Wilde, and John Howard Payne. None of the songwriters survive—George Pope Morris, Samuel Woodworth, Thomas Dunn English, Phoebe Cary, Stephen Foster. The selections from D. G. Mitchell's *Reveries of a Bachelor* disappear, along with the orations of John C. Calhoun and Daniel Webster. The historians of the mid-nineteenth century, W. H. Prescott, John Lothrop Motley, and Francis Parkman vanish, as do the southern writers (Simms, Timrod, Paul Hamilton Hayne) and the antislavery writers—Whittier and Stowe. Gone are Abraham Lincoln and all the songs and ballads of the Civil War. Out of six western humorists, only Twain survives; of the "transition poets"—Bayard Taylor, Edmund Clarence Stedman, Thomas Bailey Aldrich, Sidney Lanier, Thomas Buchanan Read, George Henry Boker, Richard Henry Stoddard, and Celia Leighton Thaxter—not one. Of the late nineteenth-century nature writers, not one. Out of a dozen poets of the same period, only Crane and Dickinson. The local colorists—Bret Harte,

General Lewis Wallace, Edward Eggleston, John Hay, Joaquin Miller, Helen Hunt Jackson, Henry Grady, Hamlin Garland, George Washington Cable, Joel Chandler Harris, Sarah Orne Jewett, Mary Wilkins Freeman, Mary Noialles Murfree, Charles Dudley Warner—cede their places to Henry James, Henry Adams, and Theodore Dreiser. The critics are wiped out in toto, along with Edward Everett Hale, Ambrose Bierce, Henry Cuyler Bunner, and Frank Stockton. The "feminine novelists" of the twentieth century whom Pattee added to his 1932 edition—Willa Cather and Edith Wharton—give way to Faulkner, Fitzgerald, and Hemingway, and, with the exception of Frost, all of the twentieth-century poets—Robinson, Lindsay, Masters, Sandburg, Lowell, Sterling, and Millay disappear. [. . .]

I have listed these excisions and revisions at length because they show in a detailed and striking manner that "literature" is not a stable entity, but a category whose outlines and contents are variable. The anthologies of the 1930s, midway between Pattee and Miller, show unmistakably that this variability is a function of the political and social circumstances within which anthologists work.[6] The thirties' anthologies include items that had not appeared in such collections before and have seldom appeared there since—cowboy songs, Negro spirituals, railroad songs, southwestern yams, and, in translation, the songs and prayers of Native Americans. They include letters, extracts from journals, passages from travel literature, and a large number of political speeches—Woodrow Wilson's "Address to Newly Naturalized Citizens," Lee's "Farewell to the Army of Northern Virginia." There are essays by Margaret Fuller and Sophia Ripley from *The Dial*, excerpts from Henry George's *Progress and Poverty* and *Social Problems*, William James' "What Pragmatism Means," and John Fiske's "Darwinism Verified." There are descriptions of America written by European writers, and a great deal of writing about, as well as by, Abraham Lincoln. One anthology, prepared by teachers from New York City, even turns the last forty pages into a sort of "melting pot" selection from the literatures of Europe and the Orient— passages from the Egyptian Book of the Dead; the sayings of Confucius and Gautama Buddha; an excerpt from Lady Murasaki; Greek, Hebrew, and Latin poetry; and translations from the literatures of Germany, Scandinavia, France, Spain, Italy, and Russia.[7] [. . .]

Yet even though anthologists characterize their projects differently, and although the contents of their volumes vary drastically, the one element that, ironically, remains unchanged throughout them all is the anthologists' claim that their main criterion of selection has been literary excellence. But, as has by now become abundantly clear, while the term "literary excellence" or "literary value" remains constant over time, its meaning—what literary excellence turns out to be in each case—does not. Contrary to what Miller believed, great literature does not exert its force over and against time, but changes with the changing currents of social and political life.

Still, someone might object that Miller's theory, whatever its abstract merits, justifies itself on practical grounds. Surely the authors represented in his anthology *are* the major writers of America, give or take a few names, while the works the anthology excludes are minor works at best. Most educated people today, if asked to say which was better, a poem by Stedman or a poem by Dickinson, would choose the latter without hesitation. And this fact would seem to bear out the rightness of

Miller's intuitions about which writers ought to be considered great. But our conviction that Miller's choice was correct does not prove anything about the intrinsic superiority of the texts he chose; it proves only that we were introduced to American literature through the medium of anthologies similar to his. The general agreement about which writers are great and which are minor that exists at any particular moment in the culture creates the impression that these judgments are obvious and self-evident. But their obviousness is not a natural fact; it is constantly being produced and maintained by cultural activity: by literary anthologies, by course syllabi, book reviews, magazine articles, book club selections, radio and television programs, and even such apparently peripheral phenomena as the issuing of commemorative stamps in honor of Hawthorne and Dickinson, or literary bus tours of New England stopping at Salem and Amherst. The choice between Stedman and Dickinson, Stowe and Hawthorne, is never made in a vacuum, but from within a particular perspective that determines in advance which literary works will seem "good."

In saying that judgments of literary value are always perspectival, and not objective or disinterested, I do not wish to be understood as claiming that there is no such thing as value or that value judgments cannot or should not be made. We are always making choices, and hence value judgments, about which books to read, teach, write about, recommend, or have on our shelves. The point is not that these discriminations are baseless; the point is that the grounds on which we make them are not absolute and unchanging but contingent and variable. As Barbara Smith has recently argued, our tastes, emphases, preferences, and priorities, literary or otherwise, do not exist in isolation, but emerge from within a dynamic system of values which determines what, at a given moment, will be considered best.[8] Thus, for example, when the anthology editors of the late fifties and early sixties decided to limit their selection of American writers to a handful, they did so within a framework of critical beliefs that were themselves embedded in a larger cultural context. The notion that fullness and depth of representation are preferable to variety was already implicit in the New Critical insistence on studying "wholes" rather than "parts"; and that insistence, in turn, was implicit in the premium that formalism placed on making judgments about the aesthetic as opposed to the historical significance of works of art. Moreover, the formalist doctrines that stood behind the exclusivity of these anthologies did not take shape in isolation either, but were themselves implicated in a web of political, legislative, demographic, and institutional circumstances, and of disciplinary rivalries, that affected the way critics articulated and carried out their aims.

The New Critics' emphasis on the formal properties of literary discourse was part of a struggle that literary academicians had been waging for some time to establish literary language as a special mode of knowledge, so that criticism could compete on an equal basis with other disciplines, and particularly with the natural sciences, for institutional support. That struggle, whose nature had been determined by the growing prestige of science in the twentieth century, was intensified in the fifties by the arms race and especially by the launching of Sputnik, which added impetus to the rivalry between the sciences and the humanities and urgency to the claims that critics made for the primacy of form in understanding "how poems mean."[9] At the same time, the emphasis on formal properties accommodated another feature of the academic scene in the 1950s, namely, the tripling of the college population, brought about by the GI Bill, postwar affluence, and an increasing demand for people with advanced degrees.[10] The theory of literature that posited a unique interrelation of form and content justified close reading as an analytic technique that lent itself successfully to teaching literature on a mass scale. These connections between the contents of literary anthologies and historical phenomena such as the Depression, the GI Bill, and the arms race, show that *literary* judgments of value do not depend on literary considerations alone, since the notion of what is literary is defined by and nested within changing historical conditions of the kind I have outlined here. Thus, the emphasis on "major" writers did not come about in response to a sudden perception of the greatness of a few literary geniuses; it emerged from a series of interconnected circumstances that moved the theory, teaching, and criticism of literature in a certain direction. [. . .]

You will recall that the entire argument thus far has been a response to the question "but is it any good?" which implies that the works I have discussed are not really literary and are therefore not worth discussing. My tactic has been to show that the assumptions behind this question—namely, that literary values are fixed, independent, and demonstrably present in certain masterworks—are mistaken, and I have used the evidence of the literary anthologies to challenge these notions one by one. But at this point someone might observe that despite changes in the contents of the anthologies, there are some authors and some works that do persist from one decade to the next and that therefore, although the perimeters of the canon may vary, its core remains unchanged, a testimony to the enduring merits of a few great masterpieces. To this objection I would reply that the evidence of the anthologies demonstrates not only that works of art are not selected according to any unalterable standard, but that their very essence is always changing in accordance with the systems of description and evaluation that are in force. Even when the "same" text keeps turning up in collection after collection, it is not really the same text at all. [. . .]

It is important to recognize that criticism creates American literature in its own image because American literature gives the American people a conception of themselves and of their history. As a spectacular example of this phenomenon, consider F. O. Matthiessen's *American Renaissance*, of which perhaps the most important sentences are these:

> The half-decade of 1850–55 saw the appearance of *Representative Men* (1850), *The Scarlet Letter* (1850), *The House of the Seven Gables* (1851), *Moby-Dick* (1851), *Pierre* (1852), *Walden* (1854), and *Leaves of Grass* (1855). You might search all the rest of American literature without being able to collect a group of books equal to these in imaginative vitality.[11]

With this list Matthiessen determined the books that students would read and critics would write about for decades to come. More important, he influenced our assumptions about what kind of person can be a literary genius, what kinds of subjects great literature can discuss, our notions about who can be a hero and who cannot, notions of what constitutes heroic behavior, significant activity, central issues. Matthiessen, who believed that criticism should "be for the good and enlightenment of all the people, and not for the pampering of a class," believed that the books he had chosen were truly representative of the American people, for these works, more than any others, called "the whole soul of man into activity."[12]

But from the perspective that has ruled this study, Matthiessen's list is exclusive and class-bound in the extreme. If you look at it carefully, you will see that in certain fundamental ways the list does not represent what most men and women were thinking about between 1850 and 1855, but embodies the views of a very small, socially, culturally, geographically, sexually, and racially restricted elite. None of the works that Matthiessen names is by an orthodox Christian, although that is what most Americans in the 1850s were, and although religious issues pervaded the cultural discourse of the period. None deals explicitly with the issues of abolition and temperance which preoccupied the country in this period, and gave rise to such popular works as *Uncle Tom's Cabin* and T. S. Arthur's *Ten Nights in a Barroom*. None of the works on the list achieved great popular success, although this six-year period saw the emergence of the first American best-sellers. The list includes no works by women, although women at that time dominated the literary marketplace. The list includes no works by males not of Anglo-Saxon origin, and indeed, no works by writers living south of New York, north of Boston, or west of Stockbridge, Massachusetts. From the point of view that has governed the foregoing chapters, these exclusions are a more important indicator of the representativeness of literary works than their power to engage "the whole soul of man."

What I want to stress is that the present study and Matthiessen's are competing attempts to constitute American literature. This book makes a case for the value of certain novels that Matthiessen's modernist critical principles had set at a discount. Instead of seeing such novels as mere entertainment, or as works of art interpretable apart from their context, which derive their value from "imaginative vitality" and address themselves to transhistorical entities such as the "soul of man," I see them as doing a certain kind of cultural work within a specific historical situation, and value them for that reason. I see their plots and characters as providing society with a means of thinking about itself, defining certain aspects of a social reality which the authors and their readers shared, dramatizing its conflicts, and recommending solutions. It is the notion of literary texts as doing work, expressing and shaping the social context that produced them, that I wish to substitute finally for the critical perspective that sees them as attempts to achieve a timeless, universal ideal of truth and formal coherence. The American Renaissance, as we now know it, provides people with an image of themselves and of their history, with conceptions of justice and of human nature, attitudes towards race, class, sex, and nationality. The literary canon, as codified by a cultural elite, has power to influence the way the country thinks across a broad range of issues. The struggle now being waged in the professoriate over which writers deserve canonical status is not just a struggle over the relative merits of literary geniuses; it is a struggle among contending factions for the right to be represented in the picture America draws of itself.

Notes

1. Carolyn Karcher first called my attention to the way literary anthologies reflect the shifting currents of social and political life by referring me to Bruce Franklin's helpful discussion of

this phenomenon in *The Victim as Criminal and Artist: Literature from the American Prison* (New York: Oxford University Press, 1978), pp. xiii–xxii. For another informative account of American literary anthologies and their relation to social and cultural issues, see Paul Lauter, "Race and Gender in the Shaping of the American Canon: A Case Study from the Twenties," *Feminist Studies*, 9, No.3 (Fall 1983), pp. 432–63.

2. *Century Readings for a Course in American Literature*, ed. Fred Lewis Pattee, 1st ed. (New York: The Century Co., 1919); *Major Writers of America*, ed. Perry Miller et al. (New York: Harcourt, Brace & World, 1962), I.

3. *Masters of American Literature*, ed. Gordon N. Ray et al. (Boston: Houghton Mifflin, 1959); *Twelve American Writers*, ed. William M. Gibson and George Arms (New York: Macmillan, 1962); *Eight American Writers*, ed. Norman Foerster and Robert P. Falk (New York: W. W. Norton & Co., 1963).

4. *Twelve American Writers*, ed. Gibson, p. vii.

5. *Major American Writers*, ed. Howard Mumford Jones and Ernest Leisy (New York: Harcourt Brace and Co., 1935), p. v. Lauter points to the even more glaring contrast between the nine writers Foerster selected to represent American prose in 1916 in *The Chief American Prose Writers*, ed. Norman Foerster (Cambridge, MA: The Riverside Press, 1919), and the eight he chose in 1963 in *Eight American Writers*. On the two lists, only three names are the same.

6. *See*, for example, *American Literature*, ed. Thomas H. Briggs et al. (Boston: Houghton Mifflin, 1933); *American Poetry and Prose*, ed. Robert Morss Lovett and Norman Foerster (Boston: Houghton Mifflin, 1934); *Major American Writers*, ed. Jones (New York, 1935). *American Life in Literature*, ed. Jay Hubbell (New York: Harper and Brothers, 1930); and *A College Book of American Literature*, ed. Milton Ellis et al. (New York: American Book Company, 1939). The thirties also saw the appearance of new types of specialized anthologies such as *Proletarian Literature*, ed. Granville Hicks et al. (New York: International Publishers, 1935).

7. This was *American Literature*, ed. Briggs.

8. Barbara Herrnstein Smith, "Contingencies of Value," *Critical Inquiry*, 10, No.1 (September 1983), pp. 1–35.

9. In 1958, eleven months after the launching of Sputnik, the president appointed a special assistant for science and technology, and the government passed the National Defense Education Act, which increased grants given to students of mathematics, the natural and social sciences, and modern languages. See Daniel Snowman, *America Since 1920* (New York: Harper & Row, 1968), p. 128.

10. Paul A. Carter, *Another Part of the Fifties* (New York: Columbia University Press, 1983), p. 169.

11. F. O. Matthiessen, *American Renaissance: Art and Expression in the Age of Emerson and Whitman* (New York: Oxford University Press, 1941), p. vii.

12. Matthiessen, p. xi, actually claims that "successive generations of common readers who make the decisions have agreed that the authors of the pre-Civil War era who bulk the largest in stature are the five who are my subject." But in the period Matthiessen delimits, 1850 to 1855, common readers were engrossed by the works of Susan Warner, Harriet Beecher Stowe, Fanny Fern, Grace Greenwood, Caroline Lee Hentz, Mary Jane Holmes, Augusta Jane Evans, Maria Cummins, D. G. Mitchell, T. S. Arthur, and Sylvanus Cobb, Jr. See James D. Hart, *The Popular Book: A History of America's Literary Taste* (Berkeley: University of California Press, 1950). With the exception of Emerson, none of the authors Matthiessen names was read by the common reader, nor did common readers have a hand in assuring their survival.

Matthiessen, who had been active in leftist politics during the thirties, needed to believe that the works he had chosen represented "all the people," at the same time that, because of his formalist critical commitments, he needed to believe that they met the "enduring requirements for great art." As Jonathan Arac has shown, in "F. O. Matthiessen, Authorizing an American Renaissance," *The American Renaissance Reconsidered*, Selected Papers from the English Institute, 1982–1983, ed. Walter Benn Michaels and Donald E. Pease (Baltimore: The Johns Hopkins University Press, 1985), pp. 90–112, because of the policy of alliance-building adopted by the Popular Front in the late thirties, Matthiessen was able to combine his Christianity, his leftist politics, and his formalist critical allegiance through a strategy of "reconciliation," which emphasized the continuity of the present with the "great tradition" of American literature.

Chapter 20

Martin Bernal (1937–)
from "Volume I: Introduction,"
Black Athena: The Afroasiatic
Roots of Classical Civilization (1987)

These volumes are concerned with two models of Greek history: one viewing Greece as essentially European or Aryan, and the other seeing it as Levantine, on the periphery of the Egyptian and Semitic cultural area. I call them the "Aryan" and the "Ancient" models. The "Ancient Model" was the conventional view among Greeks in the Classical and Hellenistic ages. According to it, Greek culture had arisen as the result of colonization, around 1500 BC, by Egyptians and Phoenicians who had civilized the native inhabitants. Furthermore, Greeks had continued to borrow heavily from Near Eastern cultures.

Most people are surprised to learn that the Aryan Model, which most of us have been brought up to believe, developed only during the first half of the 19th century. In its earlier or "Broad" form, the new model denied the truth of the Egyptian settlements and questioned those of the Phoenicians. What I call the "Extreme" Aryan Model, which flourished during the twin peaks of anti-Semitism in the 1890s and again in the 1920s and 30s, denied even the Phoenician cultural influence. According to the Aryan Model, there had been an invasion from the north— unreported in ancient tradition—which had overwhelmed the local "Aegean" or "Pre-Hellenic" culture. Greek civilization is seen as the result of the mixture of the Indo- European-speaking Hellenes and their indigenous subjects. It is from the construction of this Aryan Model that I call this volume *The Fabrication of Ancient Greece 1785–1985*.

I believe that we should return to the Ancient Model, but with some revisions; hence I call what I advocate in Volume 2 of *Black Athena* the "Revised Ancient Model." This accepts that there is a real basis to the stories of Egyptian and

Professor of Government at Cornell University, Martin Bernal argues in *Black Athena* that Greek civilization finds its roots in southwest Asia and North Africa.

Phoenician colonization of Greece set out in the Ancient Model. However, it sees them as beginning somewhat earlier, in the first half of the 2nd millennium BC. It also agrees with the latter that Greek civilization is the result of the cultural mixtures created by these colonizations and later borrowings from across the East Mediterranean. On the other hand, it tentatively accepts the Aryan Model's hypothesis of invasions—or infiltrations—from the north by Indo-European speakers sometime during the 4th or 3rd millennium BC. However, the Revised Ancient Model maintains that the earlier population was speaking a related Indo-Hittite language which left little trace in Greek. In any event, it cannot be used to explain the many non-Indo-European elements in the later language.

If I am right in urging the overthrow of the Aryan Model and its replacement by the Revised Ancient one, it will be necessary not only to rethink the fundamental bases of "Western Civilization" but also to recognize the penetration of racism and "continental chauvinism" into all our historiography, or philosophy of writing history. The Ancient Model had no major "internal" deficiencies, or weaknesses in explanatory power. It was overthrown for external reasons. For 18th- and 19th-century Romantics and racists it was simply intolerable for Greece, which was seen not merely as the epitome of Europe, but also as its pure childhood, to have been the result of the mixture of native Europeans and colonizing Africans and Semites. Therefore the Ancient Model had to be overthrown and replaced by something more acceptable.

Chapter 21

Allan Bloom (1936–1992)
"The Student and the University," *The Closing of the American Mind* (1987)

Liberal Education

What image does a first-rank college or university present today to a teen-ager leaving home for the first time, off to the adventure of a liberal education? He has four years of freedom to discover himself—a space between the intellectual wasteland he has left behind and the inevitable dreary professional training that awaits him after the baccalaureate. In this short time he must learn that there is a great world beyond the little one he knows, experience the exhilaration of it and digest enough of it to sustain himself in the intellectual deserts he is destined to traverse. He must do this, that is, if he is to have any hope of a higher life. These are the charmed years when he can, if he so chooses, become anything he wishes and when he has the opportunity to survey his alternatives, not merely those current in his time or provided by careers, but those available to him as a human being. The importance of these years for an American cannot be overestimated. They are civilization's only chance to get to him.

In looking at him we are forced to reflect on what he should learn if he is to be called educated; we must speculate on what the human potential to be fulfilled is. In the specialties we can avoid such speculation, and the avoidance of them is one of specialization's charms. But here it is a simple duty. What are we to teach this person? The answer may not be evident, but to attempt to answer the question is already to philosophize and to begin to educate. Such a concern in itself poses the question of the unity of man and the unity of the sciences. It is childishness to say, as some do, that everyone must be allowed to develop freely, that it is authoritarian to impose a point of view on the student. In that case, why have a university? If the response is

Political scientist Allan Bloom taught at Cornell University, the University of Toronto, and the University of Chicago. In this selection from his best-selling book, *The Closing of the American Mind*, Bloom defends an education in the Great Books as a counter to what he sees as a fragmentation of the curriculum in post-secondary schooling.

"to provide an atmosphere for learning," we come back to our original questions at the second remove. Which atmosphere? Choices and reflection on the reasons for those choices are unavoidable. The university has to stand for something. The practical effects of unwillingness to think positively about the contents of a liberal education are, on the one hand, to ensure that all the vulgarities of the world outside the university will flourish within it, and, on the other, to impose a much harsher and more illiberal necessity on the student—the one given by the imperial and imperious demands of the specialized disciplines unfiltered by unifying thought.

The university now offers no distinctive visage to the young person. He finds a democracy of the disciplines—which are there either because they are autochthonous or because they wandered in recently to perform some job that was demanded of the university. This democracy is really an anarchy, because there are no recognized rules for citizenship and no legitimate titles to rule. In short there is no vision, nor is there a set of competing visions, of what an educated human being is. The question has disappeared, for to pose it would be a threat to the peace. There is no organization of the sciences, no tree of knowledge. Out of chaos emerges dispiritedness, because it is impossible to make a reasonable choice. Better to give up on liberal education and get on with a specialty in which there is at least a prescribed curriculum and a prospective career. On the way the student can pick up in elective courses a little of whatever is thought to make one cultured. The student gets no intimation that great mysteries might be revealed to him, that new and higher motives of action might be discovered within him, that a different and more human way of life can be harmoniously constructed by what he is going to learn.

Simply, the university is not distinctive. Equality for us seems to culminate in the unwillingness and incapacity to make claims of superiority, particularly in the domains in which such claims have always been made—art, religion and philosophy. When Weber found that he could not choose between certain high opposites—reason vs. revelation, Buddha vs. Jesus—he did not conclude that all things are equally good, that the distinction between high and low disappears. As a matter of fact he intended to revitalize the consideration of these great alternatives in showing the gravity and danger involved in choosing among them; they were to be heightened in contrast to the trivial considerations of modern life that threatened to overgrow and render indistinguishable the profound problems the confrontation with which makes the bow of the soul taut. The serious intellectual life was for him the battleground of the great decisions, all of which are spiritual or "value" choices. One can no longer present this or that particular view of the educated or civilized man as authoritative; therefore one must say that education consists in knowing, really knowing, the small number of such views in their integrity. This distinction between profound and superficial—which takes the place of good and bad, true and false—provided a focus for serious study, but it hardly held out against the naturally relaxed democratic tendency to say, "Oh, what's the use?" The first university disruptions at Berkeley were explicitly directed against the multiversity smorgasbord and, I must confess, momentarily and partially engaged my sympathies. It may have even been the case that there was some small element of longing for an education in the motivation of those students. But nothing was done to guide or inform their energy, and the result was merely to add multilife-styles to multidisciplines, the diversity of perversity to the diversity of specialization. What we see so often happening in general happened here too; the insistent demand for greater community ended in greater isolation. Old agreements, old habits, old traditions were not so easily replaced.

Thus, when a student arrives at the university, he finds a bewildering variety of departments and a bewildering variety of courses. And there is no official guidance, no university-wide agreement, about what he *should* study. Nor does he usually find readily available examples, either among students or professors, of a unified use of the university's resources. It is easiest simply to make a career choice and go about getting prepared for that career. The programs designed for those having made such a choice render their students immune to charms that might lead them out of the conventionally respectable. The sirens sing *sotto voce* these days, and the young already have enough wax in their ears to pass them by without danger. These specialties can provide enough courses to take up most of their time for four years in preparation for the inevitable graduate study. With the few remaining courses they can do what they please, taking a bit of this and a bit of that. No public career these days—not doctor nor lawyer nor politician nor journalist nor businessman nor entertainer—has much to do with humane learning. An education, other than purely professional or technical, can even seem to be an impediment. That is why a countervailing atmosphere in the university would be necessary for the students to gain a taste for intellectual pleasures and learn that they are viable.

The real problem is those students who come hoping to find out what career they want to have, or are simply looking for an adventure with themselves. There are plenty of things for them to do—courses and disciplines enough to spend many a lifetime on. Each department or great division of the university makes a pitch for itself, and each offers a course of study that will make the student an initiate. But how to choose among them? How do they relate to one another? The fact is they do not address one another. They are competing and contradictory, without being aware of it. The problem of the whole is urgently indicated by the very existence of the specialties, but it is never systematically posed. The net effect of the student's encounter with the college catalogue is bewilderment and very often demoralization. It is just a matter of chance whether he finds one or two professors who can give him an insight into one of the great visions of education that have been the distinguishing part of every civilized nation. Most professors are specialists, concerned only with their own fields, interested in the advancement of those fields in their own terms, or in their own personal advancement in a world where all the rewards are on the side of professional distinction. They have been entirely emancipated from the old structure of the university, which at least helped to indicate that they are incomplete, only parts of an unexamined and undiscovered whole. So the student must navigate among a collection of carnival barkers, each trying to lure him into a particular sideshow. This undecided student is an embarrassment to most universities, because he seems to be saying, "I am a whole human being. Help me to form myself, in my wholeness and let me develop my real potential," and he is the one to whom they have nothing to say.

Cornell was, as in so many other things, in advance of its time on this issue. The six-year Ph.D. program, richly supported by the Ford Foundation, was directed specifically to high school students who had already made "a firm career choice" and was intended to rush them through to the start of those careers. A sop was given to desolate humanists in the form of money to fund seminars that these young careerists could take on their way through the College of Arts and Sciences. For the rest, the educators could devote their energies to arranging and packaging the program without having to provide it with any substance. That kept them busy enough to avoid thinking about the nothingness of their endeavor. This has been the preferred mode of not looking the Beast in the Jungle in the face—structure, not content. The Cornell plan for dealing with the problem of liberal education was to suppress the students' longing for liberal education by encouraging their professionalism and their avarice, providing money and all the prestige the university had available to make careerism the centerpiece of the university.

The Cornell plan dared not state the radical truth, a well-kept secret: the colleges do not have enough to teach their students, not enough to justify keeping them four years, probably not even three years. If the focus is careers, there is hardly one specialty, outside the hardest of the hard natural sciences, which requires more than two years of preparatory training prior to graduate studies. The rest is just wasted time, or a period of ripening until the students are old enough for graduate studies. For many graduate careers, even less is really necessary. It is amazing how many undergraduates are poking around for courses to take, without any plan or question to ask, just filling up their college years. In fact, with rare exceptions, the courses are parts of specialties and not designed for general cultivation, or to investigate questions important for human beings as such. The so-called knowledge explosion and increasing specialization have not filled up the college years but emptied them. Those years are impediments; one wants to get beyond them. And in general the persons one finds in the professions need not have gone to college, if one is to judge by their tastes, their fund of learning or their interests. They might as well have spent their college years in the Peace Corps or the like. These great universities—which can split the atom, find cures for the most terrible diseases, conduct surveys of whole populations and produce massive dictionaries of lost languages—cannot generate a modest program of general education for undergraduate students. This is a parable for our times.

There are attempts to fill the vacuum painlessly with various kinds of fancy packaging of what is already there—study abroad options, individualized majors, etc. Then there are Black Studies and Women's or Gender Studies, along with Learn Another Culture. Peace Studies are on their way to a similar prevalence. All this is designed to show that the university is with it and has something in addition to its traditional specialties. The latest item is computer literacy, the full cheapness of which is evident only to those who think a bit about what literacy might mean. It would make some sense to promote literacy literacy, inasmuch as most high school graduates nowadays have difficulty reading and writing. And some institutions are quietly undertaking this worthwhile task. But they do not trumpet the fact, because this is merely a high school function that our current sad state of educational affairs has thrust upon them, about which they are not inclined to boast.

Now that the distractions of the sixties are over, and undergraduate education has become more important again (because the graduate departments, aside from the professional schools, are in trouble due to the shortage of academic jobs), university officials have had somehow to deal with the undeniable fact that the students who enter are uncivilized, and that the universities have some responsibility for civilizing them. If one were to give a base interpretation of the schools' motives, one could allege that their concern stems from shame and self-interest. It is becoming all too evident that liberal education—which is what the small band of prestigious institutions are supposed to provide, in contrast to the big state schools, which are thought simply to prepare specialists to meet the practical demands of a complex society—has no content, that a certain kind of fraud is being perpetrated. For a time the great moral consciousness alleged to have been fostered in students by the great universities, especially their vocation as gladiators who fight war and racism, seemed to fulfill the demands of the collective university conscience. They were doing something other than offering preliminary training for doctors and lawyers. Concern and compassion were thought to be the indefinable X that pervaded all the parts of the Arts and Sciences campus. But when that evanescent mist dissipated during the seventies, and the faculties found themselves face to face with ill-educated young people with no intellectual tastes—unaware that there even are such things, obsessed with getting on with their careers before having looked at life—and the universities offered no counterpoise, no alternative goals, a reaction set in.

Liberal education—since it has for so long been ill-defined, has none of the crisp clarity or institutionalized prestige of the professions, but nevertheless perseveres and has money and respectability connected with it—has always been a battleground for those who are somewhat eccentric in relation to the specialties. It is in something like the condition of churches as opposed to, say, hospitals. Nobody is quite certain of what the religious institutions are supposed to do anymore, but they do have some kind of role either responding to a real human need or as the vestige of what was once a need, and they invite the exploitation of quacks, adventurers, cranks and fanatics. But they also solicit the warmest and most valiant efforts of persons of peculiar gravity and depth. In liberal education, too, the worst and the best fight it out, fakers vs. authentics, sophists vs. philosophers, for the favor of public opinion and for control over the study of man in our times. The most conspicuous participants in the struggle are administrators who are formally responsible for presenting some kind of public image of the education their colleges offer, persons with a political agenda or vulgarizers of what the specialties know, and real teachers of the humane disciplines who actually see their relation to the whole and urgently wish to preserve the awareness of it in their students' consciousness.

So, just as in the sixties universities were devoted to removing requirements, in the eighties they are busy with attempts to put them back in, a much more difficult task. The word of the day is "core." It is generally agreed that "we went a bit far in the sixties," and that a little fine-tuning has now become clearly necessary.

There are two typical responses to the problem. The easiest and most administratively satisfying solution is to make use of what is already there in the autonomous departments and simply force the students to cover the fields, i.e., take one or more courses in each of the general divisions of the university: natural science, social science and the humanities. The reigning ideology here is *breadth*, as was *openness* in the age of laxity. The courses are almost always the already existing introductory courses, which are of least interest to the major professors and merely assume the worth and reality of that which is to be studied. It is general education, in the sense in which a jack-of-all-trades is a generalist. He knows a bit of everything and is inferior to the specialist in each area. Students may wish to sample a variety of fields, and it may be good to encourage them to look around and see if there is something that attracts them in one of which they have no experience. But this is not a liberal education and does not satisfy any longing they have for one. It just teaches that there is no high-level generalism, and that what they are doing is preliminary to the real stuff and part of the childhood they are leaving behind. Thus they desire to get it over with and get on with what their professors do seriously. Without recognition of important questions of common concern, there cannot be serious liberal education, and attempts to establish it will be but failed gestures.

It is a more or less precise awareness of the inadequacy of this approach to core curricula that motivates the second approach, which consists of what one might call composite courses. These are constructions developed especially for general-education purposes and usually require collaboration of professors drawn from several departments. These courses have titles like "Man in Nature," "War and Moral Responsibility," "The Arts and Creativity," "Culture and the Individual." Everything, of course, depends upon who plans them and who teaches them. They have the clear advantage of requiring some reflection on the general needs of students and force specialized professors to broaden their perspectives, at least for a moment. The dangers are trendiness, mere popularization and lack of substantive rigor. In general, the natural scientists do not collaborate in such endeavors, and hence these courses tend to be unbalanced. In short, they do not point beyond themselves and do not provide the student with independent means to pursue permanent questions independently, as, for example, the study of Aristotle or Kant as wholes once did. They tend to be bits of this and that. Liberal education should give the student the sense that learning must and can be both synoptic and precise. For this, a very small, detailed problem can be the best way, if it is framed so as to open out on the whole. Unless the course has the specific intention to lead to the permanent questions, to make the student aware of them and give him some competence in the important works that treat of them, it tends to be a pleasant diversion and a dead end—because it has nothing to do with any program of further study he can imagine. If such programs engage the best energies of the best people in the university, they can be beneficial and provide some of the missing intellectual excitement for both professors and students. But they rarely do, and they are too cut off from the top, from what the various faculties see as their real business. Where the power is determines the life of the whole body. And the intellectual problems unresolved at the top cannot be resolved administratively below. The problem is the lack of any unity of the sciences and the loss of the will or the means even to discuss the issue. The illness above is the cause of the illness below, to which all the good-willed efforts of honest liberal educationists can at best be palliatives.

Of course, the only serious solution is the one that is almost universally rejected: the good old Great Books approach, in which a liberal education means reading certain generally recognized classic texts, just reading them, letting them dictate what the questions are and the method of approaching them—not forcing them into categories we make up, not treating them as historical products, but trying to read them as their authors wished them to be read. I am perfectly well aware of, and actually agree with, the objections to the Great Books cult. It is amateurish; it encourages an autodidact's self-assurance without competence; one cannot read all of the Great Books carefully; if one only reads Great Books, one can never know what a great, as opposed to an ordinary, book is; there is no way of determining who is to decide what a Great Book or what the canon is; books are made the ends and not the means; the whole movement has a certain coarse evangelistic tone that is the opposite of good taste; it engenders a spurious intimacy with greatness; and so forth. But one thing is certain: wherever the Great Books make up a central part of the curriculum, the students are excited and satisfied, feel they are doing something that is independent and fulfilling, getting something from the university they cannot get elsewhere. The very fact of this special experience, which leads nowhere beyond itself, provides them with a new alternative and a respect for study itself. The advantage they get is an awareness of the classic—particularly important for our innocents; an acquaintance with what big questions were when there were still big questions; models, at the very least, of how to go about answering them; and, perhaps most important of all, a fund of shared experiences and thoughts on which to ground their friendships with one another. Programs based upon judicious use of great texts provide the royal road to students' hearts. Their gratitude at learning of Achilles or the categorical imperative is boundless. Alexandre Koyré, the late historian of science, told me that his appreciation for America was great when—in the first course he taught at the University of Chicago, in 1940 at the beginning of his exile—a student spoke in his paper of Mr. Aristotle, unaware that he was not a contemporary. Koyré said that only an American could have the naive profundity to take Aristotle as living thought, unthinkable for most scholars. A good program of liberal education feeds the student's love of truth and passion to live a good life. It is the easiest thing in the world to devise courses of study, adapted to the particular conditions of each university, which thrill those who take them. The difficulty is in getting them accepted by the faculty.

None of the three great parts of the contemporary university is enthusiastic about the Great Books approach to education. The natural scientists are benevolent toward other fields and toward liberal education, if it does not steal away their students and does not take too much time from their preparatory studies. But they themselves are interested primarily in the solution of the questions now important in their disciplines and are not particularly concerned with discussions of their foundations, inasmuch as they are so evidently successful. They are indifferent to Newton's conception of time or his disputes with Leibniz about calculus; Aristotle's teleology is an absurdity beneath consideration. Scientific progress, they believe, no longer depends on the kind of comprehensive reflection given to the nature of science by men like Bacon, Descartes, Hume, Kant and Marx. This is merely historical study, and for a long time now, even the greatest scientists have given up thinking about Galileo and Newton. Progress is undoubted. The difficulties about the truth of science raised by positivism, and those about the goodness of science raised by Rousseau and Nietzsche, have not really penetrated to the center of scientific consciousness. Hence, no Great Books, but incremental progress, is the theme for them.

Social scientists are in general hostile, because the classic texts tend to deal with the human things the social sciences deal with, and they are very proud of having freed themselves from the shackles of such earlier thought to become truly scientific. And, unlike the natural scientists, they are insecure enough about their achievement to feel threatened by the works of earlier thinkers, perhaps a bit afraid that students will be seduced and fall back into the bad old ways. Moreover, with the possible exception of Weber and Freud, there are no social science books that can be said to be classic. This may be interpreted favorably to the social sciences by comparing them to the natural sciences, which can be said to be a living organism developing by the addition of little cells, a veritable body of knowledge proving itself to be such by the very fact of this almost unconscious growth, with thousands of parts oblivious to the whole, nevertheless contributing to it. This is in opposition to a work of imagination or of philosophy, where a single creator makes and surveys an artificial whole. But whether one interprets the absence of the classic in the social sciences in ways flattering or unflattering to them, the fact causes social scientists discomfort. I remember the professor who taught the introductory graduate courses in social science methodology, a famous historian, responding scornfully and angrily to a question I naively put to him about Thucydides with "Thucydides was a fool!"

More difficult to explain is the tepid reaction of humanists to Great Books education, inasmuch as these books now belong almost exclusively to what are called the humanities. One would think that high esteem for the classic would reinforce the spiritual power of the humanities, at a time when their temporal power is at its lowest. And it is true that the most active proponents of liberal education and the study of classic texts are indeed usually humanists. But there is division among them. Some humanities disciplines are just crusty specialties that, although they depend on the status of classic books for their existence, are not really interested in them in their natural state—much philology, for example, is concerned with the languages but not what is said in them—and will and can do nothing to support their own infrastructure. Some humanities disciplines are eager to join the real sciences and transcend their roots in the now overcome mythic past. Some humanists make the legitimate complaints about lack of competence in the teaching and learning of Great Books, although their criticism is frequently undermined by the fact that they are only defending recent scholarly interpretation of the classics rather than a vital, authentic understanding. In their reaction there is a strong element of specialist's jealousy and narrowness. Finally, a large part of the story is just the general debilitation of the humanities, which is both symptom and cause of our present condition.

To repeat, the crisis of liberal education is a reflection of a crisis at the peaks of learning, an incoherence and incompatibility among the first principles with which we interpret the world, an intellectual crisis of the greatest magnitude, which constitutes the crisis of our civilization. But perhaps it would be true to say that the crisis consists not so much in this incoherence but in our incapacity to discuss or even recognize it. Liberal education flourished when it prepared the way for the discussion of a unified view of nature and man's place in it, which the best minds debated on the highest level. It decayed when what lay beyond it were only specialties, the premises of which do not lead to any such vision. The highest is the partial intellect; there is no synopsis.

Chapter 22

E. D. Hirsch, Jr. (1928–)
from "Rise of the Fragmented
Curriculum," *Cultural Literacy: What
Every American Needs to Know* (1987)

The decline of American literacy and the fragmentation of the American school curriculum have been chiefly caused by the ever growing dominance of romantic formalism in educational theory during the past half century. We have too readily blamed shortcomings in American education on social changes (the disorientation of the American family or the impact of television) or incompetent teachers or structural flaws in our school systems. But the chief blame should fall on faulty theories promulgated in our schools of education and accepted by educational policymakers.

Consider William Raspberry's comments on test results which show that black students score 35 to 45 percent lower than white students in standardized achievement tests.

> The news hits like a series of bombshells as one suburban school district after another reveals that black children are significantly behind their white counterparts on standardized achievement tests. . . . Whose fault is it that blacks tend to get lower scores? I don't know all the answers to that one. Surely a part of it is the simple fact that those children who come to school already knowing a good deal of what the society deems important to know tend to find it easier to learn more of it. The more you know, the more you can learn.[1]

We know that Raspberry is right, not only about achievement tests but also about reading and writing "skills." "The more you know, the more you can learn." What does his judgment imply for schooling? Some children, he says, enter school already

---

In this selection from *Cultural Literacy*, E. D. Hirsch, Professor Emeritus of Education and Humanities at the University of Virginia, argues against what he calls "a gradual disintegration of cultural memory" in school curricula. Since the publication of *Cultural Literacy*, Hirsch's Core Knowledge Foundation has published a series of prominent books on what school children need to know, arranged by grade.

possessing the information needed to make further advances in the literate culture, while others come to school lacking that information.

Why have we failed to give them the information they lack? Chiefly because of educational formalism, which encourages us to ignore the fact that identifying and imparting the information a child is missing is most important in the earliest grades, when the task is most manageable. At age six, when a child must acquire knowledge critical for continuing development, the total quantity of missing information is not huge. As Dr. Jeanne Chall has pointed out, the technical reading skills of disadvantaged children at age six are still on a par with those of children from literate families. Yet only a year or so later, their reading skills begin to diverge according to socioeconomic status, chiefly because low-income pupils lack elementary cultural knowledge.[2] Supplying missing knowledge to children early is of tremendous importance for enhancing their motivation and intellectual self-confidence, not to mention their subsequent ability to learn new materials. Yet schools will never systematically impart missing background information as long as they continue to accept the formalistic principle that specific information is irrelevant to "language arts skills."

Educational formalism holds that reading and writing are like baseball and skating; formalism conceives of literacy as a set of techniques that can be developed by proper coaching and practice. The following definition of reading as a skill is typical of those in many education textbooks:

> Reading is a recoding and decoding process, in contrast with speaking and writing which involve encoding. One aspect of decoding is to relate the printed word to oral language meaning, which includes changing the print to sound and meaning.

(The characteristic title of this book is *Language Skills in Elementary Education*.)[3]

The skill idea becomes an oversimplification as soon as students start reading for meaning rather than for cracking the alphabetic code. If reading were just hitting the right alphabetic ball with the right phonic bat, or just learning "text strategies," it would be like baseball. We could teach everybody to groove their swings and watch for the seven different types of pitches. The trouble is that reading for meaning is a different sort of game entirely. It is different every time, depending on what the piece of writing is about. Every text, even the most elementary, implies information that it takes for granted and doesn't explain. Knowing such information is the decisive skill of reading.

Educational formalism assumes that the specific contents used to teach "language arts" do not matter so long as they are closely tied to what the child already knows, but this developmental approach ignores Raspberry's important point that different children know different things. Current schoolbooks in language arts pay little systematic attention to conveying a body of culturally significant information from grade to grade. Their "developmental" approach contrasts sharply with textbooks from earlier decades, which consciously aimed to impart cultural literacy. Texts now used to teach reading and writing are screened not for the information they convey but for their readability scores and their fit with the sequence of abstract skills that a child is expected to acquire. On this skills model, the ideal method of language arts instruction would be to adjust the reading curriculum to the interests and competencies of each child. In fact, just this formalistic approach is recommended by

P. S. Anderson as the "self-selection" method of reading instruction, according to which the child selects his or her own curriculum for mastering decoding skills.

> A special method of individualizing instruction within a classroom has been identified by the terms *self-selection* and *language approach*. The self-selection program allows each child to seek whatever reading material stimulates him and work at his own rate with what he has chosen.[4]

I cannot claim to have studied all the recent textbooks intended to train teachers or educate children in the language arts, but those I have consulted represent learning to read as a neutral, technical process of skill acquisition that is better served by up-to-date "imaginative literature" than by traditional and factual material. "Reading in the content areas" is typically regarded as inferior to reading up-to-date fictions. Current language arts textbooks not only overlook the fundamental acculturative aims of our schools but go wrong even on technical grounds, for it has become clear that effective reading depends on acquiring factual and traditional schemata. By the same token, teachers are not expected to have mastered particular factual and traditional information or any special academic discipline; they are trained only to impart skills.

To miss the opportunity of teaching young (and older) children the traditional materials of literate culture is a tragically wasteful mistake that deprives them of information they would continue to find useful in later life. The inevitable effect of this fundamental educational mistake has been a gradual disintegration of cultural memory, causing a gradual decline in our ability to communicate. The mistake has therefore been a chief cause of illiteracy, which is a subcategory of the inability to communicate. [. . .]

The American school curriculum is fragmented both horizontally across subjects and vertically within subjects.[5] For one student in grade nine, social studies may focus on family relations; for another, the focus may be on ancient history. In an American history course in one school, students may focus on industrial America, but in another school, the focus may be on westward expansion.[6]

How such fragmentation arose can be grasped in broad terms by looking at two decisive moments in American education represented by two historic documents, one published in 1893, the other in 1918. The earlier one is the *Report of the Committee of Ten on Secondary School Studies*; the later report is *Cardinal Principles of Secondary Education*.[7] The contrasts between them bring into clear relief the change in educational theory that occurred in the first quarter of this century.[8]

Between the 1920s and the 1970s, the history of American secondary education has been one of the increasing dominance of the *Cardinal Principles* and the corresponding decline of the principles of the Committee of Ten which recommended a traditional humanistic curriculum.[9] The earlier report assumed that all students would take the same humanistic subjects and recommended giving a new emphasis to natural sciences. It took for granted that secondary school offerings would continue to consist of just the traditional areas that its subcommittees had been formed to consider—Latin, Greek, English, other modern languages, mathematics, physics, chemistry and astronomy, natural history, botany, zoology, physiology, history, civil government, economy, and geography. [. . .]

In deliberate contrast to the 1893 report, *Cardinal Principles of Secondary Education* explicitly rejected the earlier focus on subject matter. Instead, it stressed the seven fundamental aims of education in a democracy: "1. Health. 2. Command of fundamental processes. 3. Worthy home membership. 4. Vocation. 5. Citizenship. 6. Worthy use of leisure. 7. Ethical character."[10] The shift from subject matter to social adjustment was a deliberate challenge to the 1893 report and to conservative school practices generally. American education should take a new direction. Henceforth it should stress utility and the direct application of knowledge, with the goal of producing good, productive, and happy citizens. [. . .]

But the ideal of accommodating individual differences by offering different types of courses became institutionalized as "tracking" and "grouping"—systems that put bright students in one class, average students in another, and poor students in a third. The principle of adjusting to diverse capacities produced, in effect, three academic castes, each of which received different kinds of information. In language arts courses, for instance, where content became an arbitrary vehicle to inculcate the "fundamental processes" of reading and writing, the tracking system led to Shakespeare for some, sports and fantasy stories for others.

The fragmentation produced by tracking reinforced the fragmentation that was being introduced by vocational courses. Schools were to be directly useful in preparing children for citizenship, work, and leisure. Girls should be taught courses in home-making and boys courses in shop or other vocational subjects. The introduction of such courses, like the introduction of the tracking system, further multiplied the different sorts of courses offered in school, leading ultimately to the shopping mall school with its extreme horizontal fragmentation.[11] [. . .]

The private schools were less influenced by the new theories, and many of them continued to provide traditional content. As a consequence, their graduates were, on the whole, more literate than those who came from the public schools. Such differences exacerbated the class stratifications that the educators of 1893 had sought to avoid.

Notes

1. *The Washington Post* (September 16, 1985), A-19.
2. J. S. Chall, C. Snow, et al., *Families and Literacy*, Final Report to the National Institute of Education, 1982. Also, J. S. Chall, "Afterword," in R. C. Anderson et al., *Becoming a Nation of Readers: The Report of the Commission on Reading* (Washington, DC: National Institute of Education, 1985), pp. 123–24.
3. P. S. Anderson, *Language Skills in Elementary Education*, 2d ed., (New York: Macmillan, 1972), pp. 209–210.
4. Ibid., p. 220.
5. My understanding of this historical process is based chiefly on the following historical works to which the reader is referred for detailed accounts and interpretations: L. A. Cremin, *The Transformation of the School: Progressivism in American Education, 1876–1957* (New York: Knopf, 1964); Patricia A. Graham, *Progressive Education: From Arcady to Academe* (New York: Teachers College Press, Columbia University, 1967); E. A. Krug, *The Shaping of the American High School*, Volume I: 1880–1920; Volume II,

1920–1941 (Madison: University of Wisconsin Press, 1969, 1972); Diane Ravitch, *The Troubled Crusade* (New York: Basic Books, 1983); David K. Cohen, "Origins," in *The Shopping Mall High School* ed. A. G. Powell, E. Farrar, and David K. Cohen (Boston: Houghton Mifflin, 1985), pp. 233–308.

6. Powell, Farrar, and Cohen, *The Shopping Mall High School*, 3–7.
7. National Education Association, *Report of the Committee of Ten on Secondary School Studies* (Washington, DC: Government Printing Office, 1893), hereafter cited as Eliot, *Report of the Committee of Ten*; C. D. Kingsley, ed., *Cardinal Principles of Secondary Education: A Report of the Commission on the Reorganization of Secondary Education, Appointed by the National Education Association*, Bulletin, 1918, No. 35 (Washington, DC: Department of the Interior, Bureau of Education, Government Printing Office, 1918).
8. "The *Cardinal Principles* quickly became the bible of school reformers. It continues to attract enthusiasts today. It helped to popularize and rationalize the new studies and standards because it identified them with a great democratic advance." Cohen, "Origins," 256.
9. See Cremin, *The Transformation of the School*. The process after 1957 is traced in Ravitch, *The Troubled Crusade*, especially chapter 2. A concise overview of the process (just 16 pages long) can be found in "The Continuing Crisis: Fashions in Education," the third chapter in Diane Ravitch, *The Schools We Deserve: Reflections on the Educational Crises of our Times* (New York: Basic Books, 1985).
10. Kingsley, *Cardinal Principles*, 7.
11. Powell, Farrar, and Cohen, *The Shopping Mall High School*, 1–7.

Chapter 23

Frank Kermode (1919–)
from "Canon and Period,"
History and Value (1988)

How does literary history fare in these circumstances? For the past quarter-century or so a rumour has circulated to the effect that it can't any longer be written. Causal connections between works chosen for attention must be spurious, and special interests, not strictly literary at all, guide the historian's hand. Yet, as Hans Robert Jauss remarks, it used to be thought that the crowning achievement of the philologist was to write the history of his national literature, to reveal its origins with pride, and to trace its stately and inevitable development. These interests in origins and development received a great fillip at the Renaissance, and they flourished well into the present century. But it then began to seem obvious that something was wrong, and that historians of literature were actually writing histories not of literature but of other things—treating literature as a set of illustrative documents, smuggling in notions of cause and connection from social and political history.[1] [. . .]

I can best start this section on canon by reading an item from the US *Chronicle of Higher Education* dated 4 September 1985. This journal is widely circulated in American institutions of higher education. On this occasion, at the beginning of a new academic year, it ran a symposium in which twenty-two authorities in various fields told readers what developments to expect over the next few years. This is the forecast for literary studies:

> The dominant concern of literary studies during the rest of the nineteen-eighties will be literary theory. Especially important will be the use of theory informed by the work of the French philosopher Jacques Derrida to gain insights into the cultures of blacks and women.
>
> In fact the convergence of feminist and Afro-American theoretical formulations offers the most challenging nexus for scholarship in the coming years. Specifically the most exciting and insightful accounts of expressive culture in general and creative writing in particular will derive from efforts that employ feminist and Afro-American

A Fellow of the British Academy, English literary critic Sir Frank Kermode has taught at several universities, including Manchester, Cambridge, and Harvard. In this essay, he addresses the authority implied by calling literary history a "canon."

approaches to the study of texts by Afro-American writers such as Zora Neale Hurston, Sonia Sanchez, Gloria Naylor and Toni Morrison.

Among the promising areas for analysis is the examination of the concerns and metaphorical patterns that are common to past and present black women writers.

Such theoretical accounts of the cultural products of race and gender will help to undermine the half-truths that white males have established as constituting American culture as a whole. One aspect of that development will be the continued reshaping of the literary canon as forgotten, neglected or suppressed texts are re-discovered.

Literary theory is also full of disruptive and deeply political potential, which Afro-American and feminist critics will labor to release in coming years.

This manifesto, for such it appears to be, was written by the Professor "of English and of Human Relations" at the University of Pennsylvania. It proposes what could well be called a radical deconstruction of the canon, putting in the place of the false elements foisted into it by white males a list of black females. These will be studied by methods specifically Afro-American. The writer points out the political implications of these developments, for he knows that the changes he prophesies will not come to pass without alterations in more than the syllabus. He assumes that the literary canon is a load-bearing element of the existing power structure, and believes that by imposing radical change on the canon you can help to dismantle the power structure.

What interests me most about this programme is not its cunning alliance of three forces that might be thought to be in principle hostile to the idea of the canon—Feminism, Afro-Americanism, and Deconstruction—so much as its tacit admission that there is such a thing as literature and that there ought to be such a thing as a canon; the opinions of the powerful about the contents of these categories may be challenged, but the concepts of themselves remain in place. Indeed the whole revolutionary enterprise simply assumes their continuance. The canon is what the insurgents mean to occupy as the reward of success in the struggle for power.

In short, what we have here is not a plan to abolish the canon but one to capture it. The association of canon with authority is deeply ingrained in us, and one can see simple reasons why it should be so. It is a highly selective instrument, and one reason why we need to use it is that we haven't enough memory to process everything. The only other option is not a universal reception of the past and its literature but a Dadaist destruction of it. It must therefore be protected by those who have it and coveted by those who don't.

Authority has invented many myths for the protection of the canon. Religious canons can be effectively closed, even at the cost of retaining within them books of which the importance is later difficult to discern, like some of the briefer New Testament letters. They can be heavily protected, credited for example with literal inspiration, so that it is forbidden to alter one jot or tittle of them, diacritical signs, instructions to cantilators, even manifest errors. And every word, every letter, is subject to minute commentary. Whatever is included is sure to have its effect on the world. Suppose, for instance, that Revelation had not got into the Christian canon, as it almost didn't; it would have been just one more lost or apocryphal apocalypse, the province only of specialist scholarship; instead it has had vast effects on social and political behaviour over many ages, and continues to do so. The Fourth Gospel was at one time under suspicion; had it not become so central a document for Christian theology millions of people would have been required to believe something quite different from the orthodox faith, and quite a lot of them might have escaped burning if not burnt for some other reason.

So canons are complicit with power; and canons are useful in that they enable us to handle otherwise unmanageable historical deposits. They do this by affirming that some works are more valuable than others, more worthy of minute attention. Whether their value is wholly dependent on their being singled out in this way is a contested issue. There is in any case a quite unmistakable difference of status between canonical and uncanonical books, however they got into the canon. But once they are in certain changes come over them. First, they are completely locked into their times, their texts as near frozen as devout scholarship can make them, their very language more and more remote. Secondly, they are, paradoxically, by this fact, set free of time. Thirdly, the separate constituents become not only books in their own right but part of a larger whole—a whole because it is so treated. Fourthly, that whole, with all its interrelated parts, can be thought to have an inexhaustible potential of meaning, so that what happens in the course of time—as the original context and language of the collection grows more and more distant—is that new meanings accrue (they may be deemed, by a fiction characteristic of this way of thinking, to be original meanings) and these meanings constantly change though their source remains unchangeable. Since all the books can now be thought of as one large book, new echoes and repetitions are discovered in remote parts of the whole. The best commentary on any verse is another verse, possibly placed very far away from it. This was a rabbinical doctrine: "I join passages from the Torah with passages from the Hagiographa, and the words of the Torah glow as the day they were given at Sinai."[2]

The temporal gap between text and comment or application ensures that in practice something like the Gadamer-Jauss hermeneutics, whether formalized or not, is always needed. The mutual influence of one canonical text on another, intemporal in itself, appearing in time only by means of commentary, is the essence of Eliot's idea of a canon, expressed in that famous passage in the essay "Tradition and the Individual Talent"—"the whole of the literature of Europe . . . has a simultaneous existence and composes a simultaneous order," though he provides, as a secular canonist must, for additions to that order: "The existing monuments form an ideal order among themselves, which is modified by the introduction of the new (the really new) work of art among them."[3] By this means "order"—timeless order—"persists after the supervention of novelty," and it does so by adjusting itself to the new. Here the idea of canon is used in the service of an order which can be discerned in history but actually transcends it, and makes everything timeless and modern.

In this, as in the formulae of hermeneutics, in the rabbinical methodology and in the Marxist aspirations toward a theory of fruitful discrepancy, there is a clear purpose of making a usable past, a past which is not simply past but also always new. The object of all such thinking about the canonical monuments, then, is to make them modern. Indeed variants of this view are found in more than one writer of the period we now think of as "Modernist." At the same time there was a rival kind of Modernism that professed a desire to destroy the monuments, to destroy the past. But the ghost of canonicity haunts even these iconoclasts. And whether one thinks of canons as objectionable because formed at random or to serve some interests at the expense of others, or whether one supposes that the contents of canons are providentially chosen, there can be no doubt that we have not found ways of ordering our thoughts about the history of literature and art without recourse to them. That is why the minorities who want to be rid of what they regard as a reactionary canon can think of no way of doing so without putting a radical one in its place. [. . .]

There is one aspect of the question I haven't sufficiently mentioned, namely that authoritative choices, although conceivably the first motion may come from an individual, normally require a consensus, and the consensus of a relatively small number of people. In the case of literature we may identify these persons with what Fish calls an "interpretive community." There must be institutional control of interpretation, as I've argued at length elsewhere,[4] and self-perpetuating institutions resist not only those they think of as incompetent for reasons of ignorance, but also the charismatic outsider. They are bound to be reactionary in some sense; the young are trained to make certain kinds of interpretation of the favoured texts, they become senior and have themselves an interest in an inherited, if modified, set of procedures, and have to make a large investment in the canon. This does not mean that there are no sects and no discontent within the institution—anybody, today, looking at schools and faculties of literature, can see sectarian discontent. There is always a possibility that within a large and not particularly centralized institution there may develop subcanons and revisions of periodization, to suit, say, feminists and Afro-Americans or Derrideans, or even feminist Afro-American Derrideans. What is certain is that revolutionary revisions would require transfers of powers, a reign of literary terror the prospect of which many of us enjoy less than the Professor of English and Human Relations. And the business of valuing selected moments and selected books, saved from the indiscriminate mass of historical fact, would in any case continue. So would the inherited methods of analysis and evaluation (for example the "metaphorical patterns that are common to past and present black women writers"). Absolute justice and perfection of conscience are unlikely to be more available under that new dispensation than they are now.

All the same, we do accept change, in the ways we conceive as open to us. But so long as we seek value in works of the past we shall be forced to submit the show of history to the desires of the mind—whoever "we" may be. And in order to do that we shall invent new grids and impose them on the past—rewrite the past to suit our modern wishes, as the past has always been rewritten. Yet valuations are handed on, and constantly redefined; so that in the end the question is not whether they are unfairly selective, but whether we want to break the only strong link we have with the past—our ability to identify with the interests of our predecessors, to qualify their judgements without necessarily overthrowing them, to converse with them in a transhistorical dimension. Though inevitably tainted with privilege and injustice, that still seems a valuable inheritance; some catastrophe might conceivably destroy it, but the destruction should not be encouraged by members of the rather small community that cares about writing or about art in general. Some workable notion of canon, some examined idea of history, though like most human arrangements they may be represented as unjust and self-serving, are necessary to any concept of past value with the least chance of survival, necessary even to the desired rehabilitation of the unfairly neglected. So the tradition of value, flawed as it is, remains valuable. Certainly it should be constantly scrutinized, so that the past, already diminished by our necessary selective manipulations, is not reduced even further by unnecessary compliance with fashion or prejudice.

Notes

1. See R. Wellek and A. Warren, *Theory of Literature* (1949), p. 263, and R. Wellek, "The Fall of Literary History," in The Attack on Literature (1982), pp. 64–77, for a survey of the present situation.
2. D. Patte, *Early Jewish Hermeneutic in Palestine* (Missoula, MT: Scholars Press, 1975), p. 44.
3. [T. S. Eliot] *Selected Prose of T. S. Eliot*, edited with an introduction by Frank Kermode (London: Faber, 1975), p. 38.
4. Kermode, *Essays on Fiction* (in US *The Art of Telling: Essays on Fiction*). (Cambridge: Harvard University Press, 1983).

Chapter 24

Barbara Herrnstein Smith (1932–) from "Contingencies of Value," *Contingencies of Value* (1988)

The Dynamics of Endurance

When we consider the cultural re-production of value on a larger timescale, the model of evaluative dynamics outlined above suggests that both (a) the "survival" or "endurance" of a text and, it may be, (b) its achievement of high canonical status not only as a "work of literature" but as a "classic" are the product neither of the objectively (in the Marxist sense) conspiratorial force of establishment institutions nor of the continuous appreciation of the timeless virtues of a fixed object by succeeding generations of isolated readers but, rather, of a series of continuous interactions among a variably constituted object, emergent conditions, and mechanisms of cultural selection and transmission. These interactions are, in certain respects, analogous to those by virtue of which biological species evolve and survive and also analogous to those through which artistic choices evolve and are found "fit" or fitting by the individual artist. The operation of these cultural-historic dynamics may be briefly indicated here in quite general terms.

At a given time and under the contemporary conditions of available materials and technology or techniques, a particular object—let us say a verbal artifact or text—may perform certain desired/able[1] functions quite well for some set of subjects. It will do so by virtue of certain of its "properties" as they have been specifically constituted—framed, foregrounded, and configured—by those subjects under those conditions and in accord with their particular needs, interests, and resources—and also perhaps largely as pre-figured by the artist who, as described earlier, in the very process of producing the work and continuously evaluating its fitness and adjusting it accordingly, will have multiply and variably constituted it.

In this essay, Barbara Herrnstein Smith, Braxton Craven Professor of Comparative Literature and English at Duke University, addresses why now-seemingly timeless works have survived.

Two related points need emphasis here. One is that the current value of a work—that is, its effectiveness in performing desired/able functions for some set of subjects—is by no means independent of *authorial* design, labor, and skill. To be sure, the artist does not have absolute control over that value, nor can its dimensions be simply equated with the dimensions of his artistic skill or genius. But the common anxiety that attention to the cultural determinants of aesthetic value makes the artist or artistic labor *irrelevant* is simply unfounded. The second point is that what may be spoken of as the "properties" of a work—its "structure," "features," "qualities," and of course its "meanings"—are not fixed, given, or inherent in the work "itself" but are at every point the variable products of particular *subjects'* interactions with it. Thus, it is never "the *same* Homer."[2] This is not to deny that some aspect, or perhaps many aspects, of a work may be constituted in similar ways by numerous different subjects, *among whom we may include the author*: to the extent that this duplication occurs, however, it will be because the subjects who do the constituting are themselves similar, not only or simply in being human creatures (and thereby, as it is commonly supposed, "sharing an underlying humanity" and so on) but in occupying a particular universe that may be, for them, in many respects recurrent or relatively continuous and stable, and/or in inheriting from one another, through mechanisms of cultural transmission, certain ways of interacting with texts and "works of literature."

To continue, however, the account of the cultural-historical dynamics of endurance. An object or artifact that performs certain desired/able functions particularly well at a given time for some community of subjects, being perhaps not only "fit" but exemplary—that is, "the best of its kind"—under those conditions, will have an immediate survival advantage; for, relative to (or in competition with) other comparable objects or artifacts available at that time, it will not only be better protected from physical deterioration but will also be more frequently used or widely exhibited and, if it is a text or verbal artifact, more frequently read or recited, copied or reprinted, translated, imitated, cited, commented upon, and so forth—in short, culturally re-produced—and thus will be more readily available to perform those or other functions for other subjects at a subsequent time. [. . .]

Nothing endures like endurance.

To the extent that we develop within and are formed by a culture that is itself constituted in part *by* canonical texts, it is not surprising that those texts seem, as Hans-Georg Gadamer puts it, to "speak" to us "directly" and even "specially": "The classical is what is preserved precisely because it signifies and interprets itself; [that is,] that which speaks in such a way that it is not a statement about what is past, as mere testimony to something that needs to be interpreted, but says something to the present as if it were said specially to us . . . This is just what the word 'classical' means, that the duration of the power of a work to speak directly is fundamentally unlimited."[3] It is hardly, however, as Gadamer implies here, because such texts are uniquely self-mediated or unmediated and hence not needful of interpretation but, rather, because they have already been so thoroughly mediated—evaluated as well as interpreted—*for* us by the very culture and cultural institutions through which they have been preserved and by which we ourselves have been formed.

What is commonly referred to as "the test of time" (Gadamer, for example, characterizes "the classical" as "a notable mode of 'being historical,'" that historical process of preservation that through the constant proving of itself sets before us something that is true")[4] is not, as the figure implies, an impersonal and impartial mechanism; for the cultural institutions through which it operates (schools, libraries, theaters, museums, publishing and printing houses, editorial boards, prize-awarding commissions, state censors, and so forth) are, of course, all managed by *persons* (who, by definition, are those with cultural power and commonly other forms of power as well); and, since the texts that are selected and preserved by "time" will always tend to be those which "fit" (and, indeed, have often been *designed* to fit) *their* characteristic needs, interests, resources, and purposes, that testing mechanism has its own built-in partialities accumulated in and thus intensified by time. For example, the character-istic resources of the culturally dominant members of a community include access to specific training and the opportunity and occasion to develop not only competence in a large number of cultural codes but also a large number of diverse (or "cosmopol-itan") interests. The works that are differentially reproduced, therefore, will tend to be those that gratify the exercise of such competencies and engage interests of that kind: specifically, works that are structurally complex and, in the technical sense, information-rich—and which, by virtue of those qualities, may be especially amenable to multiple reconfiguration, more likely to enter into relation with the emergent interests of various subjects, and thus more readily adaptable to emergent conditions. Also, as is often remarked, since those with cultural power tend to be members of socially, economically, and politically established classes (or to serve them and identify their own interests with theirs), the texts that survive will tend to be those that appear to reflect and reinforce establishment ideologies. However much canonical works may be seen to "question" secular vanities such as wealth, social position, and political power, "remind" their readers of more elevated values and virtues, and oblige them to "confront" such hard truths and harsh realities as their own mortality and the hidden griefs of obscure people, they would not be found to please long and well if they were seen radically to undercut establishment interests or *effectively* to subvert the ideologies that support them. (Construing them to the latter ends, of course, is one of the characteristic ways in which those with anti-establishment interests participate in the cultural re-production of canonical texts and thus in their endurance as well.) [. . .]

As the preceding discussion suggests, the value of a literary work is continuously produced and re-produced by the very acts of implicit and explicit evaluation that are frequently invoked as "reflecting" its value and therefore as being evidence of it. In other words, what are commonly taken to be the signs of literary value are, in effect, its *springs*. The endurance of a classical canonical author such as Homer, then, owes not to the alleged transcultural or universal value of his works but, on the contrary, to the continuity of their circulation in a particular culture. Repeatedly cited and recited, translated, taught and imitated, and thoroughly enmeshed in the network of intertextuality that continuously *constitutes* the high culture of the orthodoxly educated population of the West (and the Western-educated population of the rest of the world), that highly variable entity we refer to as "Homer" recurrently enters our experience in relation to a large number and variety of our interests and thus can perform a large number of various functions for us and obviously has performed them for many of us over a good bit of the history of our culture. It is well to recall, however, that there are many people in the world who are not—or are not yet, or choose not to be—among the orthodoxly educated population of the West: people who do not encounter Western classics at all or who encounter them under cultural and institutional conditions very different from those of American and European college professors and their students. The fact that Homer, Dante, and Shakespeare do not figure significantly in the personal economies of these people, do not perform individual or social functions that gratify their interests, *do not have value for them*, might properly be taken as qualifying the claims of transcendent universal value made for such works. As we know, however, it is routinely taken instead as evidence or confirmation of the cultural deficiency—or, more piously, "deprivation"—of such people. The fact that other verbal artifacts (not necessarily "works of literature" or even "texts") and other objects and events (not necessarily "works of art" or even artifacts) have performed and do perform for them the various functions that Homer, Dante, and Shakespeare perform for us and, moreover, that the possibility of performing the totality of such functions is always distributed over the totality of texts, artifacts, objects, and events—a possibility continuously realized and thus a value continuously "appreciated"—commonly cannot be grasped or acknowledged by the custodians of the Western canon.

Notes

1. Here and throughout this study the term "desired/able" indicates that the valued effect in question need not have been specifically desired (sought, wanted, imagined, or intended) as such by any subject. In other words, its value for certain subjects may have emerged independent of any specific human intention or agency and, indeed, may have been altogether a product of the chances of history or, as we say, a matter of luck.
2. For a careful neo-Marxist analysis of the continuous historical "rewriting" of the Homeric texts, see John Frow, *Marxism and Literary History* (Cambridge, MA: 1986), pp.172–82.
3. Hans-Georg Gadamer, *Truth and Method* (orig. pub. as *Wahrheit und Methode*, Tübingen, 1960; trans. Joel Weinsheimer and Donald G. Marshall (New York: 1982), pp. 257–58.
4. Ibid., p. 255.

Chapter 25

Arnold Krupat (1941–)
from "The Concept of the Canon,"
The Voice in the Margin (1989)

i

The concept of a literary canon is generally understood in either of two ways, each very much opposed to the other. Let me state them in their most extreme form: on the one hand, the canon is conceived of as a body of texts having the authority of perennial classics. These texts, "the great books" (as at least one American college has institutionalized them in a course of instruction), are, as they always have been and always will be, nothing less than the very best that has been thought and said. To understand their content—to have isolated for further meditation their themes or ideas—is to gain or make some nearer approach to timeless wisdom; to apprehend their form is to experience the beautiful or at the least to perceive a significant order. Sympathetic contact with these texts cannot help but make one a better person, or—the phrase is a curious one on inspection, to be sure—more human.

On the other hand, however, the canon is taken simply as the name for that body of texts which best performs in the sphere of culture the work of legitimating the prevailing social order: canonical texts are, as they always have been, the most useful for such a purpose (although the modality of their usefulness may, of course, alter with time). To understand their content is largely to accept the world view of the socially dominant class; to apprehend their form is to fail to perceive that acceptance as such. Sympathetic contact with these texts tends mostly—although not always or exclusively—to contribute to that ideological conditioning, the production of that consciousness, necessary to conform one willingly to one's—usually subordinate—class position in society. [. . .]

Arnold Krupat is Professor of Global Studies, specializing in Native American Literature, at Sarah Lawrence College. In this essay, Krupat proposes a "secular heterodoxy" for the history of American Literature.

There are difficulties with both these views of the canon, as there are attractions in both as well. The transcendental-essentialist view, for example, like all full-blown idealisms, can only be accepted as an act of faith; it requires that one engage in what Edward Said has called "religious" criticism in contradistinction to the secular criticism he so eloquently affirms. Appeals to inherent greatness, to unseen yet dimly felt orders of great power, and the like are not conducive to this-worldly understanding—unless one chooses to say with Saint Augustine *credo ut intellegam*. The traditional view of canonical authority can be maintained only by rigidly separating literary value from value of other kinds—and by positing literary judgment as a first-order judgment, something that can be formed prior to or independently of our judgments about other kinds of value. Once the introduction of empirical (not to say statistical) evidence is permitted and the condition of cognitive responsibility is accepted,[1] the transcendental-theoretical conception of the canon tends to lose a good deal of its force.

Yet a purely instrumental or pragmatic view of the canon, for all its attractions, has its problems, too. For, at least in its extreme version, it tends to see the canon as formed *exclusively* by power relations: the canonical texts are the surviving victors on the battlefield calling for due praise. If they are an obnoxious group, we can never join them and so we must get out there and beat them. But this is to adopt a Thrasymachean perspective, the world view of primitive capitalism, a crude form of social Darwinism, or indeed an equally crude version of Marxist class struggle. It is to adopt Foucault's bleak vision (somewhat modified in the last work) of discourse as power and power as everywhere, so that even to fight and win is only to become oppressor in one's turn—to force people to read *our* books now, not *theirs* (and, of course, they will fight back, conflict unending). And this sort of Hobbesian war of all against all forever is very far from what most instrumental pragmatists desire.

To avoid it, three positions in relation to the canon have, thus far, seemed open to the instrumentalist. The first is simply to declare a universal peace, as it were, one in which everyone is conceded equal greatness—or, rather, one in which any claim at all to greatness or superiority is automatically suspect. This means, in effect, that one simply dispenses entirely with the concept of the canon—of *any* canon of authorized texts—so that what one teaches and writes about is simply books that happen to be interesting, or useful for one sort of demonstration or another. Not evaluative but only functional criteria are admitted as determining text selection. Cognitive responsibility here consists in a willingness to provide arguments for justifying the functional criteria as generally reasonable ones. [. . .]

A second way to avoid endless ideological struggle for the canon of great or central books is not to jettison the canon entirely but to propose a canon whose authority is merely statistical. Wayne Charles Miller, for example, has suggested revising the canon of American literature on principles—so it seems to me—of virtually statistical representation.[2] This position on the canon I shall call the ethical–ontological reform. It claims that we should, *in fairness*, read a proportionate number of authors who *are* actually available. Since it is indisputable that America is not almost exclusively made up of white, male, eastern WASPS, courses in American literature should not almost exclusively be made up of texts by white, male, eastern WASPS. With this observation I agree wholeheartedly. But even if one were to accept the principle of proportional revision, the ethical–ontological reform has no way of deciding which authors should finally be taught, or even which authors should represent their particular ethnic and/or racial groups. If it is easy to agree that Black Elk might well be taught instead of William Cullen Bryant (but maybe it isn't so easy to agree after all, and maybe the most famous Native American autobiographer should yield to some other Native American autobiographer) or that Whittier (if he still shows up) could be dropped to make room for Frederick Douglass (but why not Linda Brent or Harriet Wilson?) or Maxine Hong Kingston, it is less easy to know whether Kingston should be supplemented by Frank Chin, or whether (for the semester has only so many weeks) Ferris Takahashi should be taught instead of Piri Thomas. (And we have still left out immigrant Jewish writers, Chicanos, Italian Americans, Scandinavians, and a great many others as yet unmelted into general Americanness.) This inability or refusal to imagine a principle that might merge valorized difference into some collective identity (a dialectical principle, to be sure, so that common identifications do not obliterate historical differentiations) finally works against the hope of a common culture; its canon reflects and can reflect nothing but the changing demographics of each historical moment. We never approximate an *American literature* but remain instead at the level of Miller's multiethnic literatures of the United States.

It is perhaps in recognition of this truth that the third way comes about for hermeneutically suspicious pragmatists to avoid a view of the canon as inevitably oppressive and endlessly contested. Rather than an abandonment of any canon whatever or the acceptance of a demographically authorized one, the position I turn to now reconstructs the canon on the basis of a strictly experiential authority. Recognizing that the traditional canonical texts have been pretty exclusively the *phallogocentric* texts, in Jacques Derrida's term, those representing the experiences and values of a very small group of elite Western males, this position decides that the books worth promoting are not just any books, not even those that offer a kind of proportional representation of what is actually there, but, instead, those which are directly relevant to their audience's experience. This reform of the canon leads increasingly to the sense that any attempt to consider American or English literature as a whole is just too hopelessly broad, and it is bound to urge a shift more and more to courses in and books about ethnic literature, working-class literature, third world women's literature, regional literature, and so on, and more nearly in isolation from one another rather than in relation to one another.

Here, one gives up the proposal to a wide range of American students and readers even of a multiethnic canon. Instead, one offers to readers of the Plains region, Willa Cather's novels and the epics of John Neihardt; to readers of the southwest, perhaps Paul Horgan and Witter Bynner; New Yorkers may choose from Emma Goldman, Mike Gold, and Abraham Cahan, Langston Hughes and early James Baldwin, Oscar Hijuelos and Piri Thomas, among others. In New England, at least to predominantly male readers, Hawthorne, Thoreau, Melville, even Longfellow and Whittier and Bryant will continue to be taught. Thus Paul Lauter's experience in trying to present Faulkner's *The Bear* to a group of working-class students in New York would rarely be repeated (they thought the story silly; they laughed),[3] for everyone will find his personal concerns more or less closely reflected in what he or she is asked to read.

The latter two of these reactions against the essentialist view of the canon and against the instrumentalism of endless conflict do retain a measure of the authority that traditional views accorded to canonical texts, for the books they propose to us are, after all, deemed to be more important or somehow more necessary than others—demographically, or at least in regard to their audiences' own experience. But what they give up is enormous. For they surrender the possibility of attaining just that sort of perspective on our individual experience and our historical moment that a broad acquaintance with literature can provide. Instead of the attempt to define a canon that might become what Elizabeth Fox-Genovese has called a "collective auto-biography" and "to introduce some notion of collective standards," they "settle for education as personal autobiography or identity," and thus "accept the worst forms of political domination" (133). In their desire to acknowledge difference and accord it its due, they give up the very possibility of a common culture, one that may at least imply a common society. An American literary canon, I should say, is worth fighting for as the complex record of possible national identities; Crèvecoeur's question, What is an American? is unanswerable, of course. But it needs again and again to be posed.

In these regards, then, the most appealing aspect of the traditional view of the canon is its determination to believe in a common culture, a body of work that defines some part of what we all (whoever "we" are) can believe in and share. To quote Elizabeth Fox-Genovese once more, ". . . however narrow and exclusive the canon we have *inherited*, the existence of *some* canon offers our best guarantee of some common culture" (132; emphasis added), one which, as I would gloss her remarks, in its het-erodoxy expresses some part of our selves as collective selves, so that we see ourselves neither as simply accommodating or opposing but, rather, informed by others dialogically. The idea of the collective self as it is presented in the collective autobi-ography we call the canon, and the social vision implied by such a collective self, is something I will return to. These are of especial importance in any attempt to keep the concept of an American literature meaningful.

ii

[. . .] I will thus appropriate the term *heterodoxy* and use it to name exactly such a principle as "unity-in-difference," as this principle may inform an American literary canon—a canon of national literature—and an American social order. I shall also try to extend it to the international order of literature and society in taking heterodoxy as informing that *cosmopolitanism* my own discussion will take as its ultimate horizon. [. . .]

For the canon of American literature, secular heterodoxy on an empirical level means something very specific: it means that any proposed canon of American literature that does not include more than merely occasional examples of the literatures produced by red and black people as well as white people—men and women, of indigenous and African, as well as European origins—is suspect on the very face of it. The history of that national formation called the United States of America is such as to insist upon the pri-macy of Euramerican, Native American, and Afro-American literary expression in any attempt to define an American literature. In saying this, I hope it is clear, to repeat, that I am not calling for some kind of proportional representation for these groups, nor restricting the canon to texts associated with these groups, aprioristically denying that canonical American books might well be produced by people of Asiatic, or mixed origins, or of any background whatever. In these regards, it is worth noting that Spanish is now the second language of the United States (as it is the second most widely spoken language in the world), a fact sufficiently important to have provoked intense efforts on the part of S. I. Hayakawa and others to pass a regressive and repressive constitutional amendment making English the official language of the United States. As I write, on the eve of Election Day, 1988, proposals are on the ballot in Arizona, Colorado, and Florida to this effect. Given the increase in American Spanish speakers, there is no doubt in my mind that Latino literature will soon exert major pressure on the canon, a development I look forward to with enthusiasm. Nonetheless, to the present, the cultural expression of red, white, and black people seems to me to have a historically urgent claim to primary attention.

I mean, here, to assert that Afro-American and Native American literary production, when we pay attention to it, offers texts equivalently excellent to the traditional Euramerican great books. It is not only that these texts should be read in the interest of fairness or simply because they are available; nor is it because they provide charming examples of "primitive" survivals: they should be read because of their abundant capacity to teach and delight. But for that capacity to be experienced and thus for the excellence of these texts to be acknowledged, it will be necessary, as I have suggested above, to recognize that what they teach frequently runs counter to the teaching of the Western tradition, and that the ways in which they delight is different from the ways in which the Western tradition has given pleasure. [. . .]

In terms of technique, even the most recent and most complexly composed Native American works are still likely to have roots in or relations to oral traditions that differ considerably in their procedures from those of the dominant, text-based culture: if these works are indeed equivalently excellent, still it must be recognized that they are differently excellent. To the extent that we are perhaps already in what Father Walter Ong calls the "secondary orality of our electronic age" (305) to the extent that print culture is already receding from the importance it had for a full five hundred years, we may currently be producing just the conditions of possibility for such a recognition. That postmodernist fiction, poetry, and painting have found a substantial audience; that the disjointed, even spasmodic styles of "Miami Vice," "Crime Story," and MTV music video have proved popular, indicates that a wide public has lost interest in attempts to represent the world realistically in causally connected, continuous linear narrative. Ronald Reagan's popularity, matched with the unpopularity of his actual political positions, is only further evidence of a paradigm shift whose description is already possible to produce, although its evaluation remains somewhat more difficult. In any event, the material situation as I can understand it, for all that I am wary of it, nonetheless seems to me encouraging for the appreciation of Native American literature.

Notes

1. I derive my conception of cognitive responsibility from Hayden White who derives it from Stephen Pepper's *World Hypotheses*. Cognitively responsible positions are those "committed

to rational defenses of their world hypotheses" as opposed to those "not so committed" (H. White, n. 23). To the mystic and fascist positions of cognitive irresponsibility mentioned by White we might currently add varieties of postmodernism. Jurgen Habermas's attempts to define conditions of "cognitive adequacy," which I admit to knowing only secondhand, seem also pertinent here.

2. See Wayne Charles Miller, "Toward a New Literary History of the United States." I should note that Miller nowhere explicitly proposes a statistical canon.

3. See Paul Lauter, "History and the Canon."

Works Cited

Fox-Genovese, Elizabeth. "Gender, Race, Class, Canon." *Salmagundi* 72 (1986): 131–43.

Ong, Walter J., S.J. *Interfaces of the Word: Studies in the Evolution of Consciousness and Culture.* Ithaca: Cornell University Press, 1977.

Said, Edward. *The World, the Text, and the Critic.* Cambridge: Harvard University Press, 1983.

Chapter 26

Charles Altieri (1942–) from "An Idea and Ideal of Literary Canon," *Canons and Consequences* (1990)

The process of strong evaluation makes clear the interests canons might serve, primarily by fostering two basic functions. One is curatorial: literary canons preserve rich, complex contrastive frameworks, which create what I call a cultural grammar for interpreting experience. Given the nature of canonical materials, however, there is no way to treat the curatorial function as simply semantic. Canons involve values, both in what they preserve and in the principles of preserving. Thus, the other basic function of canons is necessarily normative. Because these functions are interrelated, canons need not present simple dogmas. Instead, canons serve as dialectical resources, at once articulating the differences we need for a rich contrastive language and constituting models of what we can make of ourselves as we employ that language. This interrelation, in turn, applies to two basic kinds of models, each addressing a different dimension of literary works. Canons call attention to examples of what can be done within the literary medium. The canon is a repertory of inventions and a challenge to our capacity to further develop a genre or style. But in most cases, craft is both an end in itself and a means for sharpening the texts' capacity to offer a significant stance that gives us access to some aspect of nontextual experience. So in addition to preserving examples of craft, canons also establish exemplary attitudes, often while training us to search for ways to connect the two. This means that when we reflect on the cultural roles canons can play, we must take as our representative cases not only those works that directly exhibit exemplary features of craft or wisdom, but also works that fundamentally illuminate the contrastive language we must use to describe those achievements. It matters that we read the *Aeneid* because there are strong reasons to continue valuing the tragic sense of duty the work exemplifies; it

Charles Altieri is Professor of English at the University of California, Berkeley. In this selection from his book, *Canons and Consequences*, Altieri describes the canon as a "normative circle" of judgment and interpretation.

matters that we read Thomas Kyd because of the influence he exercised on Shakespeare and Eliot; it matters much less that we read George Gascoigne or Stephen Duck, the "Thresher Poet," because they neither provided significant types of exemplifying wisdom or craft nor influenced those who we think did.

The curatorial and normative practices we develop for such bodies of texts have three possible cultural consequences. The first, and most fundamental, is the most difficult to discuss. Canons play the role of institutionalizing idealization: they provide contexts for their own development by establishing examples of what ideals can be, how people have used them as stimuli and contexts for their own self-creation, and why one can claim that present acts can address more than the present. Harold Bloom offers a compelling account of the struggle a canon elicits, but his reliance on personal strength leads him to pay scant attention to other, equally significant effects of this heritage. The very idea of a canon and the example it offers create the standards writers try to meet. Indeed, canons are largely responsible for the frame of questions that allows Bloom's "agon," and, equally important, they establish the complex practices of argument by which critical evaluations can be articulated. Canons make us want to struggle, and they give us the common questions and interests we need to ennoble that project.

We share enough literary experience to obviate any need to elaborate on these pieties. So I will proceed immediately to a second, corollary cultural consequence of canons. If ideals are to play a significant role for a culture, there must be a model of authority that empiricism cannot provide. When we offer an idealization from or about the canon, we must face the question of who will judge those features of the past as worthy to become normative models—or, who will judge the reasons we offer in our idealizations of those idealizations. We return to the dilemma of circularity. But by now I hope that our reflections on the canon will manifest some of the immanent capacities of the circle. Judges for the canon must be projections from within the canon as it develops over time. Here we can construct a normative circle, analogous to the principle of competent judgment John Stuart Mill proposes as his way of testing among competing models of happiness. Our judges for ideals must be those we admire as ideal figures or those whom these ideal figures admired. Only such an audience of judges can save us from the trap of an even smaller circle. For unless we can project audiences for our evaluations who are beyond the specific interpretive community that shares our reading habits, there is little point in giving reasons for our idealizations. Our reasons would only identify our own community; they would say nothing significant about values in general or give us the distance from ourselves requisite for both self-criticism and self-direction. Similarly, unless our audience were as capacious as the ideals we project, we would contaminate them (or ourselves) in the very process of articulating our strong evaluations.

Chapter 27

Alvin Kernan (1923–)
from "Introduction: The Death of
Literature," *Death of Literature* (1990)

Literature has in the past thirty years or so passed through a time of radical disturbances that turned the institution and its primary values topsy-turvy. Talk began in the 1960s about the death of literature, with a comparison intended to Nietzsche's announcement of the death of God, and by 1982 Leslie Fiedler, a pop-lit advocate not sorry to see high-culture literature go, could happily title a book *What Was Literature?*

Internally, the traditional romantic and modernist literary values have been completely reversed. The author, whose creative imagination had been said to be the source of literature, was declared dead or the mere assembler of various bits of language and culture into writings that were no longer works of art but simply cultural collages or "texts." The great historical tradition extending from Homer to the present has been broken up in various ways. The influence of earlier poets on their successors has been declared no longer beneficial but the source of anxiety and weakness. The literary canon has been analyzed and disintegrated, while literary history itself has been discarded as a diachronic illusion, to be replaced by a synchronic paradigm. What were once the masterpieces of literature, the plays of Shakespeare or the novels of Flaubert, are now void of meaning, or, what comes to the same thing, filled with an infinity of meanings, their language indeterminate, contradictory, without foundation; their organizational structures, grammar, logic, and rhetoric, verbal sleights of hand. Such meaning as they may have is merely provisional and conferred on them by the reader, not inherent in the text or set in place for all time by the writer's word craft. Rather than being near-sacred myths of human experience of the world and the self, the most prized possessions of culture, universal statements about an unchanging and essential human nature, literature is increasingly treated as authoritarian and destructive of human freedom, the ideology of the patriarchy

---

While sympathetic to those who fear that literature is becoming part of the past, Alvin Kernan, who has taught at Yale and Princeton, contends that it is only a particular—Romantic and Modernist—sense of literary activity that may be passing.

devised to instrument male, white hegemony over the female and the "lesser breeds." Criticism, which was once the scorned servant of literature, has declared its independence and insisted that it too is literature. Not everyone accepts all of these new views, but their reality is increasingly taken as fact, and it is as fact, not a judgment of what has happened, that they are here described in as neutral a way as possible.

Externally, political radicals, old and young, from Herbert Marcuse to Terry Eagleton, have attacked literature as elitist and repressive. Television and other forms of electronic communication have increasingly replaced the printed book, especially its idealized form, literature, as a more attractive and authoritative source of knowledge. Literacy, on which literary texts are dependent, has diminished to the point that we commonplacely speak of a "literacy crisis." Courses in composition have increasingly replaced courses in literature in the colleges and universities, where enrollments and majors in literature continue to decrease nationally. The art novel has grown increasingly involute and cryptic, poetry more opaque, gloomy, and inward, and theater more hysterical, crude, and vulgar in counterproductive attempts to assert their continued importance. What was once called "serious literature" has by now only a coterie audience, and almost no presence in the world outside university literature departments. Within the university, literary criticism, already by the 1960s Byzantine in its complexity, mountainous in its bulk, and incredible in its totality, has turned on literature and deconstructed its basic principles, declaring literature an illusory category, the poet dead, the work of art only a floating "text," language indeterminate and incapable of meaning, interpretation a matter of personal choice. Many of our best authors—Nabokov, Mailer, Malamud, and Bellow were the cases I explored in an earlier book, *The Imaginary Library*—have experienced and not recovered from a crisis of confidence in the traditional values of literature and a sense of its importance to humanity.

The disintegration of literature has become scandalous enough to produce headlines and bestsellers. In 1988, Stanford University, for example, made the front pages and the TV news programs with a debate about whether its required course in great books, including many works of literature, should drop some of the classics, all written by "dead white males," to make room for the inclusion of books by women, blacks, and Third World writers. The great books which had hitherto formed the basis of liberal education were denounced as elitist, Eurocentric, and the tools of imperialism. Under this kind of pressure, the faculty and administration agreed to replace such writers as Homer and Dickens with books like Simone de Beauvoir's *Second Sex*. William Bennett, then a conservative secretary of education, debated with the president of Stanford on national television the social question at the center of the issue: the relative importance to society at large of the traditional intellectual qualities represented by the classics of literature versus social values of equality of gender and race represented by less prestigious writings.

This particular debate was but one part of a wider cultural debate about the breakdown of education, and particularly literary education, raised in books that became surprise best-sellers, *Cultural Literacy* by E. Donald Hirsch, Jr., and *The Closing of the American Mind* by Allan Bloom. Hirsch, a professor of English literature at the University of Virginia, charging that Americans are becoming culturally illiterate from not reading the best kinds of writing, offered a test in great ideas—Darwin, Freud, Marx, for example—that could be administered at home to diagnose the seriousness of the deficiency. A follow-up volume provided in handy form the means to rectify the deficiency in case, like radon in the cellar, it should be discovered. Bloom, a University of Chicago professor of politics with a bias to literature, is a follower of Leo Strauss and his view of certain classic texts, Plato particularly, as the sacred repositories of arcane truths. He charged that a pliant relativism in university faculties, derived from Nietzsche and other German philosophers, had led education away from the classic texts and their Socratic search for the good and the true. The modern student, infected with relativism, believing that all values are only opinions, and one opinion as good as another, has entirely abandoned, according to Bloom, the great books and their quest for the best course of belief or action, to live in a daze of universal tolerance, apathy, and ignorance.

Even if Bennett, Hirsch, and Bloom are taken at something less than face value, the widespread interest in their views testifies to a general concern that book culture, of which literature is a central part, is disappearing, and with it many of our society's central values. No wonder Marxists fight feminists for the right to identify the smells arising from the literary corpse. Bad faith or phallocentrism? Hegemony or gynocide? It went so far that the religious leader of Iran, the infamous Ayatollah, could without fear of reprisal take out a contract on the life of the author of a novel found offensive to Islam. Within a month, or ere Salman Rushdie's books were old, Western publishers were holding an international book fair in Iran.

If *literature* has died, *literary activity* continues with unabated, if not increased, vigor, though it is increasingly confined to universities and colleges. Stories and poetry are written and read, plays are performed, and strenuous efforts made to write well. Publishers pay large advances for novels, literary prizes are given with increasing frequency and in larger amounts, and the literature printed, of whatever quality, continues to increase. An industrious literary criticism and scholarship, largely within the academies, overproduces both literary theory and practical criticism. There are many optimists who see a new and better literary system arising phoenixlike from the ashes of the old, no longer the "repository of known truth and received values" (Levine) but "a poetics which strives to define the conditions of meaning. . . . how we go about making sense of texts" (Culler). This redirection of literature is perceived by its supporters as a giant step for humankind, and Levine and his colleagues at a conference on the humanities spoke for the more advanced criticism when they said that it seems "particularly ironic that the humanities are receiving their most severe criticism at a moment when for many of us their significance and strength have never been greater."

What has passed, or is passing, is the romantic and modernist literature of Wordsworth and Goethe, Valéry and Joyce, that flourished in capitalistic society in the high age of print, between the mid-eighteenth century and the mid-twentieth. The death of the old literature in the grand sense, Shelley's unacknowledged legislation of the world, Arnold's timeless best that has been thought and written, Eliot's unchanging monuments of the European mind, from the rock drawings in Lascaux to *The Magic Mountain*, has seemed to people who matured intellectually in the ancien régime of high culture nothing less than the setting of the sun of the human imagination in the evening-lands of Western civilization. No more eloquent defense of the old literary order has appeared than a collection of essays, *Prose and Cons*, by Maynard Mack, and no more bitter denunciation of the new ways than one of the essays, "The Life of Learning": "We are narrowing, not enlarging our horizons. We are shucking, not assuming our responsibilities. And we communicate with fewer and fewer because it is easier to jabber in a jargon than to explain a complicated matter in the real language of men. How long can a democratic nation afford to support a narcissistic minority so transfixed by its own image?" Many others, like Mack, remember that it was only yesterday that F. R. Leavis could reject C. P. Snow's argument in *The Two Cultures* that humanists were dangerously ignorant of science with the haughty observation that a humane training based on literature was the only education worth having. Cleanth Brooks and W. K. Wimsatt only a few years ago claimed critical infallibility, as if their formalist conceptions of literature were revealed truth, anathematizing heterodoxy as "the intentional fallacy" and "the heresy of paraphrase." Not long ago at all, there seemed nothing absurd in Northrop Frye's argument in *Anatomy of Criticism* that the totality of literature formed an extensive scheme, mystical in its symbolism, but orderly in its structure, originating in the fears and desires constituting the human soul and moving through history in the form of the great literary myths, corresponding with nothing less than the seasonal cycle of the natural year.

Looking back, it seems incredible that these views could have so recently been taken as seriously as they surely were. But they are gone now, and to the often bewildered, muttering, and angry survivors of the old order, the change has seemed another treason of the clerks, who are most often identified as a group of radical critics practicing, usually in the universities, what Paul Ricoeur has aptly labeled "the hermeneutics of suspicion." Phenomenology, structuralism, deconstruction, Freudianism, Marxism, feminism have been the most clamorous voices announcing the death of the old literature in recent years. Structuralism and deconstruction, the leading demystifiers of traditional literary views for a time, were a poetics militant, attacking bourgeois society by undermining its ideology and exposing all authority, including all literary authority, as illegitimate and repressive. Feminism denounced the old literature as an instrument of male domination. Marxists, the followers of Foucault, and the new historicists treated literature as a capitalist institution and a disguised instrument of hegemony, to be exposed as mere establishment propaganda. To the new Freudians, literature was another form of the repression of instinct and revolutionary impulses, to be cured by deeper analysis.

The Anglo-Saxon tendency has been to keep literature and politics as widely separated as possible, but these recent types of radical criticism have, in the manner of the Continent and particularly of Paris, associated themselves, both in theory and in practice, with new left politics and social theory. Literature has been seen as a soft area of bourgeois society, a place at which to get at and discredit capitalist ideology, to advance Third World, minority, and feminist causes, and to advocate permissiveness, openness, and freedom in all areas, sexual, interpretive, environmental.

But the social scene in which the hermeneutics of suspicion have flourished has been much larger than the narrow setting of universities, conferences of specialists, and salon politics of the literary subculture. They, and literature itself, have been only one small part of a much more extensive and deeper cultural change. Not only the arts but all our traditional institutions, the family and the law, religion and the state, have in recent years been coming apart in startling ways. The family is probably the most desperate battlefield in this massive social change: the pill, soaring divorce rates, custody battles, poor single-parent families headed by women, right-to-life and pro-choice struggles, two-career families, surrogacy, women's rights, battered wives and murdered families, the disappearance of traditional patterns of sexual differentiation, in vitro fertilization, casual attitudes toward sex, the appearance of new venereal diseases. There is good and bad in this catalog, as well as "purposes mistook / Fall'n on the inventors heads," but all are causing enormous stress in an old institution. The death throes of the nuclear family, along with the changes in other major social institutions, make the death of romantic literature seem but a trifle here. To see what has happened to literature as a part of the social revolution sometimes loosely styled postindustrialism that has been transforming modern life in the West, and to a lesser degree in the Second and Third worlds as well, provides both a historical setting in which to understand literary change and a scale which accurately measures its interesting but limited part in what has been going on. In the larger world, in fact, the death of literature may be chiefly interesting only for the precise schematic way in which it represents changes taking place elsewhere, as in the family, in more complicated and less obvious ways. The exact reversal of literary values, for example—poets are creative geniuses / the poet is dead, literary texts are supersaturated with meaning / literary texts are empty of meaning—offers almost a laboratory example of the revolutionary, as opposed to evolutionary, model of institutional change that Thomas Kuhn called "paradigm shift." [. . .]

Chapter 28

Roger Kimball (1953–)
from "Speaking Against the Humanities,"
Tenured Radicals: How Politics
Has Corrupted Our Higher Education
(1990; rev. 1998)

In order to get a more tangible sense of what humanistic inquiry means in the environment of today's academy, let us return to the Whitney Humanities Center at Yale University to consider some of the presentations that were given at a day-long public symposium in the spring of 1986 on "The Humanities and the Public Interest." The purpose of the event, in the words of a university press release, was "to re-examine the traditional association between the study of the humanities and the guardianship of humanistic values in the context of contemporary American society." Peter Brooks, who presided over this event as well, expanded on this in the press release: "The symposium will ask whether the case for the humanities can rest on traditional assumptions, or whether a new rationale is needed if the humanities are to claim a major place in contemporary modes of thought and analysis."

The symposium opened with some introductory remarks by Professor Brooks, who noted that the original impetus for the symposium was his favorite reading material, former Secretary of Education William J. Bennett's report on higher education in the humanities, *To Reclaim a Legacy*. As we have seen, this report defends precisely those "traditional assumptions" of the humanities that Professor Brooks hoped the Yale symposium would question. For himself, Professor Brooks declared his "profound disagreement" with the conclusions and general outlook of Secretary Bennett's report, taking issue especially with what he described as its "intellectual fundamentalism." Professor Brooks's opening remarks were very brief, but they established the tenor for the day's discussion; and since he identified Secretary Bennett's report as the catalyst for the symposium, we may begin by returning to take a closer look at the report's argument.

In this essay, Roger Kimball, Managing Editor at *The New Criterion*, focuses on a 1986 discussion of the Humanities at Yale, for what Kimball sees as politicized tensions within the academy about the canon.

To Reclaim a Legacy begins by reaffirming the traditional role of the humanities as the chief instrument of our cultural self-definition. Its presiding spirit is Matthew Arnold, whose faith in the ennobling effects of high culture, of "the best that has been thought and said," is patent throughout the report. Elaborating on Arnold's famous phrase, Secretary Bennett describes the humanities as "the best that has been said, thought, written, and otherwise expressed about the human experience." The humanities are important, he writes, because

> they tell us how men and women of our own and other civilizations have grappled with life's enduring, fundamental questions: What is justice? What should be loved? What deserves to be defended? What is courage? What is noble? What is base? . . .
>
> These questions are not simply diversions for intellectuals or playthings for the idle. As a result of the ways in which these questions have been answered, civilizations have emerged, nations have developed, wars have been fought, and people have lived contentedly or miserably.

The real source of the controversy surrounding Secretary Bennett's report lies not so much in such general observations as in his prescriptions for "reclaiming" the legacy he finds threatened and, in the end, in his understanding of the substance and definition of that legacy. In the simplest terms he calls for a reshaping of undergraduate study "based on a clear vision of what constitutes an educated person." In his view, the goal of the humanities should be a "common culture" rooted in the highest ideals and aspirations of the Western tradition.

Nevertheless, it is important to note that, despite accusations to the contrary, Secretary Bennett does not advocate restoration of a previous state of affairs. He insists that the solution to the current crisis in the humanities "is not a return to an earlier time when the classical curriculum was the only curriculum and college was available to only a privileged few." Given the charges of elitism and reaction that his proposals have brought forth, especially from the most elite of our universities, it seems well to emphasize the point. "American higher education today serves far more people . . . than it did a century ago," Secretary Bennett writes.

> Its increased accessibility to women, racial and ethnic minorities, recent immigrants, and students of limited means is a positive accomplishment of which our nation is justly proud. . . . But our eagerness to assert the virtues of pluralism should not allow us to sacrifice the principle that formerly lent substance and continuity to the curriculum, namely, that each college and university should recognize and accept its vital role as a conveyor of the accumulated wisdom of our civilization.

It is of course this final affirmation that has angered Secretary Bennett's opponents. For one thing, who decides what counts as "the accumulated wisdom of our civilization"? In Arnold's terms, why should the humanities be concerned primarily with the best that has been thought and said? Does that not exclude a large portion of human experience? And does not that mass of experience deserve "equal time" in our institutions of higher education? Here again, who is to say what counts as "best"? Perhaps the Arnoldian injunction has been interpreted too narrowly, too "ideologically," too exclusively? Furthermore, why should the humanities focus so intently upon the past? Why should they not concern themselves as much with the *creation* as with the *preservation* and *transmission* of culture? Such questions are at the heart of Professor Brooks's "profound disagreement" and charge of "intellectual fundamentalism"—a charge that has been loudly echoed in the academy and that was to be advanced with great zeal that Saturday at Yale's Whitney Humanities Center.

It was not, however, until the second, and most publicized, session, "The Social Mission of the Humanities," that the subject of the humanities and the public interest really came into focus. This session featured a "dialogue" between the late A. Bartlett Giamatti, who had not yet given up the presidency of Yale University to become commissioner of baseball, and Norman Podhoretz, the conservative critic and sometime editor of *Commentary* magazine. Responding to President Giamatti and Mr. Podhoretz were Henry Rosovsky, former dean and professor of social science at Harvard, and Cornel West, a ferociously articulate black radical who was then a professor at the Yale Divinity School and is now at Princeton. It was in this session that the real issues facing the humanities in contemporary American society were most clearly set forth.

Mr. Podhoretz spoke first. The humanities, he said, cannot be justified on practical grounds. Because the knowledge and culture they represent are "good in themselves," their ultimate justification is simply their intrinsic value. From this it follows that the humanities cannot directly help us in the formulation of public policy; nor do they yield any particular political position; nor indeed does acquaintance with the humanities necessarily make us morally more upright or more humane—think only of the cultivated Nazi commandants who also savored Mozart. Echoing the sentiments expressed in Secretary Bennett's report, Mr. Podhoretz identified the chief function of the humanities to be the creation of a "common culture." Central to this view of the humanities is the idea of a more or less generally recognized canon of works that define that common culture and preserve its traditions. Mr. Podhoretz admitted that there will always be disagreement about the composition of the canon at, as it were, its edges; but he claimed that, at least until recently, there has been a widely shared consensus about the core body of works that constitute "the best that has been thought and said."

In one sense, this view of the humanities can be said to be exclusive or "elitist," since it presupposes a rigorously defined notion of what it means to be an educated person. But in another sense, it is deeply democratic for it locates authority not in any class or race or sex, but in a tradition before which all are equal. As Mr. Podhoretz observed, to the extent that the humanities are crucial to the maintenance of civilized life, it is essential that as many people as possible have the opportunity to steep themselves in the great works of the canon: only thus is high culture preserved and transmitted. Furthermore, as the transmitter of the canon, of what Mr. Podhoretz described as our "intellectual patrimony," the humanities have traditionally instilled a sense of the value of the democratic tradition we have inherited. And it is in this respect, he noted, that the humanities do have a political dimension, insofar as they rest upon a belief in the value and importance of Western culture and the civilization that gave birth to it.

With the social and political upheaval of the Sixties and early Seventies, Mr. Podhoretz continued, this entire conception of the humanities came under radical assault. Not only the idea of a common culture founded upon a recognized canon of great works, but the very notion of a politically autonomous realm of culture was dismissed as naïve, ethnocentric, or somehow repressive. Even the fundamental belief in the value of Western culture and civilization—the value, that is to say, of the whole humanistic enterprise—was undermined. And while it is true that the more extreme manifestations of this revolt have disappeared, Mr. Podhoretz maintained that the radical attitudes espoused in the Sixties and Seventies live on in attenuated form in the academy—even, or rather especially, in the humanistic disciplines, in the values and assumptions that typically inform the teaching and study of the humanities. For the most part, he said, a study of the humanities now tends at best to encourage a feeling of "mild contempt" for culture as traditionally defined and at worst to inspire outright hatred of our civilization and everything it stands for. And because of this sedimented radicalism in the academy, the humanities, however much they may still add to an individual's enlightenment and culture, no longer really contribute to the common good.

Not surprisingly, Mr. Podhoretz's diagnosis was met with great hostility. I overheard the idea of a "common culture," for example, variously described as "moribund," "imperialistic," and "fascist." It was also considered to be "sexist," I gathered, judging from the knowing looks that his use of the phrase "intellectual patrimony" occasioned. President Giamatti began by telling us that he found Mr. Podhoretz's talk "internally contradictory," for is there not a contradiction between asserting the essentially private nature of the humanities and then lamenting that they no longer conduce to the commonweal? In fact, though, President Giamatti's charge depended upon distorting Mr. Podhoretz's description of the humanities. It is one thing to say that the humanities cannot be justified on instrumental grounds, as Mr. Podhoretz did, quite another to say that they are a private affair entirely without social consequence, which no one but President Giamatti thought to propose.

The president of Yale University, who at one time was known as a scholar of Renaissance literature, also came out strongly against the idea of a canon. Instead, he thought that the humanities should encourage "modes of thinking that would discipline the imagination without pretending to direct it"—the idea being, I suppose, that it doesn't much matter what one learns so long as one learns something. President Giamatti even claimed that this was the "Greek view" of education. Perhaps he meant the view current in contemporary Greece; certainly, the idea that education should seek "to discipline the imagination without I pretending to direct it" is completely foreign to the classical ideal of *paideia*, of formative education, as well as to the teachings of Plato and Aristotle. One thinks, for example, of the quite definite ideas that Plato expressed about what should and should not be taught in his discussion of education in the third book of *The Republic*. But leaving the Greek view of education to one side, President Giamatti's reservations about the importance of the canon do help us understand his central charge against Mr. Podhoretz: that his view of the humanities is "solipsistic" and "spiritually selfish." Basically, President Giamatti presented Mr. Podhoretz as an elitist who wanted to keep culture for himself. But the real difference between them was that Mr. Podhoretz wanted the *substance* of the humanities to be as widely available as possible, whereas President Giamatti was happy with what we might call universal schooling—the substance, the content, of what was taught was for him incidental.

If nothing else, President Giamatti exemplified the strategy that Henry Rosovsky, the session's first respondent, identified as the prime imperative for academic administrators—"Be vague." Professor Rosovsky went on to suggest that the hallmark of the humanities was "an eternal dissatisfaction," that the humanities ought in fact to "engender a kind of dissatisfaction," and hence that they "should not be conservative." Against Mr. Podhoretz's vision of a "common culture," Professor Rosovsky sided with President Giamatti in questioning the desirability of adhering to a canon and in extolling as an alternative to this the ideal of a "multi-culture" nourished by disparate sources and traditions. It is worth noting that the phrase "multi-culture" and its variants have become code words for an approach to the humanities that is in effect *anti*-cultural—at least anti-high-cultural. Part of the rhetoric of "pluralism" and "diversity," the elevation of "multi-cultural" experience cloaks the abandonment of traditional humanistic culture. It belongs with prattle about the humanities instilling "dissatisfaction" and the desirability of undermining the traditional canon. Such sentiments are heard everywhere in the academy today, but it did seem odd coming from the lips of a man who in the early Seventies, when he was a dean at Harvard, had been a staunch supporter of the canon and one of the chief architects of Harvard's now dismantled core curriculum. In 1974, faced with the prospect of curricular anarchy, Professor Rosovsky publicly deplored the loss of "an older community of beliefs and values" (quoted in Gilbert Allardyce, "The Rise and Fall of the Western Civilization Course," *American Historical Review*, June 1982, pages 696–697); now he looks to the loss of those beliefs and values as a prelude to the establishment of a multicultural paradise. *Autres temps, autres moeurs.*

Chapter 29

Paul Lauter (1932–)
from "Canon Theory and Emergent Practice," *Canons and Contents* (1991)

This division between the concerns of what I have come to call "canonical criticism" and those of what is called "theory" is, I think, one fact of current literary practice in the United States. In another chapter I have attempted to trace the very differing histories of canonical and academic criticism since the late 1960s; I do not wish to pursue that story here, except to underline the fact that canon criticism was initially an effort to carry the politics of the 1960s social movements into the work socially-engaged academics actually did, especially into our classrooms. Consequently, canon criticism first influenced curriculum and thus gradually the margins of publishing and scholarship. Somewhat later, it came to affect the selection of texts about which graduate students and critics write; more slowly still, which works became sufficiently revered to find their way into footnotes, indices or other measures of academic weight. More recently, it has begun altering the "mainstream" of publishing as well as generating wide public debate. Such has been the history of books like Frederick Douglass' *Narrative*, Charlotte Perkins Gilman's "The Yellow Wallpaper," or, more recently, Harriet Jacobs' *Incidents in the Life of a Slave Girl*. In the most recent stages of this process, as I shall point out below, some of the concerns of canonical criticism have converged with those arising from academic "theory." Even so, this division is only one of those that need to be explored. If one were at a convention of educators, one might be struck by the conflict between those advocating and those denouncing the canonical proposals of William Bennett, Allan Bloom, and Lynne Cheney. If one were at the National Women's Studies Association conference, one would probably notice the differences *among* those committed to changing the existing canon. Here, I want to chart the multiple conflicts that have arisen as the question of the canon has come to play a larger role on the intellectual stage.

Paul Lauter, Allan K. and Gwendolyn Miles Smith Professor of English at Trinity College, Hartford, surveys positions in the canon debate, and argues that non-canonical works provide the best material with which to reinterpret canonical works.

An issue of *Salmagundi* concentrating on "Cultural Literacy: Canon, Class, Curriculum" (#72, Fall, 1986) provides a useful starting point. The issue contains an essay by Robert Scholes called "Aiming a Canon at the Curriculum" and a set of responses to Scholes's article. For Scholes, the problem is to resist efforts by the Bennetts to impose upon educational institutions a particular canon of great books or a recipe for "cultural literacy" like E. D. Hirsch's. Scholes's essay can be taken as representative of the general reaction of the academic community to the cultural prescriptions of Reaganism. "What I am opposing," he writes, "is the learning of a set of pious clichés about a set of sacred texts" (p. 116). Most readers are sufficiently familiar (one might even say bored) with the ideas of Bennett and Cheney on this subject so I need not restate Scholes's well-drawn critique except to say that he brings into the open the right-wing politics that fuel Bennett's analysis as well as his solution. Scholes approvingly quotes T. S. Eliot's formulation of the matter: " 'to know what we want in education we must know what we want in general, we must derive our theory of education from our philosophy of life. The problem turns out to be a religious problem' " (p. 107; the Eliot quote is from "Modern Education and the Classics"). Seen in this light, the question of the canon becomes a conflict of values, and therefore, translated into public policy, of politics.

Of course, the issue is seldom joined in these terms; neither side in this debate puts forward explicitly political criteria for choosing texts they think worth studying. Bennett claims that the "great books" simply emerge over time through some kind of consensus among the properly educated. The opposition is formulated in terms of resisting the imposition of an essentially narrow vision of "Western Culture as a single coherent object, constructed of masterpieces built by geniuses," which is how Scholes characterizes the Bennett doctrine (p. 114). Bennett would not explicitly claim that Twain, Faulkner, and Martin Luther King, Jr., whose work he includes on his list, among them sufficiently "cover" the issue of racism in America; but in practice, adoption of that list as a basis of curriculum effectively marginalizes, if it does not altogether silence, the oppositional voices of writers such as David Walker, Harriet Jacobs, William Lloyd Garrison, Frances E. W. Harper, W. E. B. DuBois, or Malcolm X. Against this effort effectively to narrow the limits of significant debate, all the commentators on Scholes's piece stand united on what amounts to a platform of liberal pluralism. Even E. D. Hirsch, whose idea of "cultural literacy" is embodied in a dictionary of words and phrases that "every American should know," essentially joins Scholes on this score. Indeed, a broad consensus exists among literary practitioners to oppose efforts like those of Bennett's successor in the National Endowment for the Humanities, Lynne Cheney, to "politicize" criteria for selecting panels and projects or to project a single, federally-approved core curriculum.[1] [. . .]

Those who were united in opposition to the Bennett or Cheney prescriptions may well part company. For it is precisely at this point that those dominantly concerned with what is called "theory" turn with a shrug back to "hermeneutical recuperation." While it might not seem to matter precisely which texts one subjects to psychoanalytic, semiotic, or deconstructive analysis—and thus even the most hardened poststructuralist might in theory join the effort to broaden the canon—it turns out in practice that few "theorists" have participated in the effort to reconstruct literary canons—at least when one looks beyond the work of some feminist and black critics. Barbara Herrnstein Smith has effectively argued in her essay "Contingencies of Value"[2] that the positivistic stance and universalizing tendencies of academic criticism make it largely irrelevant to, and its practitioners finally uninterested in, the issues of valuation that are central to the question of the canon.

Furthermore, it is sometimes argued by enthusiasts of "theory" that deconstruction, in particular the work of Foucault, introduced the issue of power into literary discourse and thus raised the problem of canon formation. But the idea that cultural structures, such as a reading list or a set of criteria for admission to higher education, embody and sustain power relationships was quite clear long before poststructuralist theory emerged on the French, much less the American, scene. It was no accident that Mao Tse-tung's final effort to overcome the institutions of class in China was called the "cultural revolution"; the core of his ideas on that subject are contained in his 1942 "Talks at the Yenan Forum on Art and Literature." Further, demands for "open admissions" and "black studies" shook American universities long before Foucault had become a household word. The notion that debate over the canon derives from poststructuralist theory expresses something of the insulation of the academy. The general lack of contribution until quite recently of most American practitioners of "theory" to the debates about the canon also suggests something of the profound conservatism of the academy, in which the citation of precedent and authority ("as Derrida proposes in *Épirons* . . . ") have weighed far more heavily in the scales of tenure than the presentation of an "unknown" like Frances E. W. Harper. Such qualities exemplify the problem Mao tried to address, badly or well: how to bring into the open, and thus into question, the power relationships embedded in the consciousness of academics as well as in the cultural institutions we inhabit. And these qualities suggest that part of the importance of the effort to widen the canon is precisely the need to counter the tendency of academics to absorb social conflict into debates, over language and form, for example, that they can more easily control.[3] But revolution is not a linguistic phenomenon. [. . .]

If the second front of the debate is about *extending* rather than constricting the canon, the third front involves the substantial conflicts among those of us who work to change the existing canon; once again, I think, different political agendas translate into differing cultural priorities. One position is that taken up by cultural pluralists: We are many in these United States and any canon, any curriculum, ought to be "representative." I do not wish to disparage this answer for in many respects I agree with it; indeed, I think that the currently fashionable attacks on cultural pluralism, exemplified by Werner Sollors' "A Critique of Pure Pluralism," are badly misplaced. I shall return to Sollors' essay and to the book in which it is found shortly, but first I want to clarify differences among those committed to changing the canon.

Formulating the case for pluralism simply by asserting the value of representation obscures what is in question. For how can any canon be fully representative? Even if one ignores the problem of room at the inn—and I would be the last to deny that in anthologies, curricula, and literary histories this *is* a problem, though not always insuperable—do not the very processes of canonization necessarily reflect the structures of social and political power and thus embed in their product an unrepresentative, if widened, set of texts organized even at best along hierarchical lines? Does it not follow, then, that the goal must be to abolish canons altogether and to substitute, rather, the authority of various individual or ethnic group experiences freed from the constraints of any official discipline? It is to this issue that the most interesting essay in *Salmagundi*, that by Elizabeth Fox-Genovese, is addressed. [. . .]

Her primary argument is with those who wish to abolish all canons. While our students, she contends, may feel "colonized in relation to that elite western culture that has constituted the backbone of our humanistic education," throwing out the canon "does not solve their problem any more than expurgating all traces of western technology solves the problem of colonial peoples." Transforming "the canon and the surveys in response to changing constituencies has less to do with rewriting the story than with reinterpreting it," she concludes (pp. 135–36).

How to accomplish this goal? Fox-Genovese's proposal is to reread canonical texts as much for what they do *not* say as for what they make explicit. The work of Hobbes and Locke, for example, can be read from a gendered perspective: "We can teach elite culture from the perspective of gender, race, and class if we are prepared to accept attention to issues of gender, race, and class as proxies for the subjective testimony of those excluded from the most exalted cultural roles" (p. 140).[4] While this strategy is important, a problem with it inheres in the word "we," and in Fox-Genovese's implicit methodology. Again: "we can transform the entire focus of conventional courses by the themes we select" (p. 141); or, shifting pronoun, but retaining the teacher-dominant tactic, "one can *present* the individual as the problem rather than the solution" (p. 141; my italics). In effect, Fox-Genovese is shifting the ground of the debate from *what* is taught to *how* it is presented.[5] Her questions (though not her objectives) are similar to those asked by certain critics sceptical of the whole project of revising the canon. They ask: Does it really matter whether students read Harriet Jacobs or Nathaniel Hawthorne, Fitzgerald or LeSueur in class, since it is not at all clear what they do and do not learn by studying any of them? Or, more generally, does the reading of canonical stalwarts necessarily imply the transmission of elitist values, or the study of working-class texts dependably yield working-class consciousness? These are interesting empirical questions perhaps mainly for educational sociologists. They evoke the studies of the late 1950s which suggested that collegiate learning affects the values, at least of the students examined, marginally, if at all.[6] Such scepticism extends the critique poststructuralists have mounted against the too-easily presumed value of revolutions, or even of revolutionary movements, organizations, and goals.

One problem with these questions is that they invite broad generalizations about all students in any circumstances. In fact, a good deal of evidence, mainly anecdotal, suggests that students of diverse class and racial backgrounds respond quite differently to books such as *The Girl, Daughter of Earth*, and *Their Eyes Were Watching God*, and that the character of such responses changes in response to different political conditions. It can be extremely important for students from marginalized backgrounds to know that the domain named "literature" belongs to them as well as to others. But the main problem with sceptical social critiques, reasonable as they may seem, is that they too easily lead to political paralysis, and they play into reactionary ideologies that, for example, deny *any* value to radical change, whether retrospectively in Cuba or prospectively in South Africa. In the context of canon debate such scepticism over the value of altering canons can hardly help but provide fuel for those indifferent or hostile to such change. The argument then becomes: "Let us await persuasive evidence that it makes a difference before we disrupt the rooted norms of academic study." It is important to call into question exaggerated claims for the effects of canon change. Still, the assumption that such change matters to what readers experience, envision, and believe seems to me a useful intellectual and pedagogical hypothesis on the basis of which to act. Only through such action do we find in social practice whether or to what extent and under what conditions that hypothesis is valid. [. . .]

I suspect that the central reason it is necessary to read noncanonical texts is that they *teach us* how to view experience through the prisms of gender, race, nationality, and other forms of marginalization. From what other source would, for example, Anglo teachers in New England begin to comprehend Southwestern Latino experience than the work of writers like Américo Paredes, Rolando Hinojosa, or Gloria Anzaldúa? Reading, I want to reiterate, is not the only way we learn about how power is structured, and how those at the margins define their own relationships to such structures; there is no substitute finally for eating the pear of social change. Still, reading is a vital way to gain power in a literate society. That is why marginalized groups have always had to struggle against usage or law to obtain access to the power of literacy—or, having obtained it, to get a hearing for their literary productions, which returns us to the question of the canon.

While I think Fox-Genovese is right about the importance of rereading the traditional canon, it also seems to me that the best lens for that rereading is provided by noncanonical works themselves. For their outlook is likely to be less constrained by the deep conservatism of academe than that of professors, however progressive. In a sense, what seems to be involved is a reversal of the usual process by which noncanonical texts are appraised from the perspective and in the terms of those long established; rather, we are learning to reread—and thus to decenter—canonical texts, such as *The Great Gatsby* and *A Farewell to Arms*, from the perspectives provided by noncanonical works, like *The Girl* and *Their Eyes Were Watching God*. I have used as the slogan of the Reconstructing American Literature project the statement "so that the work of Frederick Douglass, Mary Wilkins Freeman, Agnes Smedley, Zora Neale Hurston and others is read with the work of Nathaniel Hawthorne, Henry James, William Faulkner, Ernest Hemingway and others." I offer that not as an article of pluralist faith, but as, in the first instance, an epistemological proposition. And, in the second instance, a challenge to theory and to history: What are the theoretical conceptions, the critical practices, the historical designs, the ideas about function and audience, that must be reconstructed so *that* these works can, indeed, be read? [. . .]

The fact is that neither separation nor integration provide wholly satisfactory methods for presenting or studying marginalized cultures. On the one hand, simply to integrate the writing of Jean Toomer, Zora Neale Hurston, Langston Hughes and Nella Larsen—to speak of the most likely—with Pound, Hemingway, and Eliot misrepresents the milieu from which and in which black writers created; on the other hand, to place them solely within the context of the Harlem Renaissance underplays their qualities and influence as "modernist" authors. A separate women's studies, African-American or Chicano studies department runs the risk of falsifying by homogenizing the diverse experience of all black women and men in America; and that also may contribute to the ghettoization of female and minority studies as well as to the indifference of academics in other departments to such concerns. On the other hand, without such centers of intellectual work and hubs for political struggle, the culture and experiences of the marginalized will be marginalized even more systematically, or so the experiences of curriculum "integration" projects and American history strongly suggest. Few in this country are very comfortable with the notion of assigning jobs or housing on the basis of race or ethnicity, but without specific numerical goals and timetables, the effects of racism persist even unto a generation cured—as this is not—of racism. So it is, as I argue elsewhere, with respect to culture: The distinctive qualities of the arts of marginalized groups need to be determined and celebrated by viewing each group collectively even as we acknowledge how much male and female, as well as black, Latino, white, Indian, and Asian cultural traditions in the United States overlap. That requires processes *both* of separation and of integration. Sollors' critique of pluralism in the Bercovitch book seems to be an obstacle to this dual process.

It is not, I should say, that an emphasis on or celebration of difference is, in all circumstances, socially cohesive let alone demonstrably progressive. No one usefully could argue such a generalized position in view of the reemergence of what can only be called tribalism in parts of Eastern Europe and the Middle East. The meaning and thus the value of cultural difference depends fundamentally on how it functions in particular societies at specific historical moments. Similarly, to be sure, with respect to a canon: Its establishment can and often does serve dominantly hegemonic functions, but it can also be, at least in theory, part of a process by which a society generates and maintains a necessary level of unity, and in certain circumstances, offers the opportunity for democratic participation to its members. Part of the current debate over cultural canons in the United States concerns the question of whether what is now necessary in this country is, above all, a unity of tradition (and if so, whose) or a fuller, indeed for the first time a meaningful, recognition of the character and importance of its diversity. [. . .]

For my own part, the most interesting frontiers of canon study have to do, in theoretical terms, with the implications of the material and institutional conditions of authorship and literary study, and with the functions of canons in establishing and maintaining boundaries (as well, in pedagogical terms, with the comparative study of canonical and marginalized texts about which I comment in another chapter). In her well-known article "Why Are There No Great Women Artists?" Linda Nochlin offered an institutional analysis of the processes by which art is created and valued. "The question 'Why are there no great women artists?' " she wrote, leads

> to the conclusion that art is not a free, autonomous activity of a superendowed individual, "influenced" by previous artists, and, more vaguely and superficially, by "social forces," but rather, that art making, both in terms of the development of the art maker and the nature and quality of the work of art itself, occurs in a social situation, is an integral element of the social structure, and is mediated and determined by specific and definable social institutions, be they art academies, systems of patronage, mythologies of the divine creator and artists as he-man or social outcast. . . . By stressing the *institutional*—that is, the public—rather than the *individual* or private preconditions for achievement in the arts, we have provided a model for the investigation of other areas in the field.[7]

Literary practitioners have, on the whole, been slow to take up this analytic challenge. Nina Baym has examined the "Melodramas of Beset Manhood" that provide a kind of exclusionary mythology. Jane Tompkins and Cathy Davidson have investigated the impact of particular publishing houses and practices, as well as the effect of peer networks, on the establishment and maintenance of reputations and thus of canons in the nineteenth century. Hazel Carby has explored the relationship of specific audience and educational objective to the texts of black women writers. More recently, Richard Brodhead has looked at how specific opportunities to enter the profession of authorship carried with them certain restrictions as to audience, subject, and convention—and thus implicitly, restrictions as to access to forms of composition, including those generally defined as significant or major. I have examined the effect of the post-World War I decline of women's literary clubs, and the enormous network they informed, on the radical reshaping of an American literary canon.[8] Such works are notable for a number of reasons: They demystify the processes by which reputations and canons are constructed; they offer fundamentally altered accounts of literary history and cultural relations; they define the origins and changing characteristics of the literary practices engaged in by noncanonical writers; they bring into focus contrasting readings of American history and culture provided for us by canonical and noncanonical works; and they reveal the material supports— or their lack—that prove critical not only to literary production but also to the survival of works of art—a point to which I shall return later. In short, the work of these critics helps us to understand canon formation and change in terms of concrete institutional developments at particular historical junctures.

Perhaps the most politically explosive example of such study is that contained in Martin Bernal's *Black Athena*.[9] Bernal argues that the generally accepted understanding of the origins and nature of classical Greek civilization was, in fact, constructed in Northern Europe during the mid-nineteenth century—constructed in such a way as to deny the rich Phoenician (Semitic) and Egyptian (African) influences that profoundly shaped Hellenic culture. A narrow canon of classical study was thus shaped within a definable institutional framework by identifiable intellectuals to serve particular political ends—having to do with a racist struggle to assert the dominance of Northern European, Aryan, cultures over those of Africa and Asia. Bernal's account of classical culture—indeed, of the origin of the Greek language—if sustained by further study undermines the politically-laden but ahistorical and simplistic image of Athens as the one true "cradle" of "our" civilization. But it also offers a particularly vivid instance of a more general point: that canons are not handed down from Mount Olympus, nor yet from Mount Horeb, but are the products of historically specific conflicts over culture and values. [. . .]

Thus, I think, the canon debate leads out of a narrowly construed set of professional concerns and back into the broader social and political world. The efforts to keep noncanonical work in print and to provide for its wider distribution, the controversies over literary prizes, like that recently involving Toni Morrison, the debates over whether, or to what, the "American mind" is closed make it clear that canonical issues are not simply matters of academic dispute. Like any meaningful cultural concern, the question of the canon directly affects lives. There is a vital dialectic between the recognition of a writer like Morrison and the need to understand her predecessors, like Frances E. W. Harper, Nella Larsen, Ann Petry, and Gwendolyn Brooks; but more, between the creative aspirations of such writers and the lived experience of black women in American society. Likewise, a dialectic functions between accounts of the origin of "our" civilization and who shares power within it. What is at stake, after all—to return to the initial problem with the cultural maven of Reaganism, William Bennett—is what a society sees as important from its past to the construction of its future, who decides that, and on what basis.

To be sure, the ways to reunite canon study and political action in practice are not always self-evident. At one point, the demand to take up and to take seriously Frederick Douglass, W. E. B. DuBois, Charlotte Perkins Gilman, and Agnes Smedley could, and often did, precipitate sharp conflicts, not only in curriculum committees but in the wider communities concerned with education. Today, canon study is perhaps as popular a subject for academic disquisitions as poststructuralist theory, and the issues are publication and promotion, not black power and sexual politics. It should be no surprise that an institution like the academy could thus largely absorb a serious challenge to its assumptions and structures. After all, the *New York Review of Books* found the diagram of a molotov cocktail useful to selling magazines. But if the ideas that canons are socially constructed *by* people and *in* history, that they have always changed and can be changed, that they are deeply shaped by institutions and the material conditions under which writing is produced and consumed—if these and related ideas have essentially triumphed within the academy, they remain deeply conflicted outside scholastic walls. It is all but impossible, I am told, to convince a Congressperson of the importance of studying Zora together with—God forbid in place of—Ernest. That is no defeat; it says to me, in fact, that while the advocates of a broad, multicultural canon have consolidated our position within most educational institutions, the Bennetts and Blooms have been hard at work in the public arenas. To me the next challenge is to shift the locus of struggle precisely to such public forums, even now as the academic right-wing bemoans the triumph of heterogeneity in the university. For all the academic fascination with hermeneutics and epistemology, it is in the realms of ethics and politics that the question of the canon must now be contested.

Notes

1. As in Lynne Cheney, *50 Hours: A Core Curriculum for College Students* (Washington, D.C.: National Endowment for the Humanities, 1989).
2. Barbara Herrnstein Smith, *Contingencies of Value: Alternative Perspectives for Critical Theory* (Cambridge: Harvard University Press, 1988).
3. Many of these issues of the social functions of criticism and the political character of what was not then called the "canon" were being discussed in the late 1960s among those who contributed to the volume called *The Politics of Literature*, ed. Louis Kampf and Paul Lauter (New York: Pantheon, 1970).
4. An excellent illustration of this approach is provided in a reading of Aristotle's *Politics* provided by Elizabeth Victoria Spelman in *Inessential Woman: Problems of Exclusion in Feminist Thought* (Boston: Beacon, 1988).

5. Myra Jehlen raises a similar set of concern in "How the Curriculum is the Least of Our Problems," *ADE Bulletin* 93 (Fall 1989) pp. 5–7.

6. See, for example, Philip E. Jacobs, *Changing Values in College: An Exploratory Study of the Impact of College Teaching* (New York: Harper, 1957), and Kenneth A. Feldman and Theodore M. Newcomb, *The Impact of College on Students*, 2 Vols. (San Francisco: Jossey-Bass, 1969).

7. Linda Nochlin, *Women, Art, and Power and Other Essays* (New York: Harper and Row, 1988), pp. 158, 176.

8. Nina Baym, "Melodramas of Beset Manhood: How Theories of American Fiction Exclude Women Authors," in *The New Feminist Criticism: Essays on Women, Literature and Theory*, ed. Elaine Showalter (New York: Pantheon, 1985), pp. 63–80; Jane Tompkins, *Sensational Designs* (New York: Oxford University Press, 1985); Cathy Davidson, *Revolution and the Word: The Rise of the Novel in America* (New York: Oxford University Press, 1986); Hazel V. Carby, *Reconstructing Womanhood* (New York: Oxford University Press, 1987); Richard Brodhead, in a paper delivered at the 1988 convention of the Modern Language Association, New Orleans, La.; Paul Lauter in an unpublished paper, "Clubs and Canons: Nineteenth-Century Women's Study Groups and 'American Literature' " (1989). Other notable instances are provided in Hortense Spillers' and Marjorie Pryse's *Conjuring* (Bloomington, Ind.: Indiana University Press, 1985) and Carolyn Karcher's splendid intro-duction to Lydia Maria Child's *Hobomok and Other Writings on Indians* (New Brunswick, N.J.: Rutgers University Press, 1986).

9. Martin Bernal, *Black Athena* (New Brunswick, N.J.: Rutgers University Press, 1987).

Chapter 30

Katha Pollitt (1949–)
"Why We Read: Canon to the Right of Me . . . ," *The Nation* (1991)

For the past couple of years we've all been witness to a furious debate about the literary canon. What books should be assigned to students? What books should critics discuss? What books should the rest of us read, and who are "we" anyway? Like everyone else, I've given these questions some thought, and when an invitation came my way, I leaped to produce my own manifesto. But to my surprise, when I sat down to write—in order to discover, as E. M. Forster once said, what I really think—I found that I agreed with all sides in the debate at once.

Take the conservatives. Now, this rather dour collection of scholars and diatribists—Allan Bloom, Hilton Kramer, John Silber and so on—are not a particularly appealing group of people. They are arrogant, they are rude, they are gloomy, they do not suffer fools gladly, and everywhere they look, fools are what they see. All good reasons not to elect them to public office, as the voters of Massachusetts recently decided. But what is so terrible, really, about what they are saying? I too believe that some books are more profound, more complex, more essential to an understanding of our culture than others; I too am appalled to think of students graduating from college not having read Homer, Plato, Virgil, Milton, Tolstoy—all writers, dead white Western men though they be, whose works have meant a great deal to me. As a teacher of literature and of writing, I too have seen at first hand how ill-educated many students are, and how little aware they are of this important fact about themselves. Last year I taught a graduate seminar in the writing of poetry. None of my students had read more than a smattering of poems by anyone, male or female, published more than ten years ago. Robert Lowell was as far outside their frame of reference as Alexander Pope. When I gently suggested to one student that it might benefit her to read some poetry if she planned to spend her life writing it, she told me that yes, she knew she should read more but when she encountered a really

Poet and columnist Katha Pollitt was a recipient of the 1992 National Magazine Award for this essay, in which she argues that the urgency of the canon debate is affected by a sense that college reading lists might represent the last chance for many people to do a wide range of reading.

good poem it only made her depressed. That contemporary writing has a history which it profits us to know in some depth, that we ourselves were not born yesterday, seems too obvious even to argue.

But ah, say the liberals, the canon exalted by the conservatives is itself an artifact of history. Sure, some books are more rewarding than others, but why can't we change our minds about which books those are? The canon itself was not always as we know it today: Until the 1920s, *Moby-Dick* was shelved with the boys' adventure stories. If T. S. Eliot could single-handedly dethrone the Romantic poets in favor of the neglected Metaphysicals and place John Webster alongside Shakespeare, why can't we dip into the sea of stories and fish out Edith Wharton or Virginia Woolf? And this position too makes a great deal of sense to me. After all, alongside the many good reasons for a book to end up on the required-reading shelf are some rather suspect reasons for its exclusion: because it was written by a woman and therefore presumed to be too slight; because it was written by a black person and therefore presumed to be too unsophisticated or to reflect too special a case. By all means, say the liberals, let's have great books and a shared culture. But let's make sure that all the different kinds of greatness are represented and that the culture we share reflects the true range of human experience.

If we leave the broadening of the canon up to the conservatives, this will never happen, because to them change only means defeat. Look at the recent fuss over the latest edition of the Great Books series published by Encyclopedia Britannica, headed by that old snake-oil salesman Mortimer Adler. Four women have now been added to the series: Virginia Woolf, Willa Cather, Jane Austen and George Eliot. That's nice, I suppose, but really! Jane Austen has been a certified Great Writer for a hundred years! Lionel Trilling said so! There's something truly absurd about the conservatives earnestly sitting in judgment on the illustrious dead, as though up in Writers' Heaven Jane and George and Willa and Virginia were breathlessly waiting to hear if they'd finally made it into the club, while Henry Fielding, newly dropped from the list, howls in outer darkness and the Brontës, presumably, stamp their feet in frustration and hope for better luck in twenty years, when *Jane Eyre* and *Wuthering Heights* will suddenly turn out to have qualities of greatness never before detected in their pages. It's like Poets' Corner at Manhattan's Cathedral of St. John the Divine, where mortal men—and a woman or two—of letters actually vote on which immortals to honor with a plaque, a process no doubt complete with electoral campaigns, compromise candidates and all the rest of the underside of the literary life. "No, I'm sorry, I just can't vote for Whitman. I'm a Washington Irving man myself."

Well, a liberal is not a very exciting thing to be, as *Nation* readers know, and so we have the radicals, who attack the concepts of "greatness," "shared," "culture" and "lists." (I'm overlooking here the ultraradicals, who attack the "privileging" of "texts," as they insist on calling books, and think one might as well spend one's college years deconstructing *Leave It to Beaver*.) Who is to say, ask the radicals, what is a great book? What's so terrific about complexity, ambiguity, historical centrality and high seriousness? If *The Color Purple*, say, gets students thinking about their own experience, maybe they ought to read it and forget about—and here you can fill in the name of whatever classic work you yourself found dry and tedious and never got around to finishing. For the radicals the notion of a shared culture is a lie, because it means presenting as universally meaningful and politically neutral books that reflect the interests and experiences and values of privileged white men at the expense of those of others—women, blacks, Latinos, Asians, the working class, whoever. Why not scrap the one-list-for-everyone idea and let people connect with books that are written by people like themselves about people like themselves? It will be a more accurate reflection of a multifaceted and conflict-ridden society, and will do wonders for everyone's self-esteem, except, of course, living white men—but they have too much self-esteem already.

Now, I have to say that I dislike the radicals' vision intensely. How foolish to argue that Chekhov has nothing to say to a black woman—or, for that matter, myself— merely because he is Russian, long dead, a man. The notion that one reads to increase one's self-esteem sounds to me like more snake oil. Literature is not an aerobics class or a session at the therapist's. But then I think of myself as a child, leafing through anthologies of poetry for the names of women. I never would have admitted that I needed a role model, even if that awful term had existed back in the prehistory of which I speak, but why was I so excited to find a female name, even when, as was often the case, it was attached to a poem of no interest to me whatsoever? Anna Laetitia Barbauld, author of "Life! I know not what thou art/But know that thou and I must part!"; Lady Anne Lindsay, writer of languid ballads in incomprehensible Scots dialect; and the other minor female poets included by chivalrous Sir Arthur Quiller-Couch in the old *Oxford Book of English Verse*: I have to admit it, just by their presence in that august volume they did something for me. And although it had nothing to do with reading or writing, it was an important thing they did.

Now, what are we to make of this spluttering debate, in which charges of imperialism are met by equally passionate accusations of vandalism, in which each side hates the others, and yet each one seems to have its share of reason? Perhaps what we have here is one of those debates in which the opposing sides, unbeknownst to themselves, share a myopia that will turn out to be the most telling feature of the whole discussion: a debate, for instance, like that of our Founding Fathers over the nature of the franchise. Think of all the energy and passion spent pondering the question of property qualifications or direct versus legislative elections while all along, unmentioned and unimagined, was the fact—to us so central—that women and slaves were never considered for any kind of vote.

Something is being overlooked: the state of reading, and books, and literature in our country at this time. Why, ask yourself, is everyone so hot under the collar about what to put on the required-reading shelf? It is because while we have been arguing so fiercely about which books make the best medicine, the patient has been slipping deeper and deeper into a coma.

Let us imagine a country in which reading is a popular voluntary activity. There, parents read books for their own edification and pleasure, and are seen by their children at this silent and mysterious pastime. These parents also read to their children, give them books for presents, talk to them about books and underwrite, with their taxes, a public library system that is open all day, every day. In school— where an attractive library is invariably to be found—the children study certain books together but also have an active reading life of their own. Years later it may even be hard for them to remember if they read *Jane Eyre* at home and Judy Blume in class, or the other way around. In college young people continue to be assigned certain books, but far more important are the books they discover for themselves—browsing in the library, in bookstores, on the shelves of friends, one book leading to another, back and forth in history and across languages and cultures. After graduation they continue to read, and in the fullness of time produce a new generation of readers. Oh happy land! I wish we all lived there.

In that other country of real readers—voluntary, active, self-determined readers— a debate like the current one over the canon would not be taking place. Or if it did, it would be as a kind of parlor game: What books would *you* take to a desert island? Everyone would know that the top-ten list was merely a tiny fraction of the books one would read in a lifetime. It would not seem racist or sexist or hopelessly hide-bound to put Hawthorne on the syllabus and not Toni Morrison. It would be more like putting oatmeal and not noodles on the breakfast menu—a choice part arbitrary, part a nod to the national past, part, dare one say it, a kind of reverse affirmative action: School might frankly be the place where one read the books that are a little off-putting, that have gone a little cold, that you might pass over because they do not address, in reader-friendly contemporary fashion, the issues most immediately at stake in modern life, but that, with a little study, turn out to have a great deal to say. Being on the list wouldn't mean so much. It might even add to a writer's cachet *not* to be on the list, to be in one way or another too heady, too daring, too exciting to be ground up into institutional fodder for teenagers. Generations of high school kids have been turned off to George Eliot by being forced to read *Silas Marner* at a tender age. One can imagine a whole new readership for her if grown-ups were left to approach *Middlemarch* and *Daniel Deronda* with open minds, at their leisure.

Of course, they rarely do. In America today the assumption underlying the canon debate is that the books on the list are the only books that are going to be read, and if the list is dropped no books are going to be read. Becoming a textbook is a book's only chance; all sides take that for granted. And so all agree not to mention certain things that they themselves, as highly educated people and, one assumes, devoted readers, know perfectly well. For example, that if you read only twenty-five, or fifty, or a hundred books, you can't understand them, however well chosen they are. And that if you don't have an independent reading life—and very few students do—you won't *like* reading the books on the list and will forget them the minute you finish them. And that books have, or should have, lives beyond the syllabus—thus, the totally misguided attempt to put current literature in the classroom. How strange to think that people need professorial help to read John Updike or Alice Walker, writers people actually do read for fun. But all sides agree, if it isn't taught, it doesn't count.

Let's look at the canon question from another angle. Instead of asking what books we want others to read, let's ask why we read books ourselves. I think the canon debaters are being a little disingenuous here, are suppressing, in the interest of their own agendas, their personal experience of reading. Sure, we read to understand our American culture and history, and we also read to recover neglected masterpieces, and to learn more about the accomplishments of our subgroup and thereby, as I've admit-ted about myself, increase our self-esteem. But what about reading for the aesthetic pleasures of language, form, image? What about reading to learn something new, to have a vicarious adventure, to follow the workings of an interesting, if possibly skewed, narrow and ill-tempered mind? What about reading for the story? For an expanded sense of sheer human variety? There are a thousand reasons why a book might have a claim on our time and attention other than its canonization. I once infuriated an acquaintance by asserting that Trollope, although in many ways a lesser writer than Dickens, possessed some wonderful qualities Dickens lacked: a more realistic view of women, a more skeptical view of good intentions, a subtler sense of humor, a drier vision of life which I myself found congenial. You'd think I'd advocated throwing Dickens out and replacing him with a toaster. Because Dickens is a certified Great Writer, and Trollope is not.

Am I saying anything different from what Randall Jarrell said in his great 1953 essay "The Age of Criticism"? Not really, so I'll quote him. Speaking of the literary gatherings of the era, Jarrell wrote:

> If, at such parties, you wanted to talk about *Ulysses* or *The Castle* or *The Brothers Karamazov* or *The Great Gatsby* or Graham Greene's last novel—Important books—you were at the right place. (Though you weren't so well off if you wanted to talk about *Remembrance of Things Past*. Important, but too long.) But if you wanted to talk about Turgenev's novelettes, or *The House of the Dead*, or *Lavengro*, or *Life on the Mississippi*, or *The Old Wives' Tale*, or *The Golovlyov Family*, or Cunningham-Grahame's stories, or Saint-Simon's memoirs, or *Lost Illusions*, or *The Beggar's Opera*, or *Eugene Onegin*, or *Little Dorrit*, or the *Burnt Njal Saga*, or *Persuasion*, or *The Inspector-General*, or *Oblomov*, or *Peer Gynt*, or *Far from the Madding Crowd*, or *Out of Africa*, or the *Parallel Lives*, or *A Dreary Story*, or *Debits and Credits*, or *Arabia Deserta*, or *Elective Affinities*, or *Schweik*, or—any of a thousand good or interesting but Unimportant books, you couldn't expect a very ready knowledge or sympathy from most of the readers there. They had looked at the big sights, the current sights, hard, with guides and glasses; and those walks in the country, over unfrequented or thrice-familiar territory, all alone—those walks from which most of the joy and good of reading come—were walks that they hadn't gone on very often.

I suspect that most canon debaters have taken those solitary rambles, if only out of boredom—how many times, after all, can you reread the *Aeneid*, or *Mrs. Dalloway*, or *Cotton Comes to Harlem* (to pick one book from each column)? But those walks don't count, because of another assumption all sides hold in common, which is that the purpose of reading is none of the many varied and delicious satisfactions I've mentioned; it's medicinal. The chief end of reading is to produce a desirable kind of person and a desirable kind of society. A respectful, high-minded citizen of a unified society for the conservatives, an up-to-date and flexible sort for the liberals, a sub-group-identified, robustly confident one for the radicals. How pragmatic, how moralistic, how American! The culture debaters turn out to share a secret suspicion of culture itself, as well as the antipornographer's belief that there is a simple, one-to-one correlation between books and behavior. Read the conservatives' list and produce a nation of sexists and racists—or a nation of philosopher kings. Read the liberals' list and produce a nation of spineless relativists—or a nation of open-minded world citizens. Read the radicals' list and produce a nation of psychobabblers and ancestor-worshipers—or a nation of stalwart proud-to-be-me pluralists.

But is there any list of a few dozen books that can have such a magical effect, for good or for ill? Of course not. It's like arguing that a perfectly nutritional breakfast cereal is enough food for the whole day. And so the canon debate is really an argument about what books to cram down the resistant throats of a resentful captive populace of students; and the trick is never to mention the fact that, in such circumstances, one book is as good, or as bad, as another. Because, as the debaters know from their own experience as readers, books are not pills that produce health when ingested in measured doses. Books do not shape character in any simple way—if, indeed, they do so at all—or the most literate would be the most virtuous instead of just the ordinary run of humanity with larger vocabularies. Books cannot mold a common national purpose when, in fact, people are honestly divided about what kind of country they want—and are divided, moreover, for very good and practical reasons, as they always have been.

For these burly and energetic purposes, books are all but useless. The way books affect us is an altogether more subtle, delicate, wayward and individual, not to say private, affair. And that reading is being made to bear such an inappropriate and simplistic burden speaks to the poverty both of culture and of frank political discussion in our time.

On his deathbed, Dr. Johnson—once canonical, now more admired than read— is supposed to have said to a friend who was energetically rearranging his bedclothes, "Thank you, this will do all that a pillow can do." One might say that the canon debaters are all asking of their handful of chosen books that they do a great deal more than any handful of books can do.

Chapter 31

Henry Louis Gates, Jr. (1950–) from "The Master's Pieces: On Canon Formation and the African-American Tradition," *Loose Canons: Notes on the Culture Wars* (1992)

As writers, teachers, or intellectuals, most of us would like to claim greater efficacy for our labors than we're entitled to. These days, literary criticism likes to think of itself as "war by other means." But it should start to wonder: Have its victories come too easily? The recent move toward politics and history in literary studies has turned the analysis of texts into a marionette theater of the political, to which we bring all the passions of our real-world commitments. And that's why it is sometimes necessary to remind ourselves of the distance from the classroom to the streets. Academic critics write essays, "readings" of literature, where the bad guys (for example, racism or patriarchy) lose, where the forces of oppression are subverted by the boundless powers of irony and allegory that no prison can contain, and we glow with hard-won triumph. We pay homage to the marginalized and demonized, and it feels almost as if we've righted a real-world injustice. I always think of the folktale about the fellow who killed seven with one blow.

Ours was the generation that took over buildings in the late sixties and demanded the creation of black and women's studies programs, and now, like the return of the repressed, has come back to challenge the traditional curriculum. And some of us are even attempting to redefine the canon by editing anthologies. Yet it sometimes seems that blacks are doing better in the college curriculum than they are in the streets.

This is not a defeatist moan. Just an acknowledgment that the relation between our critical postures and the social struggles they reflect upon is far from transparent.

Henry Louis Gates is Director of W. E. B. DuBois Institute for African and African-American Research, and W. E. B. DuBois Professor of the Humanities at Harvard University. In this essay, Gates defines a canon as "commonplace book of our shared culture," and goes on to describe editing the *Norton Anthology of African-American Literature*.

That doesn't mean there's no relation, of course, only that it's a highly mediated one. In any event, I do think we should be clear about when we've swatted a fly and when we've toppled a giant. [. . .]

I think back to why I went into the study of literature in the first place. I suppose the literary canon is, in no very grand sense, the commonplace book of our shared culture, in which we have written down the texts and titles that we want to remember, that had some special meaning for us. How else did those of us who teach literature fall in love with our subject than through our own commonplace books, in which we inscribed, secretly and privately, as we might do in a diary, those passages of books that named for us what we had for so long deeply felt, but could not say? I kept mine from the age of twelve, turning to it to repeat those marvelous passages that named myself in some private way. From H. H. Munro and O. Henry—I mean, some of the popular literature we had on the shelves at home—to Dickens and Austen, to Hugo and de Maupassant, I found resonant passages that I used to inscribe in my book. Finding James Baldwin and writing him down at an Episcopal church camp during the Watts riots in 1965 (I was fifteen) probably determined the direction of my intellectual life more than did any other single factor. I wrote and rewrote verbatim his elegantly framed paragraphs, full of sentences that were at once somehow Henry Jamesian and King Jamesian, yet clothed in the cadences and figures of the spirituals. I try to remind my graduate students that each of us turned to literature through literal or figurative commonplace books, a fact that we tend to forget once we adopt the alienating strategies of formal analysis. The passages in my commonplace book formed my own canon, just as I imagine each of yours did for you. And a canon, as it has functioned in every literary tradition, has served as the commonplace book of our shared culture. [. . .]

• • •

I have been thinking about these strains in black canon formation because a group of us will be editing still another anthology, which will constitute still another attempt at canon formation: W. W. Norton will be publishing the *Norton Anthology of African-American Literature*. The editing of this anthology has been a great dream of mine for a long time. After a year of readers' reports, market surveys, and draft proposals, Norton has enthusiastically embarked upon the publishing of our anthology.

I think that I am most excited about the fact that we will have at our disposal the means to edit an anthology that will define a canon of African-American literature for instructors and students at any institution which desires to teach a course in African-American literature. Once our anthology is published, no one will ever again be able to use the unavailability of black texts as an excuse not to teach our literature. A well-marked anthology functions in the academy to *create* a tradition, as well as to define and preserve it. A Norton anthology opens up a literary tradition as simply as opening the cover of a carefully edited and ample book.

I am not unaware of the politics and ironies of canon formation. The canon that we define will be "our" canon, one possible set of selections among several possible sets of selections. In part to be as eclectic and as democratically "representative" as possible, most other editors of black anthologies have tried to include as many authors and selections (especially excerpts) as possible, in order to preserve and "resurrect" the tradition. I call this the Sears and Roebuck approach, the "dream book" of black literature.

We have all benefited from this approach to collection. Indeed, many of our authors have managed to survive only because an enterprising editor was determined to marshal as much evidence as she or he could to show that the black literary tradition existed. While we must be deeply appreciative of that approach and its results, our task will be a different one.

Our task will be to bring together the "essential" texts of the canon, the "crucially central" authors, those whom we feel to be indispensable to an understanding of the shape, and shaping, of the tradition. A canon is often represented as the "essence" of the tradition, indeed, as the marrow of tradition: the connection between the texts of the canon is meant to reveal the tradition's inherent, or veiled, logic, its internal rationale.

None of us is naive enough to believe that "the canonical" is self-evident, absolute, or neutral. It is a commonplace of contemporary criticism to say that scholars make canons. But, just as often, writers make canons, too, both by critical revaluation and by reclamation through revision. Keenly aware of this—and, quite frankly, aware of my own biases—I have attempted to bring together a group of scholar-critics whose notions of the black canon might not necessarily agree with my own, or with each others'. I have tried to bring together a diverse array of ideological, methodological, and theoretical perspectives, so that we together might produce an anthology that most fully represents the various definitions of what it means to speak of an African-American literary tradition, and what it means to *teach* that tradition. And while we are at the earliest stages of organization, I can say that my own biases toward canon formation are to stress the formal relationships that obtain among texts in the black tradition—relations of revision, echo, call and response, antiphony, what have you—and to stress the vernacular roots of the tradition. For the vernacular, or oral literature, in our tradition, has a canon of its own.

But my pursuit of this project has required me to negotiate a position between, on the one hand, William Bennett, who claims that black people can have no canon, no masterpieces, and, on the other hand, those on the critical left who wonder why we want to establish the existence of a canon, any canon, in the first place. On the right hand, we face the outraged reactions of those custodians of Western culture who protest that the canon, that transparent decanter of Western values, may become—breathe the word—*politicized*. But the only way to answer the charge of "politics" is with an emphatic *tu quoque*. That people can maintain a straight face while they protest the irruption of politics into something that has always been political from the beginning—well, it says something about how remarkably successful official literary histories have been in presenting themselves as natural and neutral objects, untainted by worldly interests.

I agree with those conservatives who have raised the alarm about our students' ignorance of history. But part of the history we need to teach has to be the history of the idea of the "canon," which involves (though it's hardly exhausted by) the history of literary pedagogy and of the institution of the school. Once we understand how they arose, we no longer see literary canons as *objets trouvés* washed up on the beach of history. And we can begin to appreciate their ever-changing configuration in relation to a distinctive institutional history.

Universal education in this country was justified by the argument that schooling made good citizens, good American citizens; and when American literature started to be taught in our schools, part of the aim was to show what it was to be an American. As Richard Brodhead, a leading scholar of American literature, has observed, "no past lives without cultural mediation. The past, however worthy, does not survive by its own intrinsic power." One function of "literary history" is, then, to disguise that mediation, to conceal all connections between institutionalized interests and the literature we remember. Pay no attention to the man behind the curtain, booms the Great Oz of literary history.

Cynthia Ozick once chastised feminists by warning that *strategies become institutions*. But isn't that really another way of warning that their strategies, heaven forfend, may *succeed*? Here we approach the scruples of those on the cultural left, who worry about, well, the price of success. "Who's co-opting whom?" might be their slogan. To them, the very idea of the canon is hierarchical, patriarchal, and otherwise politically suspect. They'd like us to disavow it altogether.

But history and its institutions are not just something we study, they're also something we live, and live through. And how effective and how durable our interventions in contemporary cultural politics will be depends upon our ability to mobilize the institutions that buttress and reproduce that culture. The choice isn't between institutions and no institutions. The choice is always: What kind of institutions shall there be? Fearing that our strategies will become institutions, we could seclude ourselves from the real world and keep our hands clean, free from the taint of history. But that is to pay obeisance to the status quo, to the entrenched arsenal of sexual and racial authority, to say that they shouldn't change, become something other, and, let's hope, better than they are now.

Indeed, this is one case where we've got to borrow a leaf from the right, which is exemplarily aware of the role of education in the reproduction of values. We must engage in this sort of canon deformation precisely because Mr. Bennett is correct: the teaching of literature *is* the teaching of values; not inherently, no, but contingently, yes; it is—it has become—the teaching of an aesthetic and political order, in which no women or people of color were ever able to discover the reflection or representation of their images, or hear the resonances of their cultural voices. The return of "the" canon, the high canon of Western masterpieces, represents the return of an order in which my people were the subjugated, the voiceless, the invisible, the unrepresented, and the unrepresentable. [. . .]

Let me be specific. Those of us working in my own tradition confront the hegemony of the Western tradition, generally, and of the larger American tradition, specifically, as we set about theorizing about our tradition, and engaging in attempts at canon formation. Long after white American literature has been anthologized and canonized, and recanonized, our attempts to define a black American canon, foregrounded on its own against a white backdrop, are often decried as racist, separatist, nationalist, or "essentialist." Attempts to derive theories about our literary tradition from the black tradition—a tradition, I might add, that must include black vernacular forms as well as written literary forms—are often greeted by our colleagues in traditional literature departments as misguided attempts to secede from a union which only recently, and with considerable kicking and screaming, has been forged. What is *wrong* with you people, our friends ask us in genuine passion and concern; after all, aren't we all just citizens of literature here?

Well, yes and no. It is clear that every black American text must confess to a complex ancestry, one high and low (literary and vernacular), but also one white and black. There can be no doubt that white texts inform and influence black texts (and vice versa), so that a thoroughly integrated canon of American literature is not only politically sound, it is *intellectually* sound as well. But the attempts of scholars such as Arnold Rampersad, Houston Baker, M. H. Washington, Nellie McKay, and others to define a black American canon, and to pursue literary interpretation from within this canon, are not meant to refute the soundness of these gestures of integration. Rather, it is a question of perspective, a question of emphasis. Just as we can and must cite a black text within the larger American tradition, we can and must cite it within its own tradition, a tradition not defined by a pseudoscience of racial biology, or a mystically shared essence called blackness, but by the repetition and revision of shared themes, topoi, and tropes, a process that binds the signal texts of the black tradition into a canon just as surely as separate links bind together into a chain. It is no more, or less, essentialist to make this claim than it is to claim the existence of French, English, German, Russian, or American literature—as long as we proceed inductively, from the texts to the theory. For nationalism has always been the dwarf in the critical, canonical chess machine. For anyone to deny us the right to engage in attempts to constitute ourselves as discursive subjects is for them to engage in the double privileging of categories that happen to be preconstituted. [. . .]

Recently at Cornell, I was listening to Hortense Spillers, the great black feminist critic, read her important essay, "Mama's Baby, Papa's Maybe." Her delivery, as usual, was flawless, compelling, inimitable. And although I had read this essay as a manuscript, I had never before felt—or heard—the following lines:

> The African-American male has been touched, therefore, by the *mother*, handled by her in ways that he cannot escape, and in ways that the white American male is allowed to temporize by a fatherly reprieve. This human and historic development—the text that has been inscribed on the benighted heart of the continent—takes us to the center of an inexorable difference in the depths of American women's community: the African-American woman, the mother, the daughter, becomes historically the powerful and shadowy evocation of a cultural synthesis long evaporated—the law of the Mother—only and precisely because legal enslavement removed the African-American male not so much from sight as from *mimetic* view as a partner in the prevailing social fiction of the Father's name, the Father's law.
>
> Therefore, the female, in this order of things, breaks in upon the imagination with a forcefulness that marks both a denial and an "illegitimacy." Because of this peculiar American denial, the black American male embodies the *only* American community of males which has had the specific occasion to learn *who* the female is within itself, the infant child who bears the life against the could-be fateful gamble, against the odds of pulverization and murder, including her own. It is the heritage of the *mother* that the African-American male must regain as an aspect of his own personhood—the power of "yes" to the "female" within.

How curious a figure—men, black men, gaining their voices through the black mother. Precisely when some committed feminists or some committed black nationalists would essentialize all "others" out of their critical endeavor, Hortense Spillers rejects that glib and easy solution, calling for a revoicing of the "master's" discourse in the cadences and timbres of the Black Mother's voice. [. . .]

For me, I realized as Hortense Spillers spoke, much of my scholarly and critical work has been an attempt to learn how to speak in the strong, compelling cadences of my mother's voice. To reform core curricula, to account for the comparable eloquence of the African, the Asian, and the Middle Eastern traditions, is to begin to prepare our students for their roles as citizens of a world culture, educated through a truly human notion of "the humanities," rather than—as Bennett and Bloom would have it—as guardians at the last frontier outpost of white male Western culture, the Keepers of the Master's Pieces. And for us as scholar-critics, learning to speak in the voice of the black female is perhaps the ultimate challenge of producing a discourse of the critical Other.

Chapter 32

Gerald Graff (1937–)
from "Introduction: Conflict in America,"
Beyond the Culture Wars: How Teaching the Conflicts Can Revitalize American Education
(1992)

If we believe what we have been reading lately, American higher education is in a disastrous state. As pictured in a stream of best sellers, commission reports, polemical articles, and editorials, the academic humanities in particular look like a once-respectable old neighborhood gone bad. The stately old buildings have been defaced with spray paint, hideous accumulations of trash litter the ground, and omnipresent thought police control the turf, speaking in barbarous, unintelligible tongues while enforcing an intolerant code of political correctness on the terrorized inhabitants.

Having lived in this neighborhood for the last thirty years as a teacher of literature, a department chair, a lecturer at numerous universities, and a curriculum consultant at some, I find it hard to square these lurid accounts with my experience. As my first chapter illustrates, there is something truly astonishing about the degree of exaggeration, patent falsehood, and plain hysteria attained by the more prominent of these accounts. Much of the hysteria comes from simple fear of change, but much of it comes from the mysterious nature of certain precincts of the academic world to both other academics and the public. When a country is little known, fabulous and monstrous tales readily circulate about it, and any abuse can be passed off as typical.

It is true that staggering changes have occurred in the climate of academic life over the course of my professional life. I started graduate school in 1959, so my career spans the abyss between today's feverish struggles over books and the days when the literary canon—the body of literature thought to be worth teaching—seemed so uncontroversial that you rarely heard the word "canon." It is also true that social and cultural change has brought difficult new problems in the areas of admissions, hiring,

---

Gerald Graff, Professor of English at the University of Illinois, Chicago, sees the canon debate as a sign of the "vitality" of today's university, and recommends that colleges and universities "teach the controversies."

and campus life. In my view, however, these are the problems of success, a consequence of the vast superiority of today's university in intellectual reach and cultural
diversity to the relatively restricted campus culture of a generation ago. That today's
university is rocked by unprecedented conflicts is a measure of its vitality, not its
decline. As I see it, the challenge is to turn these very conflicts to positive account, by
transforming a scene of hatred and anger into one of educationally productive debate.
How this can be done is the subject of this book.

Though I am sympathetic to feminism, multiculturalism, and other new theories
and practices that have divided the academy, I do not argue that these movements
have the final word about culture—only that the questions they are raising deserve to
be taken seriously. Yet one would never guess from the overheated and ill-informed
accounts given by today's popular critics that the issues in the battle over education
are ones on which reasonable people might legitimately disagree. Arguments that at
the very least are worthy of debate—like the argument that political factors such
as race, class, gender, and nationality have influenced art and criticism far more than
education has traditionally acknowledged—have been reduced by their opponents to
their crudest and most strident form and thus dismissed without a hearing (see
Chapters 2 and 8). A complex set of issues that cry out for serious debate has been
turned into a clear-cut choice—as one prominent conservative puts it—"between
culture and barbarism."[1]

No doubt it pleases such critics to think of themselves as last-ditch defenders of
civilization against the invasion of barbarian relativists and terrorists. But if the goal
is constructive educational reform, then such apocalyptic posturing is a dead end.
One does not have to be a tenured radical to see that what has taken over the
educational world today is not barbarism and unreason but, simply, conflict. The first
step in dealing productively with today's conflicts is to recognize their legitimacy.

This book asks us to rethink the premise that the eruption of fundamental conflict
in education has to mean educational and cultural paralysis. My argument is that
conflict has to mean paralysis only as long as we fail to take positive advantage of it.

Acknowledging the legitimacy of social conflict, however, is not an easy thing even
for Americans of goodwill. We may not hesitate to embrace cultural diversity, but
when diversity leads to clashes of interests, as it naturally will, we find ourselves at a
loss. Such conflict seems vaguely un-American, a legacy of the less abundant societies
of the Old World. President Bush echoed an old American tradition when he recently
declared that class conflict is "for European democracies . . . it isn't for the United
States of America. We are not going to be divided by class."[2]

In fact, there is little reason to think we Americans are any less divided by class than
other nations, but we are certainly better at concealing it from ourselves. A combination
of affluence and geography has enabled more fortunate Americans to avoid noticing
unpleasant social conflicts by the simple device of moving away from them. In times past
there was the frontier, settled mostly by conquerable Indian tribes, to which Americans
could flee when urban conflict became too intense. More recently the flight has been out
to suburbs and malls, on freeways which let us drive past our social problems, or into
high rises from which those problems need be viewed only from a distance.

In our mass-produced fantasies we are virtually obsessed with conflict, but of a
stylized, unreal, or commercially trivialized kind. Our popular films and television

programs often deal with the sorts of conflicts that can be resolved by a fistfight, a car chase, or a shoot-out at the OK Corral. Our TV commercials stage endless disputations between partisans of old and new improved brands of soap, toothpaste, or deodorant, including the great debate over whether we should drink a particular beer because it is Less Filling or because it Tastes Great. Other commercials present modern life as a conflict-free utopia in which races freely intermingle and the world's ethnic groups join hands on a hillside to hymn their desire to buy the world a Coke. Our national obsession with athletic contests (one I fully share) is at least partly explained by the fact that conflict in sports, unlike in real life, is safe and satisfying, with clear-cut winners and losers.

Lately, however, conflicts over race, gender, and ethnicity have become so frequent and conspicuous that we seem to be getting more accustomed to dealing with them. This is seen in the public fascination with social-conflict films like *Do the Right Thing, Thelma and Louise,* and *Dances with Wolves* and with events like the Clarence Thomas-Anita Hill sexual harassment hearings. Yet the same conflicts that we have begun to accept in society still stick in our craw when they appear in education. The race, class, and gender conflicts that national newsmagazines treat as understandable and legitimate in films and public hearings have been depicted by these same magazines as a catastrophe for education.

Clearly, we still long to think of education as a conflict-free ivory tower, and the university *tries* to live up to this vision. While it welcomes diversity and innovation, it neutralizes the conflicts which result from them. This it does by keeping warring parties in noncommunicating courses and departments and by basing the curriculum on a principle of live and let live: I won't try to prevent you from teaching and studying what you want if you don't try to prevent me from teaching and studying what I want.

The effects of this amiable rule of laissez-faire have by no means been all bad, and it would be a serious mistake to try to abolish it entirely. It has enabled the American curriculum to relieve the increasingly conflicting pressures placed on it by painlessly expanding its frontiers, adding new subjects, courses, and programs without asking those in control of the already established ones to change their ways. It is only by such peaceful coexistence that the university could have achieved the improbable feat of becoming modern society's chief patron of cultural innovation without ceasing to stand for staunchly traditional values.[3]

As I point out in Chapter 7, the modern university has from the beginning rested on a deeply contradictory mission. The university is expected to preserve, transmit, and honor our traditions, yet at the same time it is supposed to produce new knowledge, which means questioning received ideas and perpetually revising traditional ways of thinking. The smooth functioning of the modern university has depended on a silent agreement to minimize this conflict between old and new, and times of relative affluence have afforded the room to pursue both missions without fatal collisions. This explains how it is possible for both the left and the right to believe with some reason that the opposing party is in charge.

Today, however, we see the end of the growth economy that for so long enabled the university to cushion its conflicts by indefinitely expanding the departmental and curricular playing field. Meanwhile, the contradictions that have accumulated as the academy has diversified have become so deep, antagonistic, and openly political that it has become impossible to prevent them from becoming visible to outsiders. In no other American institution do we find such a mind-boggling juxtaposition of clashing ideologies: corporate managers side by side with third world Marxists; free market economists with free-form sculptors; mandarin classical scholars with post-modern performance artists; football coaches with deconstructive feminists. Peaceful coexistence is increasingly strained, and it is harder to hold the conflicts at bay by the silent agreement not to wash dirty linen in public.

The result is today's educational crisis. It is a sign of the university's vitality that the crisis is happening so openly there. The academic curriculum has become a prominent arena of cultural conflict because it is a microcosm, as it should be, of the clash of cultures and values in America as a whole. As the democratization of culture has brought heretofore excluded groups into the educational citadel, with them have come the social conflicts that their exclusion once kept safely distant. A generation ago decisions about what was worth teaching and what counted as "culture" were still circumscribed by a relatively homogeneous class with a relatively common back-ground. Today new constituencies—women, blacks, gays, and immigrant groups from Asia and Latin America in particular—demand a say in how culture will be defined. And even more offensive to those who are used to having their way without controversy, these upstarts are now often in a position to put their ideas into practice. A less "canonical" faculty and student body implies a less canonical curriculum, dramatizing the fact that culture itself is a debate, not a monologue.[4]

Never comfortable with conflict to begin with, we are naturally prone to interpret these challenges as symptoms of disintegration. Many of the well-publicized horror stories about intolerant political correctness on campus—when they have not been shown to be simply bogus[5]—seem to me a symptom not of left-wing McCarthyism, as has been charged, but of fear in the face of controversy. In some cases, at least, I believe that what teachers have perceived as "harassment" is simply the novel experience of being in a minority and having to argue for one's beliefs instead of taking them for granted. Some overzealous proponents of cultural diversity have indeed behaved obnoxiously in attempting to sensitize their student and faculty colleagues whether they wish to be sensitized or not. But it is not necessarily a symptom of intolerance if a feminist student challenges a teacher's interpretation of Henry James for acceding to a stereotype about women or if a black student asks why a slave diary has not been assigned in a course on the Civil War.

I suspect that the teachers who have reacted to such criticisms by canceling their courses and offering themselves to the media as helpless victims of political correctness would have done better to stay and argue the issues with their students. Good teachers, after all, *want* their students to talk back. They know that student docility is a far more pervasive problem than student intransigence. Good students for their part appreciate teachers who take strong positions on controversial questions—though they do not appreciate brainwashing.

If the public furor over political correctness has shed more heat than light, it has at least proved that the gap between American culture and the ivory tower has closed. There is an old joke that academic disputes are especially poisonous because so little is at stake in them. But the stakes are no longer so trivial: Today's academic disputes over which texts should be taught in the humanities, over the competing claims of

Western and non-Western culture, and over the pros and cons of affirmative action and codes regulating hate speech mirror broader social conflicts over race, ethnicity, and privilege. Even the quarrels sparked by esoteric literary theories about the pertinence of gender questions to the study of Shakespeare echo debates over sex roles in the larger society provoked by feminists, gay activists, and the entry of women into the professional work force. At a moment when many in our society are questioning traditional assumptions about romantic love, heterosexuality, the nuclear family, abortion, aging, free speech, and the American flag, we should not be too outraged that decisions about which books to assign no longer go without saying.

But if outrage is not a helpful response to the conflicts occasioned by new interests, ideas, and constituencies, neither is the liberal complacency that has been content to say, "Sure, we can handle that innovation; we'll just add a new course on it." The current educational crisis has exposed the limitations of the live-and-let-live philosophy of curriculum, enabling conservatives to take the lead in the education debate and attract many disillusioned liberals to their side. The conservatives speak powerfully because they recognize the incoherence of a curriculum that is content to go on endlessly multiplying courses and subjects like boutiques at a mall. Unfortunately the conservatives' only prescription for curing this incoherence is to superimpose a higher order on the curriculum, an order that they like to call the "common culture" but that is really only *their* idea of order, one contender among several competing ones.

The history of modern American education has pitted the liberal pluralist solution (everyone do his or her own thing) against the conservative solution (everyone do the conservatives' thing). What is happening, today, I believe, is that both the liberal pluralist and the conservative solutions have outlived their usefulness. Everyone doing his or her own thing has made a mess of the curriculum, but cleaning up the mess by reverting to a narrowly defined traditional curriculum can only make a far worse mess. Since such a traditional curriculum would mean cutting away the vast areas of the world of knowledge and culture that do not fit the conservative vision of reality, it could be institutionalized only by forcing it down the throats of dissenting teachers and students. But then the same holds for the extreme radical vision.

Antagonistic as they are in most respects, the liberal pluralist and the conservative solutions are actually two sides of the same coin; neither is able to imagine any positive role for cultural conflict. Liberal pluralists are content to let cultural and intellectual diversity proliferate without addressing the conflicts and contradictions that result, whereas conservatives would exclude or shut down those conflicts. Neither strategy works in a world in which cultural and philosophical conflicts, increasingly, can no longer be evaded or shut down. A combination of changing demographic patterns in the wider culture, making student bodies and faculties more diverse, and of unsettling new ideas in the academic disciplines, challenging traditional disciplinary axioms, has created conflicts that cannot be successfully coped with by the traditional educational philosophies and curricular structures, much less by shaking our fists, shouting about relativism and lost standards, and calling for a return to the eternal verities. A solution, however, is latent in the problem itself, if only we can stop listening to those who tell us that controversy is a symptom of barbarism and that education was better in the past because it was calmer.

Where the university *has* failed—and here is the point on which many on the right, left, and center should be able to agree—is in making a focused curriculum out of its lively state of contention. Too much of the current debate is simply irrelevant to the educational problem as it is experienced by the struggling student. The most neglected fact about the culture war is that its issues are clearer and more meaningful to the contending parties than they are to that student. It is not the conflicts dividing the university that should worry us but the fact that students are not playing a more active role in them.

As I argue in Chapters 2 through 6, it won't matter much whose list of books wins the canon debate if students remain disaffected from the life of books and intellectual discussion, as too many have been since long before any canon revisionists arrived on the academic scene. It is easy to forget that for most American students the problem has usually been how to deal with *books* in general, regardless of which faction is drawing up the reading list. Here educators are wasting a major opportunity, for the conflicts that are now adding to the confusions of students have the potential to help them make better sense of their education and their lives. There is really no other choice. These conflicts are not going to go away, and students need to learn to deal with them in the culturally diverse world in which they already live and will live after graduation.

In this book I argue that the best solution to today's conflicts over culture is to teach the conflicts themselves, making them part of our object of study and using them as a new kind of organizing principle to give the curriculum the clarity and focus that almost all sides now agree it lacks. In a sense this solution constitutes a compromise, for it is one that conflicting parties can agree on. But it is really a way of avoiding the evasive compromise represented by the pluralist cafeteria counter curriculum, which leaves it up to students to connect what their teachers do not.

In an important sense, academic institutions are *already* teaching the conflicts every time a student goes from one course or department to another, but they are doing it badly. As I point out in Chapter 6, students typically experience a great clash of values, philosophies, and pedagogical methods among their various professors, but they are denied a view of the interactions and interrelations that give each subject meaning. They are exposed to the *results* of their professors' conflicts but not to the process of discussion and debate they need to see in order to become something more than passive spectators to their education. Students are expected to join an intellectual community that they see only in disconnected glimpses. This is what has passed for "traditional" education, but a curriculum that screens students from the controversies between texts and ideas serves the traditional goals of education as poorly as it serves those of reformers.

Nobody wants to turn the curriculum into a shouting match, of course. But the curriculum is already a shouting match, and one that will only become more angry and polarized if ways are not found to exploit rather than avoid its philosophical differences. When teachers in rival camps do not engage one another in their classrooms, all sides get comfortable preaching to the already converted. We get clashing forms of political correctness that become ever more entrenched the less they are forced to speak to one another. In a vicious circle, opposing viewpoints are so rarely debated that on the rare occasions when they are, the discussion is naturally hostile and confused, and this result then seems to prove that reasoned debate is not possible. Here, as I see it, is the essence of the problems of "Balkanization," separatism, and particularism that have come so to worry us: not the lack of agreement but of the respectful disagreement that supposedly is the strength of democracies and educational institutions.

That is why, however admirable the intention, adding courses in non-Western culture to existing general education requirements (as is now being done or contemplated at many schools and colleges) will only once more postpone the debate that has always been avoided in the past. It is not that non-Western courses are inherently separatist, as so many charge, but that *the established curriculum is separatist*, with each subject and course being an island with little regular connection to other subjects and courses. It is important to bring heretofore excluded cultures into the curriculum, but unless they are put in dialogue with traditional courses, students will continue to struggle with a disconnected curriculum, and suspicion and resentment will continue to increase. For the same reasons, the new field of "cultural studies" should be an open debate about culture and not the euphemism for various kinds of leftist studies that it has become. At the least, cultural studies and women's studies courses should be in dialogue with traditional ones.

In addition to being educationally defective, a disconnected curriculum in which one hand never knows what the other is doing is also very expensive. The cafeteria counter curriculum evolved, after all, during a period of affluence when universities had the luxury to hire specialists in almost everything and encourage them to go their separate ways. Such an ill-coordinated mode of organization would put a commercial firm out of business in a few months, and it may now put many universities out of business if they do not find ways to make teaching more collaborative. I argue in Chapters 6 and 9 that for reasons of both economy and pedagogy we need to rethink what I call the course fetish and the myth of the great teacher, which rest on the notion that by some law of nature teaching must be a solo performance.

I grant that making harsh disagreements productive for education is not easy. How will departments and colleges agree on what to disagree about? Who will determine the agenda of debate and decide which voices are included and excluded, and how will the inequalities between students and teachers, the tenured and the untenured, the eminent and the obscure be overcome? Some will see the introduction of non-Western texts into traditional introductory courses not as a debatable issue but as a capitulation to political pressure. Others will rightly be offended by proposals to debate questions like "Did the Holocaust really happen?" or "Is homosexuality a disease?" where no reputable scholar considers the question open or where it is framed in a way that puts one group on the defensive.

Numerous teachers, departments, and colleges have managed to overcome these obstacles, however, as we shall see in Chapter 9. They have recognized that students need to see the connections between the different interpretations, ideas, and values in the curriculum if they are to enter actively into academic discussions. The point was made best by an instructor who had joined with several colleagues to teach an introductory literature course: "Our students were able to argue with us because they saw us arguing with each other."

These teachers and institutions pick up at the very point at which today's disputes have become deadlocked. They assume that there is something unreal about the either/or choice we have been offered between teaching Western or non-Western culture, that in a culturally diverse society, a wide range of cultures and values should be and will be taught. But they also see that teaching different cultures and values implies teaching them in relation to one another so that the differences and points of intersection become comprehensible. I find it much easier to clarify the traditional idea that great literature is universal when I teach feminist critiques of that idea. Opposing texts and theories need one another to become intelligible to students. As one of my students put it after our class had read Joseph Conrad's *Heart of Darkness* alongside the very different treatment of Africa by the Nigerian novelist Chinua Achebe, *Things Fall Apart*, she thought she better understood the Europeanness of Conrad because she now had something to compare it with.

Teaching the conflicts has nothing to do with relativism or denying the existence of truth. The best way to make relativists of students is to expose them to an endless series of different positions which are *not* debated before their eyes. Acknowledging that culture is a debate rather than a monologue does not prevent us from energetically fighting for the truth of our own convictions. On the contrary, when truth is disputed, we can seek it only by entering the debate—as Socrates knew when he taught the conflicts two millennia ago.

Notes

1. Roger Kimball, "Tenured Radicals: A Postscript," *New Criterion*, 9, No. 5 (January 1991) p. 13.
2. Quoted in Benjamin De Mott, "The Myth of Classlessness," *New York Times* (October 10, 1990), p. A23.
3. I have traced these processes of conflict avoidance in the field of literary studies in my earlier book *Professing Literature: An Institutional History* (Chicago: University of Chicago Press, 1987), from which I draw at times in the present book.
4. I am indebted for this point to Susan Lowry of the University of Wisconsin at Milwaukee.
5. Recent articles have adduced evidence that at least one of the most sensational "PC stories" was baseless or grossly exaggerated. See Jon Weiner's account of the Stephan Thernstrom case at Harvard, "What Happened at Harvard," *Nation*, 253, No. 10 (September 30, 1991) pp. 384–88; see also Rosa Ehrenreich, "What Campus Radicals?" *Harper's*, 283, No. 1699 (December 1991) pp. 57–61. These and other pieces challenging the reliability of widely circulated PC stories have recently been collected in *Beyond PC: Toward a Politics of Understanding*, ed. Patricia Aufderheide (St. Paul, MN.: Graywolf Press, 1992), pp. 92–121.

Chapter 33

John Guillory
from "Preface" and "Canonical and Noncanonical: The Current Debate,"
Cultural Capital: The Problem of Literary Canon Formation (1993)

Preface

Symbolic struggles are always much more effective (and therefore realistic) than objectivist economists think, and much less so than pure social marginalists think. The relationship between distributions and representations is both the product and the stake of a permanent struggle between those who, because of the position they occupy within the distributions, have an interest in subverting them by modifying the classifications in which they are expressed and legitimated, and those who have an interest in perpetuating misrecognition, an alienated cognition that looks at the world through categories the world imposes.

—Bourdieu, *The Logic of Practice*

The largest thesis of this book is that the debate about the canon has been misconceived from the start, and that its true significance is one of which the contestants are not generally aware. The most interesting question raised by the debate is not the familiar one of which texts or authors will be included in the literary canon, but the question of why the debate represents a crisis in literary study. [. . .] I will argue that evaluative judgments are the necessary but not sufficient condition for the process of canon formation, and that it is only by understanding the social function and institutional protocols of the school that we will understand how works are preserved, reproduced, and disseminated over successive generations and centuries. Similarly, where the debate speaks about the canon as representing or failing to represent

---

In this essay, John Guillory, Professor of English at New York University, considers the canon with reference to Pierre Bourdieu's idea of "cultural capital."

particular social groups, I will speak of the school's historical function of distributing, or regulating access to, the forms of cultural capital. By insisting on the interrelation between representation and distribution, I hope to move beyond a certain confusion which both founds and vitiates the liberal pluralist critique of the canon, a confusion between representation in the political sense—the relation of a representative to a constituency—and representation in the rather different sense of the relation between an image and what the image represents. The collapse of the latter sense into the former has had the unfortunate effect of allowing the participants in the "symbolic struggle" over representation in the canon to overestimate the political effects of this struggle, at the same time that the participants have remained relatively blind to the social and institutional conditions of symbolic struggles. I will argue that the concept of cultural capital can provide the basis for a new historical account of both the process of canon formation and the immediate social conditions giving rise to the debate about the canon. For while the debate seems to its participants to be about the contents of the literary canon, its significance goes well beyond the effects of any new consensus about a truly "representative" canon. The canon debate signifiesnothing less than a crisis in the form of cultural capital we call "literature."

The concept of "cultural capital" is derived from the work of Pierre Bourdieu, where it facilitates a revisionary sociology of great depth and complexity. The purpose of importing the term into the debate about the canon is not to endorse Bourdieu's project in its totality (I have dissented on occasion from particular conclusions of Bourdieu's) but to introduce an entirely different theoretical perspective into the present debate. The theory of cultural capital implies that the proper social context for analyzing the school and its literary curriculum is *class*. Yet the argument of this book is not simply, on that account, "Marxist." For Bourdieu the concept of class is preeminently a sociological concept, and one which is, as Marxists know, undertheorized in Marx himself. If there exists a form of capital which is specifically symbolic or *cultural*, the production, exchange, distribution, and consumption of this capital presupposes the division of society into groups that can be called classes. Bourdieu's sociology assumes such a division, but it does not assume that an economic account of classes is sufficient in itself.[1] Such an account would omit precisely what in Bourdieu's theory is "cultural." The theory of cultural capital belongs to the general field of what in France goes by the name of "post-Marxist" thought; but this is an affiliation which is much harder to claim in our own country, where there is no indigenous Marxist tradition to overthrow or move beyond. Without aspiring either to a consistent Marxism or post-Marxism, I have sought rather to make visible the relative absence of class as a working category of analysis in the canon debate. This may seem surprising to participants in the debate, who have always argued that exclusions from the canon are determined by the race, gender, or class identities of authors. But the argument of this book is that one cannot infer a process of exclusion from the canon by setting out from the category of class, a fact which explains why examples of excluded authors always happen to be those whose identities are marked by race or gender. The fact of class determines whether and how individuals gain access to the means of literary production, and the system regulating such access is a much more efficient mechanism of social exclusion than acts of judgment. By foregrounding the question of the relation between social groups and the means of literary production,

I have thus attempted to resist the easy assumption that whatever one says about race and gender goes without saying for class too. [. . .]

The largest context for analyzing the school as an institution is therefore the *reproduction* of the social order, with all of its various inequities. The particular authors who happen to be canonical have a minor role in this system of reproduction, but the far larger role belongs to the school itself, which regulates access to literary production by regulating access to literacy, to the practices of reading and writing. The literary syllabus is the institutional form by means of which this knowledge is disseminated, and it constitutes capital in two senses: First, it is *linguistic* capital, the means by which one attains to a socially credentialed and therefore valued speech, otherwise known as "Standard English." And second, it is *symbolic* capital, a kind of knowledge-capital whose possession can be displayed upon request and which thereby entitles its possessor to the cultural and material rewards of the well-educated person. For reasons to be argued more fully within the chapters of this book, I regard these two kinds of capital as ultimately more socially significant in their effects than the "ideological" content of literary works, a content which the critics of the canon see as reinforcing the exclusion of minority authors from the canon by expressing the same values which determine exclusionary judgments. Literary works must be seen rather as the vector of ideological notions which do not inhere in the works themselves but in the context of their institutional presentation, or more simply, in the way which they are taught.

For the purposes of a sociologically informed history of canon formation, it is the category of "literature" which invites the closest scrutiny. That category organizes the literary curriculum in such a way as to create the illusion of a fixed and exclusive "canon," an illusion which is belied by the real history of literary curricula in the schools. For that very reason, calling the canon into question has failed to inaugurate a historico-critical inquiry into the category of literature, even while it has registered a crisis in the cultural capital so denominated. The overarching project of the present study is an inquiry into just this crisis, one which attempts to explain why the category of literature has come to seem institutionally dysfunctional, a circumstance which I will relate to the emergence of a technically trained "New Class," or "professional-managerial class."[2] To put this thesis in its briefest form, the category of "literature" names the cultural capital of the old bourgeoisie, a form of capital increasingly marginal to the social function of the present educational system. From this perspective the issue of "canonicity" will seem less important than the historical crisis of literature, since it is this crisis—the long-term decline in the cultural capital of literature—which gives rise to the canon debate. The category of literature remains the *impensé* of the debate, in spite of what passes on the left as a critique of that category's transcendent value, and on the right as a mythological "death of literature." [. . .]

It is nevertheless an interesting consequence of the canon debate that it has called every act of judgment into question, not simply because judgment is always historical, local, or institutional, but more profoundly because it is exercised at all. The latter position is expressed unequivocally by a participant in the debate over the Stanford "Western Culture" course:

> The *notion* of a core list is inherently flawed, regardless of what kinds of works it
> includes or excludes. It is flawed because such a list undermines the critical stance that

we wish students to take toward the materials they read. . . . A course with such readings creates two sets of books, those privileged by being on the list and those not worthy of inclusion. Regardless of the good intentions of those who create such lists, the students have not viewed and will not view these separate categories as equal.[3]

It is difficult to see how the logic of such an argument would allow *any* works to be taught, since every syllabus of study selects some works rather than others. The curious logic of this argument conflates the syllabus, a selection of texts for study in a particular institutional context, with the canon itself—the sum total of works supposed to be "great." A syllabus will necessarily be limited by the constraints of a particular class and its rubric, even by the irreducibly material constraint that only so much can be read or studied in a given class. In no classroom is the "canon" itself the object of study. Where does it appear, then? It would be better to say that the canon is an *imaginary* totality of works. No one has access to the canon as a totality. This fact is true in the trivial sense that no one ever reads every canonical work; no one can, because the works invoked as canonical change continually according to many different occasions of judgment or contestation. What this means is that the canon is never other than an imaginary list; it never appears as a complete and uncontested list in any particular time and place, not even in the form of the omnibus anthology, which remains a selection from a larger list which does not itself appear anywhere in the anthology's table of contents. In this context, the distinction between the canonical and the noncanonical can be seen not as the form in which judgments are actually made about individual works, but as an effect of the syllabus as an institutional instrument, the fact that works not included on a given syllabus appear to have no status at all.[4] The historical condition of literature is that of a complex continuum of major works, minor works, works read primarily in research contexts, works as yet simply shelved in the archive. Anyone who studies historical literatures knows that the archive contains an indefinite number of works of manifest cultural interest and accomplishment. While these works might be regarded as "noncanonical" in some pedagogic contexts—for example, the context of the "great works" survey—their noncanonical status is not necessarily equivalent in anyone's judgment to a zero-degree of interest or value. The fact that we conventionally recognize as "the canon" only those works included in such survey courses or anthologies as the Norton or the Oxford suggests to what extent the debate about the canon has been driven by institutional agendas, for which the discourse of the "masterpiece" provides such a loud accompaniment. The merest familiarity with historical context brings the continuum of cultural works back into focus and demonstrates that the field of writing does not contain only two kinds of works, either great or of no interest at all. For this reason the category of the "noncanonical" is entirely inadequate to describe the status of works which do not appear in a given syllabus of study. [. . .]

What one would like to comprehend with a finer set of terms is the relation between the material constraints of the syllabus, as an instrument of pedagogy, and the various imaginary totalities projected out of historical curricula. The syllabus has the form of a list, but the items on the list are given a specious unity by reference to a whole from which they are supposed to be a representative selection. This specious unity indeed characterizes not only the canon but the syllabi we call English literature, Romantic literature, women's literature, Afro-American literature. The canon achieves its imaginary totality, then, not by embodying itself in a really existing list, but by retroactively constructing its individual texts as a *tradition*, to which works may be added or subtracted without altering the impression of totality or cultural homogeneity. A tradition is "real," of course, but only in the sense in which the imaginary is real.[5] A tradition always retroactively unifies disparate cultural productions (and this is no less true for the tradition of women writers or the tradition of Afro-American writers); while such historical fictions are perhaps impossible to dispense with, one should always bear in mind that the concept of a given tradition is much more revealing about the immediate context in which that tradition is defined than it is about the works retroactively so organized. Also, and perhaps more interestingly, the larger and more disparate the body of works to be retroactively unified, the more urgent and totalizing the concept of tradition is likely to be. If a principle of specious unity is implicit in the construction of any syllabus, this means that the form of the syllabus sets up the conditions within which it is possible to forget that the syllabus is just a list, that there is no concrete cultural totality of which it is the expression. The confusion of the syllabus with the canon thus inaugurates a pedagogy of misreading, wherein a given text's historical specificity is effaced as it is absorbed into the unity of the syllabus/canon. [. . .]

From the perspective of long-term developments in the educational system, the canon debate itself may seem oddly beside the point. Bennett and his associates already acknowledged in their 1984 document that the "crisis of the humanities" refers to the fact that fewer undergraduates choose to major in traditional humanities than in the past. One has the impression in surveying the musings of the right-wing pundits that this fact is the result of nothing less than abdication by the professors of their duty to teach the traditional texts.[6] Nothing could be further from the truth—these texts still constitute the vastly greater part of the humanities curriculum—and in that sense the complaint of the New Right is simply fraudulent. A welcome reality check is provided by Patrick Brantlinger in his analysis of the "crisis":

> Tradition gives the humanities an importance that current funding and research priorities belie. At giant public "multiversities" like the Big Ten schools, humanities courses are taken by many students only as requirements—a sort of force-feeding in writing skills, history, great books, and appropriate "values" before they select the chutes labeled "pre-professional"—pre-med, pre-law, and so forth. . . . Clearly, one doesn't need to blame the radical sixties for the current marginalization and sense of irrelevance that pervades the humanities today.[7]

The crisis of the humanities is the result not of university professors' unwillingness to teach great works (the idea is an insult especially to those teachers and graduate students who could not find employment in the recessions of the 70s and 80s) but of the decisions students themselves make in the face of economic realities. Granted the fact that the crisis is not the result of curricular decisions by humanities teachers, why is the content of the curriculum the site of such controversy? The canon debate will not go away, and it is likely to intensify as the positions of the right and of the multiculturalists are further polarized. The very strength of the reactionary backlash, its success in acquiring access to the national media and funding for its agitprop, suggests that the symptomatic importance of the debate is related in some as yet obscurely discerned way to the failure of the contestants to give an account of the general decline in the significance of the humanities in the educational system. It has proven to be much easier to quarrel about the content of the curriculum than to confront the implications of a fully emergent professional-managerial class which no longer requires the cultural capital of the old bourgeoisie. The decline of the humanities was never the result of newer noncanonical courses or texts, but of a large-scale "capital flight" in the domain of culture. The debate over what amounts to the supplementation (or modernization) of the traditional curriculum is thus a misplaced response to that capital flight, and as such the debate has been conducted largely in the realm of the pedagogic imaginary. I would propose, then, that the division now characterizing the humanities syllabus—between Western and multicultural, canonical and noncanonical, hegemonic and nonhegemonic works—is the symptom of a more historically significant split between two kinds of cultural capital, one of which is "traditional," the other organic to the constitution of the professional-managerial class. [. . .]

If this analysis is correct it does not seem the most effective strategy for the left to cede to the right the *definition* of cultural capital; but this is exactly what multiculturalism does when it yields canonical works to the right, when it accepts the right's characterization of the canonical syllabus as constitutive of a unified and monolithic Western culture. Basing its agenda upon such assumptions, a left politics of representation seems to have no other choice than to institutionalize alternative syllabi as representative images of non-Western or "counter"-cultures. This is finally why the project of legitimizing noncanonical works in the university produces an irresolvable contradiction between the presentation of these works as equal in cultural value to canonical works, and at the same time as the embodiment of countercultural values which by their very definition are intended to delegitimize the cultural values embodied in canonical works. The polarization of the debate into Western culturalism versus multiculturalism must then be seen not as a simple conflict between regressive and progressive pedagogies but as the symptom of the transformation of cultural capital in response to social conditions not yet recognized as the real and ultimately determining context of the canon debate. Both the right-wing attempt to shore up the cultural capital of the "great works" by advocating a return to a core curriculum, and the pluralist advocacy of multiculturalism respond to the same demographic circumstances, the heterogeneous constituency of the university. But neither version of culturalist politics responds to the heterogeneous constitution of cultural capital, and hence both movements are condemned to register this condition symptomatically, as a false perception of the mutual (cultural) exclusivity of canonical and noncanonical works.

Notes

1. The concept of class is as controverted as any other notion in social theory, and the class analysis I propose in this book will not wait for any definitive resolution of that controversy.

Here I can only point to the larger features on the map in order to situate my own analysis (and Bourdieu's) in something like a region of this domain. For Marx and Marxism, class is defined by position in the relations of production, and there are only two of these: capital and labor; alternatively, the bourgeoisie and the proletariat (the existence of a "petty bourgeoisie" does not alter the dualism of this structure). The rigor of this model, which apprehends class distinction in economic terms, leaves largely unresolved the question of how cultural distinctions are articulated on the template of the capital–labor structure. The solution to this problem is ultimately what is at stake in every class analysis, including Bourdieu's. For the tradition of "bourgeois" sociology (Durkheim and Weber are as important in Bourdieu's theory as the work of Marx), class is primarily a cultural concept, although the expression of many cultural class traits will in practice depend on material resources. With the concept of "cultural capital" Bourdieu undertakes a certain negotiation between the domains of the economy and culture. Without pronouncing on the success of this negotiation, I would insist here that the aporia between the cultural and the economic is the most fundamental problem confronted by bourgeois sociology, as well as by that hypothetical "Marxist" sociology which Marx himself never produced. For Bourdieu's own account of the concept of capital, see "The Forms of Capital," in *The Handbook of Theory and Research for the Sociology of Education*, ed. John G. Richardson (New York: Greenweed Press, 1983), pp. 241–258. Rather than offer here a precise definition of cultural capital, I have followed Bourdieu's own practice in constructing the concept through the contexts of its deployment.

2. The concept of the "New Class" invokes Alvin Gouldner, *The Future of Intellectuals and the Rise of the New Class* (New York: Oxford University Press, 1979). The "professional-managerial class" invokes the groundbreaking essay of Barbara and John Ehrenreich, "The Professional-Managerial Class," in *Between Labor and Capital*, ed. Pat Walker (Montreal: Black Rose Press, 1979). The controversy in the Walker volume over the question of whether the group of professional-managerial workers constitutes a class is scarcely settled in that volume. The Ehrenreichs proceed on the assumption that a class has both an economic and a cultural component, neither of which can be simply derived from, or reduced to, the other. This working assumption makes for a less than tidy theory of class, but the Ehrenreichs are not especially disturbed by this untidiness. Gouldner's conception of cultural capital casts the New Class in the role of a new bourgeoisie, the historical successors to the old. I have kept Gouldner's theory in mind, somewhat warily, preferring Bourdieu's less narrativizing mode of class analysis. For the more limited purposes of the argument I wish to make in this book, it perhaps does not matter whether the professional-managerial classes are conceived to be a distinct class, or, as Erik Olin Wright argues in his response to the Ehrenreichs in the Walker volume, a "contradictory location within class relations" (203). What matters to the present argument is that the emergence of the professional-managerial class has enormously altered the constitution and distribution of cultural capital in the school system, and that these new conditions remain the unremarked horizon of the canon debate.

3. Cited in Mary Louise Pratt, "Humanities for the Future: Reflections on the Western Culture Debate at Stanford," *South Atlantic Quarterly*, 89 (1990), p. 14.

4. This sense of the noncanonical is well exemplified in the argument of Paul Lauter, "Race and Gender in the Shaping of the American Literary Canon: A Case Study from the Twenties," in *Feminist Criticism and Social Change: Sex, Class and Race in Literature and Culture*, ed. Judith Newton and Deborah Rosenfelt (New York: Methuen, 1985), pp. 19–44, which provides an interesting and informative narrative about the construction of anthologies of American writing. Interpreting this information is not easy, however, since how one understands the narrative depends upon how adequate the paradigm of

inclusion/exclusion is to describe the survival or disappearance of literary works. While Lauter states that "Obviously, no conclave of cultural cardinals establishes a literary canon," he is concerned to show that "in the 1920s processes were set in motion that virtually eliminated black, white, female and all working-class writers from the canon." These processes were "the professionalization of the teaching of literature, the development of an aesthetic theory that privileged certain texts, and the historiographic organization of the body of literature into conventional 'periods' and 'themes'" (23). It is difficult to see how these criteria are intrinsically unfavorable to minority writers, but the ease with which they can be made to coincide with and explain the disappearance of any given minority writer from one anthology to the next is a measure of how difficult it is for us to imagine that the social identity of the author is not the *real criterion* for every judgment, no matter where or when. The processes Lauter discovers are on his own account considerably more complex than judgments based on the social identity of authors. It is perhaps time to recognize that it is only the emergence in our own time of social identity as a *positive* criterion of judgment (as the basis, in Lauter's phrase, of establishing a "more representative and accurate literary canon") that requires a revisionist history in which social identity is the major *negative* criterion of judgment.

5. On this subject, see *The Invention of Tradition*, ed. Eric Hobsbawm and Terence Ranger (Cambridge: Cambridge University Press, 1983).

6. Allan Bloom, *The Closing of the American Mind: How Higher Education Has Failed Democracy and Impoverished the Souls of Today's Students* (New York: Simon and Schuster, 1987), 352.

7. Patrick Brantlinger, *Crusoe's Footprints: Cultural Studies in Britain and America* (New York: Routledge, 1990), 7.

Chapter 34

Vassilis Lambropoulos (1953–) from "The Rites of Interpretation," *The Rise of Eurocentrism* (1993)

Readers of *Mimesis* will remember the well-prepared and touching comparison in chapter 1, where the two basic types of literary representation in Western culture are dramatically contrasted. The scene of Odysseus' recognition by his old housekeeper Euryclea in the *Odyssey* is examined in great thematic and stylistic detail, and then interpreted against a parallel reading of the sacrifice of Isaac in Genesis. The wide variety of distinct features exhibited in the two texts is organized in two correspondding sets of diametrically opposed character and tone. These sets are then seen as concise pictures of the worldview expressed in the respective works, and are used as the basis for a broad outline of the Homeric and the Biblical systems of thought. At the end of the chapter, the two types are set forth as the starting point for the investigation of European literary representation that the rest of the book conducts through the centuries, from antiquity to modern times.

All this is scrupulously explored and narrated in painstaking philological fashion. Passages are selected carefully and read thoroughly, distinctions are made with an informed eye on stylistic detail, and differences are established with discriminating attention to the particular aspects and the overall pattern of the texts. Both works are considered as epics, but their qualities are found to differ in such a fundamental way that they express (and allow for) opposing modes of understanding and of literary writing. Erich Auerbach (1892–1957) states that he chose to elaborate on this opposition because it operates at the foundations of Western literature, and therefore must be posed at the beginning of his study. But his presentation immediately raises questions. *Mimesis* (1946) does not have an introduction: there is no first, separate section to present its purpose and describe its approach. Instead, the work begins *in medias res*: "Readers of the *Odyssey* will remember the well-prepared and touching

---

Reworking the beginning of Erich Auerbach's essay, "Odysseus's Scar" (see Chapter 9), Vassilis Lambropoulos, C. P. Cavafy Chair in Modern Greek and Professor of Classical Studies and Comparative Literature at the University of Michigan, argues that Auerbach, rather than comparing and contrasting two modes of interpretation, privileges what Auerbach casts as the Biblical mode.

scene in book 19, when Odysseus has at last come home" (Auerbach 1953: 3). It begins with a first chapter which, like the rest, bears a neutrally descriptive title, "Odysseus' Scar," and immediately proceeds to conduct a close reading of a classic text. Only after several pages does it become clear that it deals with two texts, rather than one, that it seeks to establish the origins of Western mimetic modes, and that it functions as an introduction to the whole volume. Thus the title is deceptive: while it seems to promise a treatment of a Homeric passage, the chapter is as much about Abraham's sacrifice as it is about Odysseus' scar. It appears, then, that the book is introduced in a surreptitious manner. The suppression of the character of the piece and of its second major topic are closely linked: what at first glance looks like a first chapter and a discussion of the *Odyssey* proves to be an introduction and a comparison of Homer with the Old Testament.

The basic opposition, which the essay establishes but the title does not acknowledge, is posited and developed in a long series of dichotomies, purported to articulate the distinctive features of the Homeric and the Biblical style: external–internal, presence–absence, unity–disconnectedness, totality–fragmentation, illuminated–obscure, clarity–ambiguity, foreground–background, simplicity–complexity, stability–fermentation, serenity–anguish, being–becoming, legend–history. In all these binary oppositions, the first member refers to the Homeric world and the second to the Biblical, while each polarity indicates the antithesis and clash of the two worldviews and mimetic modes. Auerbach argues that the two sets of categories indicate contrasting ways of thinking and dictate contrasting ways of understanding them: each has to be comprehended in its own terms. Consequently, he insists: "Homer can be analyzed . . . but cannot be interpreted" (13), while "the text of the Biblical narrative . . . is so greatly in need of interpretation on the basis of its own content" (15). Auerbach refrains from explicitly defining his terminology; but from the basic sets of categories it may at least be inferred that analysis (which applies to the Homeric) is more of a description of simple incidents, surface meanings, and direct messages, while interpretation (which responds to the Biblical) uncovers hidden meanings, implied messages, and complex significances. This is not the place to discuss the critical validity of such a distinction.[1] It is more important to see how the approach called "interpretation" describes Auerbach's own method of reading literature.

Auerbach is faithful to his position when he reads the scene from Genesis in that he conducts an in-depth, penetrating interpretation which seeks to elucidate all its dimensions. As exemplified in this application, interpretation is the search for an ultimate explanation of both meaning and purpose. It tries to uncover the hidden, obscure, silent, ineffable, multiple meanings of a text, promising and at the same time threatening, retrievable yet always elusive, under the thick layers of language. It also tries to explain the purpose of it all, to describe the overall plan, to specify the final direction toward which everything is moving. In this part of his investigation, Auerbach is consistent. But he does not show the same consistency in his approach to Homer. For although he argues that the Greek epic allows only for analysis, his discussion exhibits all the unmistakable signs of an interpretive reading: it presents the hidden complexity of the incident with Euryclea, traverses successive layers of significance, exposes invisible assumptions, and finally builds on it a whole theory about Homeric mimesis. Interpretive understanding is again his guiding motive, since he asks persistently why everything in the text happens in this way. Auerbach violates his own epistemological principle and applies an interpretive reading to the *Odyssey*, a Biblical reading to a Homeric text. Although he argues that the two works express opposite worldviews and dictate different readings, he uses for both the approach derived from the second. He does not read Homer against the Bible, as he claims, but rather reads Homer through the Bible: his is a Biblical treatment. Thus his conclusion that Homer cannot be interpreted is an interpretive one, which results from a successful search for deep meanings in his work. Auerbach treats both works in an interpretive fashion, seeking to uncover their artistic essence behind the literary surface.

What appears to be omitted in the title of the essay is the most important element; what is not mentioned is the dominant feature; what is missing is central to what is there—the Biblical mode of mimesis and interpretation. The title promises a study on the recognition of Odysseus' scar, but the essay delivers a model of literary interpretation derived from Abraham's sacrifice; and the number above the title indicates a first chapter but refers to an introduction. These deceptive signs are part of the same tactic: while the essay identifies itself as a chapter on Homer, it is in fact an introduction to the Biblical method of reading; what seems to be an example of representation is nothing less than a model of interpretation. Thus the subtitle of the book, "The Representation of Reality in Western Literature," in order to reflect its approach, should read: "The Interpretation of the Representation of Reality in Western Literature." As the introductory chapter shows, the purpose of the whole project is not to analyze the dominant modes of this representation (i.e., present, describe, show their structure and effects), but rather interpret them (i.e., explain the secret meanings and purposes, unravel the significant pattern of their emergence and development). Auerbach's approach is exclusively Biblical: he comprehends literature according to rules that he finds dictated in the Bible, and consequently sees Western literary tradition as a (secular) Bible.

The purpose of Auerbach's book is to provide a sweeping Biblical view of literary history. His choice of texts alone is ample evidence. All his selections are canonical[2] (and often predictable), made from the revered masterpieces (Cahn 1979) of the dominant European tradition: *Satyricon, Chanson de Roland, Divina Commedia, Decameron, Gargantua et Pantagruel, Henry IV, Don Quixote, Manon Lescaut, Luise Millerin, To the Lighthouse*—to mention but a few. These choices make the book "a massive reaffirmation of the Western cultural tradition" (Said 1983: 8). Furthermore, selections are made and arranged with the Bible as a model. According to Auerbach, the Bible is the greatest canonical book, the Book of books, the absolute Book—the Book containing all the books that are worth reading and preserving. In it (and because of it), there are no other gods, no other books, no other world.[3] "The truth claim of the Bible, Auerbach says, is so imperious that reality in its sensuous or charming aspect is not dwelt upon; and the spotlight effect, which isolates major persons or happenings, is due to the same anagogical demand that excludes all other places and concerns. Bible stories do not flatter or fascinate like Homer's; they do not give us something artfully rendered; they force readers to become interpreters and to find the presence of what is absent in the fraught background, the densely layered (Auerbach uses the marvelous word *geschichtet*) narrative" (Hartman 1986: 15). As the central cultural construct of an entire tradition, it constitutes a colossal tautology and self-affirmation (and concomitantly a monument of ethnocentrism as well as censorship): the book that tells you what to read is both the single one worth reading and the privileged domain of human experience: "it seeks to overcome our reality: we are to fit our own life into its world, feel ourselves to be elements in its structure of universal history" (Auerbach 1953: 15).[4] Auerbach treats the Western literary canon in similar terms: his is a universal history of literature without references, notes, or bibliography; without any room for minor characters, neglected incidents, or marginal works. We are commanded to have no other books before it. As a historical survey, it is organized in autonomous, self-contained units, and deals with a tradition of glorious achievements from its origins through its continuous evolution to the present. The notion of the tradition itself is not discussed, and its authority is recognized unquestionably. The unity, borders, jurisdiction, arid goals of that authority are established. The driving implication is that the West has its own Bible, although a secular one, which is its literary canon. [. . .]

Mimesis is directly and extensively modeled on the Bible, and aspires to work like it: it consists of episodic stories of concentrated tension and high significance; it exhibits a discontinuous and yet evolutionary unity; it is driven by an urgent sense of universal history; it makes absolute claims of historical truth; it has a concrete, stable point of reference which makes everything involved in its sphere meaningful; it is fraught with religious, social, and political background; it employs a multi-layered, multi-dimensional narrative; finally, it seeks canonical authority. *Mimesis* aspires to be recognized as the Old Testament of exegetical philology, the Bible of literary criticism, by presenting and defending history as tradition, reading as interpretation. In its effort to cover Western literature in a definitive way, it employs two principal arguments: there is only one literature worth reading, the very canon which is its subject; and there is only one proper way of reading this canon, Biblical interpretation. By adopting the world-view and reenacting in an intensely dramatic fashion the method of its model, it attempts to achieve the same canonical status in the field of literary studies.[5]

Notes

1. Comparative philology, looking at the common oral-formulaic ground of Biblical and Homeric poetry, concludes that "on the basis of style, the Hebraic mind or worldview cannot be distinguished from the intelligence behind the *Odyssey or Iliad*" (Whallon, "Old Testament Poetry and Homeric Epic," *Comparative Literature* 18 [1966] 113–31), and that the distinctive features "Auerbach found in Homeric epic are also to be found in Old Testament poetry" (130). Thus it takes Auerbach to task for unjustifiably comparing two different literary provinces (Homeric) poetry and (Hebrew) prose, instead of comparing the two oral poetic traditions.

2. Since the special issue of *Critical Inquiry* (September 1983) on "Canons" (von Hallberg 1984), the subject has acquired great importance in many fields. Among the better discussions within literary studies, see Butler 1988; Condren 1985; Deleuze and Guattari 1986; Doyle 1989; Fiedler and Baker 1981; Fowler 1982; Fowles 1987; Gilbert 1985; Hemadi 1978 (Part II); Kermode 1985; Lauter 1985; McLaren 1988; Munich 1985; Robinson 1985; Rosenfelt 1982; von Hallberg 1984 and 1985; and West 1987. Rasula 1987 is a

postmodern novel about the authority of the text. For a parallel trend in religious studies, see Barr 1983; Beckwith 1985; Blenkinsopp 1977; Coats and Long 1977; Kermode 1987 ("The Canon"); Metzer 1987; Morgan 1990; and von Campenhausen 1972.

3. "The Bible's claim to truth is not only far more urgent than Homer's, it is tyrannical—it excludes all other claims. The world of the Scripture stories is not satisfied with claiming to be a historically true reality—it insists that it is the only real world, is destined for autocracy. . . . The Scripture stories do not, like Homer's, court our favor, they do not flatter us that they may please us and enchant us-they see to subject us, and if we refuse to be subjected we are rebels" (Auerbach 1953: 14–15). The author is reiterating a point made by Hegel in *The Philosophy of Fine Art*: "It is only the limited Jewish national god which is unable to tolerate other gods in its company for the reason that it purports as the one god to include everything, although in regard to the definition of its form it fails to pass beyond its exclusiveness wherein the god is merely the God of His own people. Such a god manifests its universality in fact only through his creation of Nature and as Lord of the heavens and the earth. For the rest he remains the god of Abraham, who led his people Israel out of Egypt, gave them laws on Sinai, and divided the land of Canaan among the Jews. And through this narrow identification of him with the Jewish nation he is in a quite peculiar way the god of this folk. . . . Consequently this austere, national god is so jealous, and ordains in his jealousy that men shall see elsewhere merely false idols. The Greeks, on the contrary, discovered their gods among other nations and accepted what was foreign among themselves" (Hegel, *On Tragedy*, ed. Anne and Henry Paolucci [New York: Harper & Row], pp. 182–83).

4. Similar claims have been made about its importance for literature: "Blake described the Bible as 'the Great Code of Art.' The suggestion is that it codifies and stabilizes rules for the production of meaning that can then be applied to individual texts read anagogically as books and chapters in a secular scripture built up, as Shelley says in the *Defence [of Poetry]* from 'the cooperating thoughts of one great mind.' " (Tilottama Rajan, "The Supplement of Reading," *New Literary History*, 17, No.3 [1986] pp. 573–94: 585).

5. Auerbach's strategy has obviously worked, since he has been appropriated by academics and turned into a "master," a model of the traditional humanist intellectual: "Auerbach is representative for those American critics and students of literature who believe in the enduring cultural importance, not just of literature, but of critical, humanistic scholarship in an age of need. Auerbach functions as a fantastic source for American critics and theorists; his primary function is not as a philological model but as a sign that in an anti-historical, anti-humanistic age of relativism, mass-cultural leveling, and the increasing irrelevance of writers and critics, it is not only possible for critics to perform opportune and important acts, to construct monumental synthetic texts in the face of massive specialization, to invent new techniques for dealing with changed cultural conditions, and to do all this out of the unique intellectual and existential experience of the individual scholar, but also, in so doing, to relegitimate culturally a certain image of the responsible and responsive authoritative critical voice" (Paul Bové, *Intellectuals in Power: A Genealogy of Critical Humanism* [New York: Columbia University Press, 1986] pp. 80–81). An investigation of the "ideological investment in Auerbach's redemptive qualities" (107) shows that the critic's "own claims and much of his own rhetoric nonetheless draw upon and emerge out of the discourse and values" (139–40) of the German mandarin tradition of anthropological humanism which he seems to oppose. Although Auerbach rejects the subservience of academics to the *Kulturstaat* and their support for the official (educational and other) policy and nationalist ideology, he never loses his faith in an elite in charge of Western tradition. Indeed, in his vision of an aristocracy of cultivation, the mission of the elite transcends national boundaries to take over (like *Mimesis*) the legacy of *Weltliteratur*. The intellectual

is called upon to play a universal role. This strategy of (self) legitimation recalls Auerbach's later condition that "the humanist who hopes to be effective in preserving or renewing humanity must experience exile and alienation (as he and Dante both did), in order to be able to transcend the traps of nationalism that threaten humanity's very existence" (175). This position of willful alienation is a source of ever greater authority: "By virtue of this total and loving exile, the philologist becomes a universal intellectual whose very 'marginality' to the forces or powers of modernity, as well as to the orthodox effects of a national or institutional tradition, empowers him to do work basic to the humanist enterprise" (177). Thus the "grand strategy of exile" (180) becomes another source of prestige for the socially displaced humanist. Auerbach's identification with Dante must be seen in the context of this effort of authorial canonization (197).

Chapter 35

Edward Said (1935–2003)
from "Connecting Empire to Secular Interpretation," *Culture and Imperialism* (1994)

From long before World War Two until the early 1970s, the main tradition of comparative-literature studies in Europe and the United States was heavily dominated by a style of scholarship that has now almost disappeared. The main feature of this older style was that it was scholarship principally, and not what we have come to call criticism. No one today is trained as were Erich Auerbach and Leo Spitzer, two of the great German comparatists who found refuge in the United States as a result of fascism: this is as much a quantitative as a qualitative fact. Whereas today's comparatist will present his or her qualifications in Romanticism between 1795 and 1830 in France, England, and Germany, yesterday's comparatist was more likely, first, to have studied an earlier period; second, to have done a long apprenticeship with various philological and scholarly experts in various universities in various fields over many years; third, to have a secure grounding in all or most of the classical languages, the early European vernaculars, and their literatures. The early-twentieth-century comparatist was a *philolog* who, as Francis Fergusson put it in a review of Auerbach's *Mimesis*, was so learned and had so much stamina as to make "our most intransigent 'scholars'—those who pretend with the straightest faces to scientific rigor and exhaustiveness—[appear to be] timid and relaxed."[1]

Behind such scholars was an even longer tradition of humanistic learning that derived from that efflorescence of secular anthropology—which included a revolution in the philological disciplines—we associate with the late eighteenth century and with such figures as Vico, Herder, Rousseau, and the brothers Schlegel. And underlying *their* work was the belief that mankind formed a marvelous, almost symphonic whole whose progress and formations, again as a whole, could be studied exclusively

Edward Said, who had been the Old Dominion Foundation Professor of the Humanities at Columbia University, proposes in this essay a "contrapuntal" reading of Western literature within "the dynamic global environment created by imperialism."

as a concerted and secular historical experience, not as an exemplification of the divine. Because "man" has made history, there was a special hermeneutical way of studying history that differed in intent as well as method from the natural sciences. These great Enlightenment insights became widespread, and were accepted in Germany, France, Italy, Russia, Switzerland, and subsequently, England.

It is not a vulgarization of history to remark that a major reason why such a view of human culture became current in Europe and America in several different forms during the two centuries between 1745 and 1945 was the striking rise of nationalism during the same period. The interrelationships between scholarship (or literature, for that matter) and the institutions of nationalism have not been as seriously studied as they should, but it is nevertheless evident that when most European thinkers celebrated humanity or culture they were principally celebrating ideas and values they ascribed to their own national culture, or to Europe as distinct from the Orient, Africa, and even the Americas. What partly animated my study of Orientalism was my critique of the way in which the alleged universalism of fields such as the classics (not to mention historiography, anthropology, and sociology) was Eurocentric in the extreme, as if other literatures and societies had either an inferior or a transcended value. (Even the comparatists trained in the dignified tradition that produced Curtius and Auerbach showed little interest in Asian, African, or Latin American texts.) And as the national and international competition between European countries increased during the nineteenth century, so too did the level of intensity in competition between one national scholarly interpretative tradition and another. Ernest Renan's polemics on Germany and the Jewish tradition are a well-known example of this.

Yet this narrow, often strident nationalism was in fact counteracted by a more generous cultural vision represented by the intellectual ancestors of Curtius and Auerbach, scholars whose ideas emerged in pre-imperial Germany (perhaps as compensation for the political unification eluding the country), and, a little later, in France. These thinkers took nationalism to be a transitory, finally secondary matter: what mattered far more was the concert of peoples and spirits that transcended the shabby political realm of bureaucracy, armies, customs barriers, and xenophobia. Out of this catholic tradition, to which European (as opposed to national) thinkers appealed in times of severe conflict, came the idea that the comparative study of literature could furnish a trans-national, even trans-human perspective on literary performance. Thus the idea of comparative literature not only expressed universality and the kind of understanding gained by philologists about language families, but also symbolized the crisis-free serenity of an almost ideal realm. Standing above small-minded political affairs were both a kind of anthropological Eden in which men and women happily produced something called literature, and a world that Matthew Arnold and his disciples designated as that of "culture," where only "the best that is thought and known" could be admitted.

Goethe's idea of *Weltliteratur*—a concept that waffled between the notion of "great books" and a vague synthesis of *all* the world's literatures—was very important to professional scholars of comparative literature in the early twentieth century. But still, as I have suggested, its practical meaning and operating ideology were that, so far as literature and culture were concerned, Europe led the way and was the main subject of interest. In the world of great scholars such as Karl Vossler and De Sanctis, it is most specifically Romania that makes intelligible and provides a center for the enormous grouping of literatures produced world-wide; Romania underpins Europe, just as (in a curiously regressive way) the Church and Holy Roman Empire guarantee the integrity of the core European literatures. At a still deeper level, it is from the Christian Incarnation that Western realistic literature as we know it emerges. This tenaciously advanced thesis explained Dante's supreme importance to Auerbach, Curtius, Vossler, and Spitzer.

To speak of comparative literature therefore was to speak of the interaction of world literatures with one another, but the field was epistemologically organized as a sort of hierarchy, with Europe and its Latin Christian literatures at the center and top. When Auerbach, in a justly famous essay entitled "Philologie der *Weltliteratur*," written after World War Two, takes note of how many "other" literary languages and literatures seemed to have emerged (as if from nowhere: he makes no mention of either colonialism or decolonization), he expresses more anguish and fear than pleasure at the prospect of what he seems so reluctant to acknowledge. Romania is under threat.

Certainly American practitioners and academic departments found this European pattern a congenial one to emulate. The first American department of comparative literature was established in 1891 at Columbia University, as was the first journal of comparative literature. Consider what George Woodberry—the department's first chaired professor—had to say about his field:

The parts of the world draw together, and with them the parts of knowledge, slowly knitting into that one intellectual state which, above the sphere of politics and with no more institutional machinery than tribunals of jurists and congresses of gentlemen, will be at last the true bond of all the world. The modern scholar shares more than other citizens in the benefits of this enlargement and intercommunication, this age equally of expansion and concentration on the vast scale, this infinitely extended and intimate commingling of nations with one another and with the past; his ordinary mental experience includes more of race-memory and of race-imagination than belonged to his predecessors, and his outlook before and after is on greater horizons; he lives in a larger world—is, in fact, born no longer to the freedom of the city merely, however noble, but to that new citizenship in the rising state which—the obscurer or brighter dream of all great scholars from Plato to Goethe—is without frontiers or race or force, but there is reason supreme. The emergence and growth of the new study known as Comparative Literature are incidental to the coming of this larger world and the entrance of scholars upon its work: the study will run its course, and together with other converging elements goes to its goal in the unity of mankind found in the spiritual unities of science, art, and love.[2]

Such rhetoric uncomplicatedly and naively resonates with the influence of Croce and De Sanctis, and also with the earlier ideas of Wilhelm von Humboldt. But there is a certain quaintness in Woodberry's "tribunals of jurists and congresses of gentlemen," more than a little belied by the actualities of life in the "larger world" he speaks of. In a time of the greatest Western imperial hegemony in history, Woodberry manages to celebrate a still higher, strictly ideal unity. He is unclear about how "the spiritual unities of science, art and love" are to deal with less pleasant realities, much less how "spiritual unities" can be expected to overcome the facts of materiality, power, and political division.

Academic work in comparative literature carried with it the notion that Europe and the United States together were the center of the world, not simply by virtue of their political positions, but also because their literatures were the ones most worth studying. When Europe succumbed to fascism and when the United States benefited so richly from the many emigré scholars who came to it, understandably little of their sense of crisis took root with them. *Mimesis*, for example, written while Auerbach was in exile from Nazi Europe in Istanbul, was not simply an exercise in textual explication, but—he says in his 1952 essay to which I have just referred—an act of civilizational survival. It had seemed to him that his mission as a comparatist was to present, perhaps for the last time, the complex evolution of European literature in all its variety from Homer to Virginia Woolf. Curtius's book on the Latin Middle Ages was composed out of the same driven fear. Yet how little of that spirit survived in the thousands of academic literary scholars who were influenced by these two books! *Mimesis* was praised for being a remarkable work of rich analysis, but the sense of its mission died in the often trivial uses made of it.[3] Finally in the late 1950s *Sputnik* came along, and transformed the study of foreign languages—and of comparative literature—into fields directly affecting national security. The National Defense Education Act[4] promoted the field and, with it, alas, an even more complacent ethnocentrism and covert Cold Warriorism than Woodberry could have imagined.

As *Mimesis* immediately reveals, however, the notion of Western literature that lies at the very core of comparative study centrally highlights, dramatizes, and celebrates a certain idea of history, and at the same time obscures the fundamental geographical and political reality empowering that idea. The idea of European or Western literary history contained in it and the other scholarly works of comparative literature is essentially idealistic and, in an unsystematic way, Hegelian. Thus the principle of development by which Romania is said to have acquired dominance is incorporative and synthetic. More and more reality is included in a literature that expands and elaborates from the medieval chronicles to the great edifices of nineteenth-century narrative fiction—in the works of Stendhal, Balzac, Zola, Dickens, Proust. Each work in the progression represents a synthesis of problematic elements that disturb the basic Christian order so memorably laid out in the *Divine Comedy*. Class, political upheavals, shifts in economic patterns and organization, war: all these subjects, for great authors like Cervantes, Shakespeare, Montaigne, as well as for a host of lesser writers, are enfolded within recurringly renewed structures, visions, stabilities, all of them attesting to the abiding dialectical order represented by Europe itself.

The salutary vision of a "world literature" that acquired a redemptive status in the twentieth century coincides with what theorists of colonial geography also articulated. In the writings of Halford Mackinder, George Chisolm, Georges Hardy, Leroy-Beaulieu, and Lucien Fevre, a much franker appraisal of the world system appears, equally metrocentric and imperial; but instead of history alone, now both empire and actual geographical space collaborate to produce a "world-empire" commanded by Europe. But in this geographically articulated vision (much of it based, as Paul Carter shows in *The Road to Botany Bay*, on the cartographic results of actual geographical exploration and conquest) there is no less strong a commitment to the belief that European pre-eminence is natural, the culmination of what Chisolm calls various "historical advantages" that allowed Europe to override the "natural advantages" of the more fertile, wealthy, and accessible regions it controlled.[5] Fevre's *La Terre et l'evolution humaine* (1922), a vigorous and integral encyclopedia, matches Woodberry for its scope and utopianism.

To their audience in the late nineteenth and early twentieth centuries, the great geographical synthesizers offered technical explanations for ready political actualities. Europe *did* command the world; the imperial map *did* license the cultural vision. To us, a century later, the coincidence or similarity between one vision of a world system and the other, between geography and literary history, seems interesting but problematic. What should we do with this similarity?

First of all, I believe, it needs *articulation* and *activation*, which can only come about if we take serious account of the present, and notably of the dismantling of the classical empires and the new independence of dozens of formerly colonized peoples and territories. We need to see that the contemporary global setting— overlapping territories, intertwined histories—was already prefigured and inscribed in the coincidences and convergences among geography, culture, and history that were so important to the pioneers of comparative literature. Then we can grasp in a new and more dynamic way both the idealist historicism which fuelled the comparatist "world literature" scheme and the concretely imperial world map of the same moment.

But that cannot be done without accepting that what is common to both is an elaboration of power. The genuinely profound scholarship of the people who believed in and practiced *Weltliteratur* implied the extraordinary privilege of an observer located in the West who could actually survey the world's literary output with a kind of sovereign detachment. Orientalists and other specialists about the non-European world—anthropologists, historians, philologists—had that power, and, as I have tried to show elsewhere, it often went hand in glove with a consciously undertaken imperial enterprise. We must articulate these various sovereign dispositions and see their common methodology.

An explicitly geographical model is provided in Gramsci's essay *Some Aspects of the Southern Questions*. Under-read and under-analyzed, this study is the only sustained piece of political and cultural analysis Gramsci wrote (although he never finished it); it addresses the geographical conundrum posed for action and analysis by his comrades as to how to think about, plan for, and study southern Italy, given that its social disintegration made it seem incomprehensible yet paradoxically crucial to an understanding of the north. Gramsci's brilliant analysis goes, I think, beyond its tactical relevance to Italian politics in 1926, for it provides a culmination to his journalism before 1926 and also a prelude to *The Prison Notebooks*, in which he gave, as his towering counterpart Lukacs did not, paramount focus to the territorial, spatial, geographical foundations of social life.

Lukacs belongs to the Hegelian tradition of Marxism, Gramsci to a Vichian, Crocean departure from it. For Lukacs the central problematic in his major work through *History and Class Consciousness* (1923) is temporality; for Gramsci, as even a cursory examination of his conceptual vocabulary immediately reveals, social history and actuality are grasped in geographical terms—such words as "terrain," "territory," "blocks," and "region" predominate. In *The Southern Question*, Gramsci not only is at pains to show that the division between the northern and southern regions of Italy is basic to the challenge of what to do politically about the national working-class movement at a moment of impasse, but also is fastidious in describing the peculiar topography of the south, remarkable, as he says, for the striking contrast between the large undifferentiated mass of peasants on the one hand, and the presence of "big" landowners, important publishing houses, and distinguished cultural formations on the other. Croce himself, a most impressive and notable figure in Italy, is seen by Gramsci with characteristic shrewdness as a southern philosopher who finds it easier to relate to Europe and to Plato than to his own crumbling meridional environment.

The problem therefore is how to connect the south, whose poverty and vast labor pool are inertly vulnerable to northern economic policies and powers, with a north that is dependent on it. Gramsci formulates the answer in ways that forecast his celebrated animadversions on the intellectual in the *Quaderni*: he considers Piero Gobetti, who as an intellectual understood the need for connecting the northern pro-letariat with the southern peasantry, a strategy that stood in stark contrast with the careers of Croce and Guistino Fortunato, and who linked north and south by virtue of his capacity for organizing culture. His work "posed the Southern question on a terrain different from the traditional one [which regarded the south simply as a back-ward region of Italy] by introducing into it the proletariat of the North."[6] But this introduction could not occur, Gramsci continues, unless one remembered that intel-lectual work is slower, works according to more extended calendars than that of any other social group. Culture cannot be looked at as an immediate fact but has to be seen (as he was to say in the *Quaderni*) *sub specie aeternitatis*. Much time elapses before new cultural formations emerge, and intellectuals, who depend on long years of preparation, action, and tradition, are necessary to the process.

Gramsci also understands that in the extended time span during which the coral-like formation of a culture occurs, one needs "breaks of an organic kind." Gobetti represents one such break, a fissure that opened up within the cultural structures that supported and occluded the north – south discrepancy for so long in Italian history. Gramsci regards Gobetti with evident warmth, appreciation, and cordiality as an individual, but his political and social significance for Gramsci's analysis of the south-ern question—and it is appropriate that the unfinished essay ends abruptly with this consideration of Gobetti—is that he accentuates the need for a social formation to develop, elaborate, build upon the break instituted by his work, and by his insistence that intellectual effort itself furnishes the link between disparate, apparently autonomous regions of human history.

What we might call the Gobetti factor functions like an animating connective that expresses and represents the relationship between the development of comparative literature and the emergence of imperial geography, and does so dynamically and organically. To say of both discourses merely that they are imperialist is to say little about where and how they take place. Above all it leaves out what makes it possible for us to articulate them *together*, as an ensemble, as having a relationship that is more than coincidental, conjunctural, mechanical. For this we must look at the domina-tion of the non-European world from the perspective of a resisting, gradually more and more challenging alternative.

Without significant exception the universalizing discourses of modern Europe and the United States assume the silence, willing or otherwise, of the non-European world. There is incorporation, there is inclusion; there is direct rule; there is coercion. But there is only infrequently an acknowledgement that the colonized people should be heard from, their ideas known.

It is possible to argue that the continued production and interpretation of Western culture itself made exactly the same assumption well on into the twentieth century, even as political resistance grew to the West's power in the "peripheral" world. Because of that, and because of where it led, it becomes possible now to reinterpret the Western cultural archive as if fractured geographically by the activated imperial divide, to do a rather different kind of reading and interpretation. In the first place, the history of fields like comparative literature, English studies, cultural analysis, anthropology can be seen as affiliated with the empire and, in a manner of speaking, even contributing to its methods for maintaining Western ascendancy over non-Western natives, especially if we are aware of the spatial consciousness exemplified in Gramsci's "southern question." And in the second place our interpretative change of perspective allows us to challenge the sovereign and unchallenged authority of the allegedly detached Western observer.

Western cultural forms can be taken out of the autonomous enclosures in which they have been protected, and placed instead in the dynamic global environment created by imperialism, itself revised as an ongoing contest between north and south, metropolis and periphery, white and native. We may thus consider imperialism as a process occurring as part of the metropolitan culture, which at times acknowledges, at other times obscures the sustained business of the empire itself. The important point—a very Gramscian one—is how the national British, French, and American cultures maintained hegemony over the peripheries. How within them was consent gained and continuously consolidated for the distant rule of native peoples and territories?

As we look back at the cultural archive, we begin to reread it not univocally but *contrapuntally*, with a simultaneous awareness both of the metropolitan history that is narrated and of those other histories against which (and together with which) the dominating discourse acts. In the counterpoint of Western classical music, various themes play off one another, with only a provisional privilege being given to any particular one; yet in the resulting polyphony there is concert and order, an organized interplay that derives from the themes, not from a rigorous melodic or formal principle outside the work. In the same way, I believe, we can read and interpret English novels, for example, whose engagement (usually suppressed for the most part) with the West Indies or India, say, is shaped and perhaps even determined by the specific history of colonization, resistance, and finally native nationalism. At this point alternative or new narratives emerge, and they become institutionalized or discursively stable entities.

It should be evident that no one overarching theoretical principle governs the whole imperialist ensemble, and it should be just as evident that the principle of domination and resistance based on the division between the West and the rest of the world—to adapt freely from the African critic Chinweizu—runs like a fissure throughout. That fissure affected all many local engagements, overlappings, interdependencies in Africa, India, and elsewhere in the peripheries, each different, each with its own density of associations and forms, its own motifs, works, institutions, and—most important from our point of view as rereaders—its own possibilities and conditions of knowledge. For each locale in which the engagement occurs, and the imperialist model is disassembled, its incorporative, universalizing, and totalizing codes rendered ineffective and inapplicable, a particular type of research and knowledge begins to build up.

An example of the new knowledge would be the study of Orientalism or Africanism and, to take a related set, the study of Englishness and Frenchness. These identities are today analyzed not as god-given essences, but as results of collaboration between African history and the study of Africa in reorganization of knowledge during the First Empire. In an important sense, we are dealing with the formation of cultural identities understood not as essentializations (although part of their enduring appeal is that they seem and are considered to be like essentializations) but as contrapuntal ensembles, for it is the case that no identity can ever exist by itself and without an array of opposites, negatives, oppositions: Greeks always require barbarians, and Europeans Africans, Orientals, etc. The opposite is certainly true as well. Even the mammoth engagements in our own time over such essentializations as "Islam," the "West," the "Orient," "Japan," or "Europe" admit to a particular knowledge and structures of attitude and reference, and those require careful analysis and research.

If one studies some of the major metropolitan cultures—England's, France's and United States', for instance—in the geographical contexts of their struggles for (and over) empires, a distinctive cultural topography becomes apparent. In using the phrase "structures of attitude and reference" I have this topography in mind, as I also have in mind Raymond Williams's seminal phrase "structures of feeling." I am talking about the way in which structures of location and geographical reference appear in the cultural languages of literature, history, or ethnography, sometimes allusively and sometimes carefully plotted, across several individual works that are not otherwise connected to one another or to an official ideology of "empire."

In British culture, for instance, one may discover a consistency of concern in Spenser, Shakespeare, Defoe, and Austen that fixes socially desirable, empowered space in metropolitan England or Europe and connects it by design, motive, and development to distant or peripheral worlds (Ireland, Venice, Africa, Jamaica), conceived of as desirable but subordinate. And with these meticulously maintained references come attitudes—about rule, control, profit and enhancement and suitability—that grow with astonishing power from the seventeenth to the end of the nineteenth century. These structures do not arise from some pre-existing (semi-conspiratorial) design that the writers then manipulate, but are bound up with the development of Britain's cultural identity, as that identity imagines itself in a geographically conceived world. Similar structures may be remarked in French and American cultures, growing for different reasons and obviously in different ways. We are not yet at the stage where we can say whether these globally integral structures are preparations for imperial control and conquest, or whether they accompany such enterprises, or whether in some reflective or careless way they are a result of empire. We are only at a stage where we must look at the astonishing frequency of geographical articulations in the three Western cultures that most dominated far-flung territories. In the second chapter of this book I explore this question and advance further arguments about it.

To the best of my ability to have read and understood these "structures of attitude and reference," there was scarcely any dissent, any departure, any demurral from them: there was virtual unanimity that subject races should be ruled, that they *are* subject races, that one race deserves and has consistently earned the right to be considered the race whose main mission is to expand beyond its own domain. (Indeed, as Seeley was to put it in 1883, about Britain—France and the United States had their own theorists—the British could only be understood as such.) It is perhaps embarrassing that sectors of the metropolitan cultures that have since become vanguards in the social contests of our time were uncomplaining members of this imperial consensus. With few exceptions, the women's as well as the working-class movement was pro-empire. And, while one must always be at great pains to show that different imaginations, sensibilities, ideas, and philosophies were at work, and that each work of literature or art is special, there was virtual unity of purpose on this score: the empire must be maintained, and it *was* maintained.

Reading and interpreting the major metropolitan cultural texts in this newly activated, reinformed way could not have been possible without the movements of resistance that occurred everywhere in the peripheries against the empire. In the third chapter of this book I make the claim that a new global consciousness connects all the various local arenas of anti-imperial contest. And today writers and scholars from the formerly colonized world have imposed their diverse histories on, have mapped their local geographies in, the great canonical texts of the European center. And from these overlapping yet discrepant interactions the new readings and knowledges are beginning to appear. One need only think of the tremendously powerful upheavals that occurred at the end of the 1980s—the breaking down of barriers, the popular insurgencies, the drift across borders, the looming problems of immigrant, refugee, and minority rights in the West—to see how obsolete are the old categories, the tight separations, and the comfortable autonomies.

It is very important, though, to assess how these entities were built, and to understand how patiently the idea of an unencumbered English culture, for example, acquired its authority and its power to impose itself across the seas. This is a tremendous task for any individual, but a whole new generation of scholars and intellectuals from the Third World is engaged on just such an undertaking.

Here is a word of caution and prudence is required. One theme I take up is the uneasy relationship between nationalism and liberation, two ideals or goals for people engaged against imperialism. In the main it is true that the creation of very many newly independent nation-states in the post-colonial world has succeeded in re-establishing the primacy of what have been called imagined communities, parodied and mocked by writers like V. S. Naipaul and Conor Cruise O'Brien, hijacked by a host of dictators and petty tyrants, enshrined in various state nationalisms. Nevertheless in general there is an oppositional quality to the consciousness of many Third World scholars and intellectuals, particularly (but not exclusively) those who are exiles, expatriates, or refugees and immigrants in the West, many of them inheritors of the work done by earlier twentieth-century expatriates like George Antonius and C. L. R. James. Their work in trying to connect experience across the imperial divide, in re-examining the great canons, in producing what in effect is a critical literature cannot be, and generally has not been, co-opted by the resurgent nationalisms, despotisms, and ungenerous ideologies that betrayed the liberationist ideal in favor of the nationalist independence actuality.

Moreover, their work should be seen as sharing important concerns with minority and "suppressed" voices within the metropolis itself: feminists, African-American writers, intellectuals, artists, among others. But here too vigilance and self-criticism are crucial, since there is an inherent danger to oppositional effort of becoming institutionalized, marginality turning into separatism, and resistance hardening into dogma. Surely the activism that reposits and reformulates the political challenges in intellectual life is safe-guarded against orthodoxy. But there is always a need to keep community before coercion, criticism before solidarity, and vigilance ahead of assent.

Since my themes here are a sort of sequel to *Orientalism*, which like this book was written in the United States, some consideration of America's cultural and political environment is warranted. The United States is no ordinary large country. The United States is the last superpower, an enormously influential, frequently interventionary power nearly everywhere in the world. Citizens and intellectuals of the United States have a particular responsibility for what goes on between the United States and the rest of the world, a responsibility that is in no way discharged or fulfilled by saying that the Soviet Union, Britain, France, or China were, or are, worse. The fact is that we are indeed responsible for, and therefore more capable of, influencing *this* country in ways that we were not for the pre-Gorbachev Soviet Union, or other countries. So we should first take scrupulous note of how in Central and Latin America—to mention the most obvious—as well as in the Middle East, Africa, and Asia, the United States has replaced the great earlier empires and is *the* dominant outside force.

Looked at honestly, the record is not a good one. United States military interventions since World War Two have occurred (and are still occurring) on nearly every continent, many of great complexity and extent, with tremendous national investment, as we are now only beginning to understand. All of this is, in William Appleman Williams's phrase, empire as a way of life. The continuing disclosures about the war in Vietnam, about the United States' support of the "contras" in Nicaragua, about the crisis in the Persian Gulf, are only part of the story of this complex of interventions. Insufficient attention is paid to the fact that United States Middle Eastern and Central American policies—whether exploiting a geo-political opening among Iranian so-called moderates, or aiding the so-called Contra Freedom Fighters in overthrowing the elected government of Nicaragua, or coming to the aid of the Saudi and Kuwaiti royal families—can only be described as imperialist.

Even if we were to allow, as many have, that United States foreign policy is principally altruistic and dedicated to such unimpeachable goals as freedom and democracy, there is considerable room for skepticism. The relevance of T.S. Eliot's remarks in "Tradition and the Individual Talent" about the historical sense are demonstrably important. Are we not as a nation repeating what France and Britain, Spain and Portugal, Holland and Germany, did before us? And yet do we not tend to regard ourselves as somehow exempt from the more sordid imperial assumption on our part that our destiny is to rule and to lead the world, a destiny that we have assigned ourselves as part of our errand into the wilderness?

In short, we face as a nation the deep, profoundly perturbed and perturbing question of our relationship to others—other cultures, states, histories, experiences, traditions, peoples, and destinies. There is no Archimedean point beyond the question from which to answer it; there is no vantage outside the actuality of relationships among cultures, among unequal imperial and non-imperial powers, among us and others; no one has the epistemological privilege of somehow judging, evaluating, and interpreting the world free from the encumbering interests and engagements of the ongoing relationships themselves. We are, so to speak, *of* the connections, not outside and beyond them. And it behooves us as intellectuals and humanists and secular critics to understand the United States in the world of nations and power from *within* the actuality, as participants in it, not detached outside observers who, like Oliver Goldsmith, in Yeats's perfect phrase, deliberately sip at the honeypots of our minds.

Contemporary travails in recent European and American anthropology reflect these conundrums and embroilments in a symptomatic and interesting way. That cultural practice and intellectual activity carry, as a major constitutive element, an unequal relationship of force between the outside Western ethnographer-observer and the primitive, or at least different, but certainly weaker and less developed non-Europeans, non-Western person. In the extraordinarily rich text of *Kim*, Kipling extrapolates the political meaning of that relationship and embodies it in the figure of Colonel Creighton, an ethnographer in charge of the Survey of India, also the head of British intelligence service in India, the "Great Game" to which young Kim belongs. Modern Western anthropology frequently repeated that problematic relationship, and in recent works of a number of theoreticians deals with the almost insuperable contradiction between a political actuality based on force, and a scientific and humane desire to understand the Other hermeneutically and sympathetically in modes not influenced by force.

Whether these efforts succeed or fail is a less interesting matter than what distinguishes them, what makes them possible: an acute and embarrassed awareness of the all-pervasive, unavoidable imperial setting. In fact, there is no way that I know of apprehending the imperial contest itself. This, I would say, is a cultural fact of extraordinary political as well as interpretive importance, yet it has not been recognized as such in cultural and literary theory, and is routinely circumvented or occluded in cultural discourses. To read most cultural deconstructionists, or Marxists, or new historicists is to read writers whose political horizon, whose historical location is within a society and culture deeply enmeshed in imperial domination. Yet little notice is taken of this horizon, few acknowledgements of the setting are advanced, little realization of the imperial closure itself is allowed for. Instead, one has the impression that interpretation of other cultures, texts, and peoples—which at bottom is what all interpretation is about—occurs in a timeless vacuum, so forgiving and permissive as to deliver the interpretation directly into a universalism free from attachment, inhibition, and interest.

We live of course in a world not only of commodities but also of representation, and representations—their production, circulation, history, and interpretation—are the very element of culture. In much recent theory the problem of representation is deemed to be central, yet rarely is it put in its full political context, a context that is primarily imperial. Instead we have on the one hand an isolated cultural sphere, believed to be freely and unconditionally available to weightless theoretical speculation and investigation, and, on the other, a debased political sphere, where the real struggle between interests is supposed to occur. To the professional student of culture—the humanist, the critic, the scholar—only one sphere is relevant, and, more to the point, it is accepted that the two spheres are separated, whereas the two are not only connected but ultimately the same.

A radical falsification has become established in this separation. Culture is exonerated of any entanglements with power, representations are considered only as apolitical images to be parsed and construed as so many grammars of exchange, and the divorce of the present from the past is assumed to be complete. And yet, far from this separation of spheres being a neutral or accidental choice, its real meaning is as an act of complicity, the humanist's choice of a disguised, denuded, systematically purged textual model over a more embattled model, whose principal features would inevitably coalesce around the continuing struggle over the question of empire itself.

Let me put this differently, using examples that will be familiar to everyone. For at least a decade, there has been a decently earnest debate in the United States over the meaning, contents, and goals of liberal education. Much but not all of this debate was stimulated in the university after the upheavals of the 1960s, when it appeared for the first time in this century that the structure, authority, and tradition of American education were challenged by marauding energies, released by socially and intellectually inspired provocations. The newer currents in the academy, and the force of what is called theory (a rubric under which were herded many new disciplines like psychoanalysis, linguistics, and Nietzschean philosophy, unhoused from the traditional fields such as philology, moral philosophy, and the natural sciences), acquired prestige and interest; they appeared to undermine the authority and the stability of established canons, well-capitalized fields, long-standing procedures of accreditation, research, and the division of intellectual labor. That all this occurred in the modest and circumscribed terrain of cultural – academic praxis simultaneously with the great wave of anti-war, anti-imperialist protest was not fortuitous but, rather, a genuine political and intellectual conjuncture.

There is considerable irony that our search in the metropolis for a newly invigorated, reclaimed tradition follows the exhaustion of modernism and is expressed variously as post-modernism or, as I said earlier, citing Lyotard, as the loss of the legitimizing power of the narratives of Western emancipation and enlightenment; simultaneously, modernism is rediscovered in the formerly colonized, peripheral world, where resistance, the logic of daring, and various investigations of age-old tradition (*al-Turath*, in the Islamic world) together set the tone.

One response in the West to the new conjunctures, then, has been profoundly reactionary: the effort to reassert old authorities and canons, the effort to reinstate ten or twenty or thirty essential Western books without which a Westerner would not be educated—these efforts are couched in the rhetoric of embattled patriotism.

But there can be another response, worth returning to here, for it offers an important theoretical opportunity. Cultural experience or indeed every cultural form is radically, quintessentially hybrid, and if it has been the practice in the West since Immanuel Kant to isolate cultural and aesthetic realms from the worldly domain, it is now time to rejoin them. This is by no means a simple matter, since—I believe—it has been the essence of experience in the West at least since the late eighteenth century not only to acquire distant domination and reinforce hegemony, but also to divide the realms of culture and experience into apparently separate spheres. Entities such as races and nations, essences such as Englishness or Orientalism, modes of production such as the Asiatic or Occidental, all of these in my opinion testify to an ideology whose cultural correlatives well precede the actual accumulation of imperial territories world-wide.

Most historians of empire speak of the "age of empire" as formally beginning around 1878, with "the scramble for Africa." A closer look at the cultural actuality reveals a much earlier, more deeply and stubbornly held view about overseas European hegemony; we can locate a coherent, fully mobilized system of ideas near the end of the eighteenth century, and there follows the set of integral developments such as the first great systematic conquests under Napoleon, the rise of nationalism and the European nation-state, the advent of large-scale industrialization, and the consolidation of power in the bourgeoisie. This is also the period in which the novel form and the new historical narrative become pre-eminent, and in which the importance of subjectivity to historical time takes hold.

Yet most cultural historians, and certainly all literary scholars, have failed to remark the *geographical* notation, the theoretical mapping and charting of territory that underlies Western fiction, historical writing, and philosophical discourse of the time. There is first the authority of the European observer—traveller, merchant, scholar, historian, novelist. Then there is the hierarchy of spaces by which the metropolitan center and, gradually, the metropolitan economy are seen as dependent upon an over-seas system of territorial control, economic exploitation, and a socio-cultural vision; without these stability and prosperity at home—"home" being a word with extremely potent resonances—would not be possible. The perfect example of what I mean is to be found in Jane Austen's *Mansfield Park*, in which Thomas Bertram's slave plantation in Antigua is mysteriously necessary to the poise and the beauty of Mansfield Park, a place described in moral and aesthetic terms well before the scramble for Africa, or before the age of empire officially began. As John Stuart Mill puts it in the *Principles of Political Economy*:

> These [outlying possessions of ours] are hardly to be looked upon as countries, . . . but more properly as outlying agricultural or manufacturing estates belonging to a larger community. Our West Indian colonies, for example, cannot be regarded as countries with a productive capital of their own . . . [but are rather] the place where England finds it convenient to carry on the production of sugar, coffee and a few other tropical commodities.[6]

Read this extraordinary passage together with Jane Austen, and a much less benign picture stands forth than the usual one of cultural formations in the pre-imperialist age. In Mill we have the ruthless proprietary tones of the white master used to effacing the reality, work, and suffering of millions of slaves, transported across the middle passage, reduced only to an incorporated status "for the benefit of the propri-etors." These colonies are, Mill says, to be considered as hardly anything more than a convenience, an attitude confirmed by Austen, who in *Mansfield Park* sublimates the agonies of Caribbean existence to a mere half dozen passing references to Antigua.

And much the same processes occur in other canonical writers of Britain and France; in short, the metropolis gets its authority to a considerable extent from the devaluation as well as the exploitation of the outlying colonial possession. (Not for nothing, then, did Walter Rodney entitle his great decolonizing treatise of 1972 *How Europe Underdeveloped Africa*.)

Lastly, the authority of the observer, and of European geographical centrality, is buttressed by a cultural discourse relegating and confining the non-European to a secondary racial, cultural, ontological status. Yet this secondariness is, paradoxically, essential to the primariness of the European; this of course is the paradox explored by Césaire, Fanon, and Memmi, and it is but one among many of the ironies of modern critical theory that it has rarely been explored by investigators of the aporias and impossibilities of reading. Perhaps this is because it places emphasis not so much on *how* to read, but rather on *what* is read and *where* it is written about and represented. It is to Conrad's enormous credit to have sounded in such a complex and riven prose the authentic imperialist note—how you supply the forces of world-wide accumulation and rule with a self-confirming ideological motor (what Marlow in *Heart of Darkness* calls efficiency with devotion to an idea at the back of it, "it" being the taking away of the earth from those with darker complexions and flatter noses) and simultaneously draw a screen across the process, saying that art and culture having nothing to do with "it."

What to read and what to do with that reading, that is the full form of the question. All the energies poured into critical theory, into novel and demystifying theoretical praxes like the new historicism and deconstruction and Marxism have avoided the major, I would say determining, political horizon of modern Western culture, namely imperialism. This massive avoidance has sustained a canonical inclusion and exclusion: you include the Rousseaus, the Nietzsches, the Wordsworths, the Dickenses, Flauberts, and so on, and at the same you exclude their relationships with the protracted, complex, and striated work of empire. But why is this a matter of what to read and about where? Very simply, because critical discourse has taken no cognizance of the enormously exciting, varied post-colonial literature produced in resistance to the imperialist expansion of Europe and the United States in the past two centuries. To read Austen without also reading Fanon and Cabral—and so on and on—is to disaffiliate modern culture from its engagements and attachments. That is a process that should be reversed.

But there is more to be done. Critical theory and literary historical scholarship have reinterpreted and revalidated major swatches of Western literature, art, and philosophy. Much of this has been exciting and powerful work, even though one often senses more an energy of elaboration and refinement than a committed engagement to what I would call secular and affiliated criticism; such criticism cannot be undertaken without a fairly strong sense of how consciously chosen historical models are relevant to social and intellectual change. Yet if you read and interpret modern European and American culture as having had something to do with imperialism, it becomes incumbent upon you also to reinterpret the canon in the light of texts whose place there has been insufficiently linked to, insufficiently weighted toward the expansion of Europe. Put differently, this procedure entails reading the canon as a polyphonic accompaniment to the expansion of Europe, giving a revised direction and valence to writers such as Conrad and Kipling, who have always been read as sports, not as writers whose manifestly imperialist subject matter has a long subterranean or implicit and proleptic life in the earlier work of writers like, say, Austen or Chateaubriand.

Second, theoretical work must begin to formulate the relationship between empire and culture. There have been a few milestones—Kiernan's work, for instance, and Martin Green's—but concern with the issue has not been intense. Things, however, are beginning to change, as I noted earlier. A whole range of work in other disciplines, a new group of often younger scholars and critics—here, in the Third World, in Europe—are beginning to embark on the theoretical and historical enterprises; many of them seem in one way or another to be converging on questions of imperialist discourse, colonialist practice, and so forth. Theoretically we are only at the stage of trying to inventory the *interpellation* of culture by empire, but the efforts so far made are only slightly more than rudimentary. And as the study of culture extends into the mass media, popular culture, micro-politics, and so forth, the focus on modes of power and hegemony grows sharper.

We should keep before us the prerogatives of the present as signposts and paradigms for the study of the past. If I have insisted on integration and connections between the past and the present, between imperializer and imperialized, between culture and imperialism, I have done so not to level or reduce differences, but rather to convey a more urgent sense of the interdependence between things. So vast and yet so detailed is imperialism as an experience with crucial cultural dimensions, that we must speak of overlapping territories, intertwined histories common to men and women, whites and non-whites, dwellers in the metropolis and on the peripheries, past as well as present and future; these territories and histories can only be seen from the perspective of the whole of secular human history.

Notes

1. Francis Fergusson, *The Human Image in Dramatic Literature* (New York: Doubleday, Anchor, 1957), pp. 205–06.
2. George E. Woodberry, "Editorial" (1903), in *Comparative Literature: The Early Years, An Anthology of Essays*, eds. Hans Joachim Schulz and Phillip K. Rein (Chapel Hill: University of North Carolina Press, 1973), p. 211. See also Harry Levin, *Grounds for Comparison* (Cambridge, MA.: Harvard University Press, 1972), pp. 57–130; Claudio Guillérn, *Entre lo uno y lo diviso: Introduccion a la literature comparada* (Barcelona: Editorial Critica, 1985), pp. 54–121.
3. Erich Auerbach, *Mimesis: The Representation of Reality in Western Literature*, trans. Willard Trask (Princeton: Princeton University Press, 1953). See also Said, "Secular Criticism," in *The World, the Text, and the Critic* (Cambridge: Harvard University Press, 1983), pp. 31–53 and 148–49.
4. The National Defense Education Act (NDEA). An act of the United States Congress passed in 1958, it authorized the expenditure of $195 million for science and languages, both deemed important for national security. Departments of Comparative Literature were among the beneficiaries of this act.

5. Cited in Smith, *Uneven Development* (Oxford: Blackwell, 1984), pp. 101–2.
6. Antonio Gramsci, "Some Aspects of the Southern Questions," in *Selections from Political Writings, 1922–1926*, trans. and ed. Quintin Hoare (London: Lawrence and Wishart, 1978), p. 461. For an unusual application of Gramsci's theories about "Southernism," see Timothy Brennan, "Literary Criticism and the Southern Question," *Cultural Critique*, No. II (Winter 1988–1989), 89–114.
7. John Stuart Mill, *Principles of Political Economy*, Vol. 3, ed. J.M. Robson (Toronto: University of Toronto Press, 1965), p. 693.

Chapter 36

Michael Bérubé (1961–)
from "Higher Education and American
Liberalism," *Public Access* (1994)

To this point I've addressed only those people who could conceivably be persuaded
that the PC attack on American academe has been conducted in a fundamentally
unethical manner; that it has created a toxic climate in which it is impossible to call
unproblematically for "debate" with conservatives on higher education and the
public sphere; and that it is intimately tied to broader attacks on leftists, liberals and
moderates in the culture at large. Many of my readers will not need convincing on
any of these counts. But others, I trust, may be inclined to believe, with John Searle,
that "long-term assaults on the integrity of the intellectual enterprise" have been
more likely to come from inside the universities themselves than from conservative
newsweeklies and Washington think tanks.[1] It is difficult for academics to remember,
moreover, that for a larger constituency spanning the cultural spectrum from
Dittoheads to Deadheads to department heads, the Cheney-D'Souza attacks are
nothing more than the just desserts of stuffed-shirt professors, snooty intellectuals
and Lexus-driving Marxists who had it coming anyway. And if the attacks play a little
fast and loose with the record, hell, that's politics. Grow up.

Yet the larger context of the culture wars debate cuts curiously across the interests
of this constituency. For it has to do with the politics of higher education, the "fail-
ings" of elementary and secondary education, and the reprivatization of culture and
industry pursued with malice aforethought during the Reagan era. The PC debates
on campus, as the overheated polemics of 1991 made clear, are of a piece with conservative assaults on public education, public television, and public funding for the
arts. In these struggles, too, college professors are often seen as liabilities by their
potential allies: where millions of ordinary citizens can see the point of rallying to
keep "Sesame Street" on the air, or creating safe and well-run public schools, or allowing "controversial" artists to put body parts on wallpaper in the name of freedom of
expression, few can find it in themselves to get worked up about the possibility that

Michael Bérubé is Paterno Family Professor in Literature at Pennsylvania State University. In this essay,
Bérubé casts the canon debates of the 1980s and early 1990s as the cultural face of a larger, conservative
political offensive.

Lynne Cheney's NEH [National Endowment for the Humanities] may not have given a fair hearing to grant proposals concerning poststructuralism or multiculturalism. Add to this the extraordinary ease with which anyone can ridicule or demonize college professors, largely because no one outside academe has any clear idea of what professors do inside or outside the classroom, and you can see that conservative criticisms of higher education have greater chances of success—particularly among self-described "liberals"—than their criticisms of public education, PBS [Public Broadcasting System] or Karen Finley's performance art.

Nevertheless, the right has begun to discover that it cannot delegitimate higher education in toto simply by casting aspersions on college professors. In an economy where college graduates stand a healthy chance of raising their standard of living in the midst of their nation's gradual economic decline, while non graduates are likely to lose ground with each passing year of their lives, "college"—as an ideological sign and as a credentializing mechanism—remains freighted with the hopes of millions of Americans who see it as their only chance for a life better than that of their parents. What results is a class schizophrenia concerning colleges and the people who run them: professors are widely distrusted and loathed for their cushy working conditions, but colleges as institutions remain widely respected as gatekeepers of the promise of better working conditions for their graduates. [. . .]

What truly endangers the future of higher education, then, are the PC wars in tandem with the growing mad-as-hell taxpayer outrage at the professional autonomy of faculty, an outrage most effectively expressed as the demand that universities curtail professorial research and require more undergraduate instruction from their employees. Now, there's nothing outrageous about that demand per se. But the emphasis on teaching at the expense of research, like the class resentment over professors' wages and workplace autonomy, overestimates the size of its target by focusing exclusively on the elite of the academic profession to the exclusion of all those who are already teaching three or four classes a semester, working part-time at $1500 to $2500 per course without benefits, or employed in non-tenure-track jobs that require them to relocate more often than most American workers.[2] These teachers, whose numbers are growing as universities continue to "downsize," must remain invisible to the public if the attacks on PC teachers and useless, tax-dollar-devouring researchers are to succeed. So far, they have been invisible indeed; and when there is no public outcry over the exploitation and betrayal of workers at Hormel, Caterpillar or General Motors, it's safe to say that no one will stop traffic at the realization that almost two-fifths of the country's college teachers are working part-time for pitiful wages. [. . .]

The answer, I suggest, is unintelligible outside the context of Cold War cultural politics—which, significantly, seem quite capable of surviving the Cold War itself. Searle's argument is the genial Western Civ counterpart to the standard conservative argument against American dissenters: because American society (Western thought) is the locus of freedom, and already enables dissent from within, it is unnecessary to engage in actual dissent. More than this, such dissent is the stuff of treason rather than patriotism, since—so the argument goes—the enemies of Western thought (the Soviet Union specifically, international Communism generally) themselves would not permit the dissent that American dissenters take for granted. In recent years the specific content of "that which is not the West" has shifted somewhat, and now that its referent cannot so clearly be Communism, "Western" will sometimes take, as its implied opposite term, "Islamic." Still, the general opposition is the same: we self-critical American-Western individuals are locked in struggle with evil empires and terrorist rings made up of groupthinkers with no traditions of self-criticism, and therefore we must suppress our capacity for self-criticism and present a united front if we are to prevail over totalitarian philosophies. [. . .]

The terrain of American cultural politics, in short, is laced with contradictions; so much so that, despite the frequency with which the directional terms "cultural right" and "cultural left" appear in this book, I want to argue that the meanings and constituencies implied by these terms are neither obvious nor stable, just as they can be neither disentangled from nor directly mapped onto the correlative terms "political right" and "political left." What I've tried to show in this section is how these contradictions shape the cultural moment of the essays in this book: American higher education maintains broad public support as the ticket to economic security, even as Americans are rightly anxious as to whether that support is justified. College professors, especially those in the humanities, do not inspire the same faith. "Politically correct" campus policies such as affirmative action and curricular revision are often defended by a coalition of traditional liberals and centrists, but literary theory and speech codes draw fire from liberals and conservatives alike; "theory," for its part, is both embraced and rejected by different factions among nonacademic feminists, gay activists, and various "minority" social movements. In the political culture at large, the very terms "liberal" and "conservative" are increasingly contested, as the Clinton White House often finds it either undesirable or impossible to break decisively with the Reagan–Bush policies of the 1980s, while Republicans, given four years to regroup, remain alternately embarrassed and energized by the "cultural war" wing of the right, which, since floating major candidates in the 1988 and 1992 elections, has come to coalesce around a social agenda advocating the repression and occasional beating of gay and lesbian citizens, the murder and terrorization of abortion doctors and family planning workers, and (underwriting these) the moral superiority of Christian culture.[3] The postwar American consensus has plainly unraveled, Cold War definitions of "American" seek a new Other on which to predicate a national Us, and American higher education (its postwar growth heretofore dependent on its usefulness to Cold War research and development) anxiously waits to discover whether its numerous intellectual investments in nonvocational training—the more speculative social sciences, the humanities, and the "studies" programs spawned by the 1960s—will find any base of public support other than those grounded on the purely economic and credentializing functions of "college." It is not an easy task to try to discover one's potential allies and opponents in such a foggy atmosphere as this (damn all these icebergs!), but it is not a dull time to be a cultural critic in the United States.

Notes

1. John Searle, "The Storm over the University," in *Debating P.C.: The Controversy over Political Correctness on College Campuses*, ed., Paul Berman (New York: Dell, 1992), p. 87. (Hereafter cited by page number.)

2. See Paul Lauter," 'Political Correctness' and the Attack on American Colleges," in *Higher Education under Fire: Politics, Economics, and the Crisis of the Humanities*, ed., Michael Bérubé and Cary Nelson (New York: Routledge, 1994).

3. The hard right resists such characterizations of itself, but is finding it increasingly enjoyable—and politically profitable—to indulge its fascist fringe. For an amazingly forthright defense of anti-gay discrimination on the grounds that it prevents "waverers" from becoming gay, see E. L. Pattullo's "Straight Talk about Gays," *Commentary*, 94, No. 6 (1992), pp. 21–24; for extremism in defense of extremism (and the moral superiority of Christians), see Pat Buchanan's 1993 speech to Pat Robertson's Christian Coalition, as reported by Thomas B. Edsall, "Buchanan Warns GOP of Schism on Abortion: Christians Told Not to Abandon 'Culture War,' " *Washington Post* (September 12, 1993), p. A8.

Chapter 37

Harold Bloom (1930–) from "An Elegy for the Canon," *The Western Canon: The Books and the School of Ages* (1994)

Originally the canon meant the choice of books in our teaching institutions, and despite the recent politics of multiculturalism, the Canon's true question remains: What shall the individual who still desires to read attempt to read, this late in history? The Biblical three-score years and ten no longer suffice to read more than a selection of the great writers in what can be called the Western tradition, let alone in all the world's traditions. Who reads must choose, since there is literally not enough time to read everything, even if one does nothing but read. Mallarmé's grand line—"the flesh is sad, alas, and I have read all the books"—has become a hyperbole. Overpopulation, Malthusian repletion, is the authentic context for canonical anxieties. Not a moment passes these days without fresh rushes of academic lemmings off the cliffs they proclaim the political responsibilities of the critic, but eventually all this moralizing will subside. Every teaching institution will have its department of cultural studies, an ox not to be gored, and an aesthetic under-ground will flourish, restoring something of the romance of reading.

Reviewing bad books, W. H. Auden once remarked, is bad for the character. Like all gifted moralists, Auden idealized despite himself, and he should have survived into the present age, wherein the new commissars tell us that reading good books is bad for the character, which I think is probably true. Reading the very best writers—let us say Homer, Dante, Shakespeare, Tolstoy—is not going to make us better citizens. Art is perfectly useless, according to the sublime Oscar Wilde, who was right about everything. He also told us that all bad poetry is sincere. Had I the power to do so, I would command that these words be engraved above every gate at every university, so that each student might ponder the splendor of the insight.

President Clinton's inaugural poem, by Maya Angelou, was praised in a *New York Times* editorial as a work of Whitmanian magnitude, and its sincerity is indeed

Harold Bloom is Sterling Professor of Humanities at Yale, and Berg Visiting Professor of English at NYU. In this essay, Bloom describes a "School of Resentment" opposed, he says, to the Canon because it rejects the strong originality of the aesthetic.

overwhelming; it joins all the other instantly canonical achievements that flood our academies. The unhappy truth is that we cannot help ourselves; we can resist, up to a point, but past that point even our own universities would feel compelled to indict us as racists and sexists. I recall one of us, doubtless with irony, telling a *New York Times* interviewer that "We are all feminist critics." That is the rhetoric suitable for an occupied country, one that expects no liberation from liberation. Institutions may hope to follow the advice of the prince in Lampedusa's *The Leopard*, who counsels his peers, "Change everything just a little so as to keep everything exactly the same."

Unfortunately, nothing ever will be the same because the art and passion of reading well and deeply, which was the foundation of our enterprise, depended upon people who were fanatical readers when they were still small children. Even devoted and solitary readers are now necessarily beleaguered, because they cannot be certain that fresh generations will rise up to prefer Shakespeare and Dante to all other writers. The shadows lengthen in our evening land, and we approach the second millennium expecting further shadowing.

I do not deplore these matters; the aesthetic is, in my view, an individual rather than a societal concern. In any case there are no culprits, though some of us would appreciate not being told that we lack the free, generous, and open societal vision of those who come after us. Literary criticism is an ancient art; its inventor, according to Bruno Snell, was Aristophanes, and I tend to agree with Heinrich Heine that "There is a God, and his name is Aristophanes." Cultural criticism is another dismal social science, but literary criticism, as an art, always was and always will be an elitist phenomenon. It was a mistake to believe that literary criticism could become a basis for democratic education or for societal improvement. When our English and other literature departments shrink to the dimensions of our current Classics departments, ceding their grosser functions to the legions of Cultural Studies, we will perhaps be able to return to the study of the inescapable, to Shakespeare and his few peers, who after all, invented all of us.

The Canon, once we view it as the relation of an individual reader and writer to what has been preserved out of what has been written, and forget the canon as a list of books for required study, will be seen as identical with the literary Art of Memory, not with the religious sense of canon. Memory is always an art, even when it works involuntarily. Emerson opposed the party of Memory to the party of Hope, but that was in a very different America. Now the party of Memory *is* the party of Hope, though the hope is diminished. But it has always been dangerous to institutionalize hope, and we no longer live in a society in which we will be allowed to institutionalize memory. We need to teach more selectively, searching for the few who have the capacity to become highly individual readers and writers. The others, who are amenable to a politicized curriculum, can be abandoned to it. Pragmatically, aesthetic value can be recognized or experienced, but it cannot be conveyed to those who are incapable of grasping its sensations and perceptions. To quarrel on its behalf is always a blunder.

What interests me more is the flight from the aesthetic among so many in my profession, some of whom at least began with the ability to experience aesthetic value. In Freud, flight is the metaphor for repression, for unconscious yet purposeful forgetting. The purpose is clear enough in my profession's flight: to assuage displaced guilt.

Forgetting, in an aesthetic context, is ruinous, for cognition, in criticism, always relies on memory. Longinus would have said that pleasure is what the resenters have forgotten. Nietzsche would have called it pain; but they would have been thinking of the same experience upon the heights. Those who descend from there, lemminglike, chant the litany that literature is best explained as a mystification promoted by bourgeois institutions.

This reduces the aesthetic to ideology, or at best to metaphysics. A poem cannot be read *as a poem*, because it is primarily a social document or, rarely yet possibly, an attempt to overcome philosophy. Against this approach I urge a stubborn resistance whose single aim is to preserve poetry as fully and purely as possible. Our legions who have deserted represent a strand in our traditions that has always been in flight from the aesthetic: Platonic moralism and Aristotelian social science. The attack on poetry either exiles it for being destructive of social well-being or allows it sufferance if it will assume the work of social catharsis under the banners of the new multiculturalism. Beneath the surfaces of academic Marxism, Feminism, and New Historicism, the ancient polemic of Platonism and the equally archaic Aritsotelian social medicine continue to course on. I suppose that the conflict between these strains and the always beleaguered supporters of the aesthetic can never end. We are losing now, and doubtless we will go on losing, and there is a sorrow in that, because many of the best students will abandon us for other disciplines and professions, an abandonment already well under way. They are justified in doing so, because we could not protect them against our profession's loss of intellectual and aesthetic standards of accomplishment and value. All that we can do now is maintain some continuity with the aesthetic and not yield to the lie that what we oppose is adventure and new interpretations.

Freud famously defined anxiety as being *Angst vor etwas*, or anxious expectations. There is always something in advance of which we are anxious, if only of expectations that we will be called upon to fulfill. Eros, presumably the most pleasurable of expectations, brings its own anxieties to the reflective consciousness, which is Freud's subject. A literary work also arouses expectations that it needs to fulfill or it will cease to be read. The deepest anxieties of literature are literary; indeed, in my view, they define the literary and become all but identical with it. A poem, novel, or play acquires all of humanity's disorders, including the fear of mortality, which in the art of literature is transmuted into the quest to be canonical, to join communal or societal memory. Even Shakespeare, in the strongest of his sonnets, hovers near this obsessive desire or drive. The rhetoric of immortality is also a psychology of survival and a cosmology.

Where did the idea of conceiving a literary work that the world would not willingly let die come from? It was not attached to the Scriptures by the Hebrews, who spoke of canonical writings as those that polluted the hands that touched them, presumably because mortal hands were not fit to hold sacred writings. Jesus replaced the Torah for Christians, and what mattered most about Jesus was the Resurrection. At what date in the history of secular writing did men begin to speak of poems or stories as being immortal? The conceit is in Petrarch and is marvelously developed by Shakespeare in his sonnets. It is already a latent element in Dante's praise of his own *Divine Comedy*. We cannot say that Dante secularized the idea, because he subsumed everything and so, in a sense, secularized nothing. For him, his poem was prophecy, as much as Isaiah was prophecy, so perhaps we can say that Dante invented our modern idea of the canonical. Ernst Robert Curtius, the eminent medieval scholar, emphasizes that Dante considered only two journeys into the beyond, before his own, to be authentic: Virgil's Aeneas in Book 6 of his epic and St. Paul's as recounted in 2 Corinthians 12: 2. Out of Aeneas came Rome; out of Paul came Gentile Christianity; out of Dante was to come, if he lived to the age of eighty-one, the fulfillment of the esoteric prophecy concealed in the *Comedy*, but Dante died at fifty-six.

Curtius, ever alert to the fortune of canonical metaphors, has an excursus upon "Poetry as Perpetuation" that traces the origin of the eternity of poetic fame to the *Iliad* (6.359) and beyond to Horace's *Odes* (4.8, 28), where we are assured that it is the Muse's eloquence and affection that allow the hero never to die. Jakob Burckhardt, in a chapter on literary fame that Curtius quotes, observes that Dante, the Italian Renaissance poet-philologist, had "the most intense consciousness that he is a distributor of fame and indeed of immortality," a consciousness that Curtius locates among the Latin poets of France as early as 1100. But at some point this consciousness was linked to the idea of a secular canonicity, so that not the hero being celebrated but the celebration itself was hailed as immortal. The secular canon, with the word meaning a catalog of approved authors, does not actually begin until the middle of the eighteenth century, during the literary period of Sensibility, Sentimentality, and the Sublime. The *Odes* of William Collins trace the Sublime canon in Sensibility's heroic precursors from the ancient Greeks through Milton and are among the earliest poems in English written to propound a secular tradition of canonicity.

The Canon, a word religious in its origins, has become a choice among texts struggling with one another for survival, whether you interpret the choice as being made by dominant social groups, institutions of education, traditions of criticism, or, as I do, by late-coming authors who feel themselves chosen by particular ancestral figures. Some recent partisans of what regards itself as academic radicalism go so far as to suggest that works join the Canon because of successful advertising and propaganda campaigns. The compeers of these skeptics sometimes go farther and question even Shakespeare, whose eminence seems to them something of an imposition. If you worship the composite god of historical process, you are fated to deny Shakespeare his palpable aesthetic supremacy, the really scandalous originality of his plays. Originality becomes a literary equivalent of such terms as individual enterprise, self-reliance, and competition, which do not gladden the hearts of Feminists, Afrocentrists, Marxists, Foucault—inspired New Historicists, or Deconstructors—of all those whom I have described as members of the School of Resentment.

One illuminating theory of canon formation is presented by Alastair Fowler in his *Kinds of Literature* (1982). In a chapter on "Hierarchies of Genres and Canons of Literature," Fowler remarks that "changes in literary taste can often be referred to revaluation of genres that the canonical works represent." In each era, some genres are regarded as more canonical than others. In the earlier decades of our time, the American prose romance was exalted as a genre, which helped to establish Faulkner, Hemingway, and Fitzgerald as our dominant twentieth-century writers of prose fiction, fit successors to Hawthorne, Melville, Mark Twain, and the aspect of Henry James that triumphed in *The Golden Bowl* and *The Wings of the Dove*. The effect of this exaltation of romance over the "realistic" novel was that visionary narratives like Faulkner's *As I Lay Dying*, Nathanael West's *Miss Lonlelyhearts*, and Thomas Pynchon's *The Crying of Lot 49* enjoyed more critical esteem than Theodore Dreiser's *Sister Carrie* and *An American Tragedy*. Now a further revision of genres has begun with the rise of the journalistic novel, such as Truman Capote's *In Cold Blood*, Norman Mailer's *The Executioner's Song*, and Tom Wolfe's *The Bonfire of the Vanities*; *An American Tragedy* has recovered much of its luster in the atmosphere of these works.

The historical novel seems to have been permanently devalued. Gore Vidal once said to me, with bitter eloquence, that his outspoken sexual orientation had denied him canonical status. What seems likelier is that Vidal's best fictions (except for the sublimely outrageous *Myra Breckenridge*) are distinguished historical novels— *Lincoln*, *Burr*, and several more—and this subgenre is no longer available for canon-ization, which helps to account for the morose fate of Norman Mailer's exuberantly inventive *Ancient Evenings*, a marvelous anatomy of humbuggery and bumbuggery that could not survive its placement in the ancient Egypt of *The Book of the Dead*. History writing and narrative fiction have come apart, and our sensibilities seem no longer able to accommodate them one to the other.

Fowler goes a long way toward expounding the question of just why all genres are not available at any one time:

> we have to allow for the fact that the complete range of genres is never equally, let alone fully, available in any one period. Each age has a fairly small repertoire of genres that its readers and critics can respond to with enthusiasm, and the repertoire easily available to its writers is smaller still: the temporary canon is fixed for all but the greatest or strongest or most arcane writers. Each age makes new deletions from the repertoire. In a weak sense, all genres perhaps exist in all ages, shadowly embodied in bizarre and freakish exceptions. . . . But the repertoire of active genres has always been small and subject to proportionately significant deletions and additions . . . some critics have been tempted to think of the generic system almost on a hydrostatic model—as if its total substance remained constant but subject to redistributions.
>
> But there is no firm basis for such speculation. We do better to treat the movements of genres simply in terms of aesthetic choice.

I myself would want to argue, partly following Fowler, that aesthetic choice has always guided every secular aspect of canon formation, but that is a difficult argu-ment to maintain at this time when the defense of the literary canon, like the assault against it, has become so heavily politicized. Ideological defenses of the Western Canon are as pernicious in regard to aesthetic values as the onslaughts of attackers who seek to destroy the Canon or "open it up," as they proclaim. Nothing is so essen-tial to the Western Canon as its principles of selectivity, which are elitist only to the extent that they are founded upon severely artistic criteria. Those who oppose the Canon insist that there is always an ideology involved in canon formation; indeed, they go farther and speak of the ideology *of* canon formation, suggesting that to make a canon (or to perpetuate one) is an ideological act *in itself.*

The hero of these anticanonizers is Antonio Gramsci, who in his *Selections from the Prison Notebooks* denies that any intellectual can be free of the dominant social group if he relies upon merely the "special qualification" that he shares with the craft of his fellows (such as other literary critics): "Since these various categories of traditional intellectuals experience through an 'esprit de corps' their uninterrupted historical qualification, they thus put themselves forward as autonomous and independent of the dominant social group."

As a literary critic in what I now regard as the worst of all times for literary criticism, I do not find Gramsci's stricture relevant. The esprit de corps of professionalism, so curiously dear to many high priests of the anticanonizers, is of no interest whatsoever to me, and I would repudiate any "uninterrupted historical continuity" with the Western academy. I desire and assert a continuity with a handful or so of critics before this century and another handful or so during the past three generations. As for "special qualification," my own, contra Gramsci, is purely personal. Even if "the dominant social group" were to be identified with the Yale Corporation, or the trustees of New York University, or of American universities in general, I can search out no *inner* connection between any social group and the specific ways in which I have spent my life reading, remembering, judging, and interpreting what we once called "imaginative literature." To discover critics in the service of a social ideology one need only regard those who wish to demystify or open up the Canon, or their opponents who have fallen into the trap of becoming what they beheld. But neither of these groups is truly *literary*.

The flight from or repression of the aesthetic is endemic in our institutions of what still purport to be higher education. Shakespeare, whose aesthetic supremacy has been confirmed by the universal judgment of four centuries, is now "historicized" into pragmatic diminishment, precisely because his uncanny aesthetic power is a scandal to any ideologue. The cardinal principle of the current School of Resentment can be stated with singular bluntness: what is called aesthetic value emanates from class struggle. This principle is so broad that it cannot be wholly refuted. I myself insist that the individual self is the only method and the whole standard for apprehending aesthetic value. But "the individual self," I unhappily grant, is defined only against society, and part of its agon with the communal inevitably partakes of the conflict between social and economic classes. Myself the son of a garment worker, I have been granted endless time to read and meditate upon my reading. The institution that sustained me, Yale University, is ineluctably part of an American Establishment, and my sustained meditation upon literature is therefore vulnerable to the most traditional Marxist analyses of class interest. All my passionate proclamations of the isolate selfhood's aesthetic value are necessarily qualified by the reminder that the leisure for meditation must be purchased from the community.

No critic, not even this one, is a hermetic Prospero working white magic upon an enchanted island. Criticism, like poetry, is (in the hermetic sense) a kind of theft from the common stock. And if the governing class, in the days of my youth, freed one to be a priest of the aesthetic, it doubtless had its own interest in such a priesthood. Yet to grant this is to grant very little. The freedom to apprehend aesthetic value may rise from class conflict, but the value is not identical with the freedom, even if it cannot be achieved without that apprehension. Aesthetic value is by definition engendered by an interaction between artists, an influencing that is always an interpretation. The freedom to be an artist, or a critic, necessarily rises out of social conflict.

But the source or origin of the freedom to perceive, while hardly irrelevant to aesthetic value, is not identical with it. There is always guilt in achieved individuality; it is a version of the guilt of being a survivor and is not productive of aesthetic value.

Without some answer to the triple question of the agon—more than, less than, equal to?—there can be no aesthetic value. That question is framed in the figurative language of the Economic, but its answer will be free of Freud's Economic Principle. There can be no poem in itself, and yet something irreducible does abide in the aesthetic. Value that cannot be altogether reduced constitutes itself through the process of inter artistic influence. Such influence contains psychological, spiritual, and social components, but its major element is aesthetic. A Marxist or Foucault-inspired historicist can insist endlessly that the *production* of the aesthetic is a question of historical forces, but production is not in itself the issue here. I cheerfully agree with the motto of Dr. Johnson—"No man but a blockhead ever wrote, except for money"—yet the undeniable economics of literature, from Pindar to the present, do not determine questions of aesthetic supremacy. And the openers-up of the Canon and the traditionalists do not disagree much on where the supremacy is to be found: in Shakespeare. Shakespeare *is* the secular canon, or even the secular scripture; forerunners and legatees alike are defined by him alone for canonical purposes. This is the dilemma that confronts partisans of resentment: either they must deny Shakespeare's unique eminence (a painful and difficult matter) or they must show why and how history and class struggle produced just those aspects of his plays that have generated his centrality in the Western Canon.

Here they confront insurmountable difficulty in Shakespeare's most idiosyncratic strength: he is always ahead of you, conceptually and imagistically, whoever and whenever you are. He renders you anachronistic because he *contains* you; you cannot subsume him. You cannot illuminate him with a new doctrine, be it Marxism or Freudianism or Demanian linguistic skepticism. Instead, he will illuminate the doctrine, not by prefiguration but by postfiguration as it were: all of Freud that matters most is there in Shakespeare already, with a persuasive critique of Freud besides. The Freudian map of the mind is Shakespeare's; Freud seems only to have prosified it. Or, to vary my point, a Shakespearean reading of Freud illuminates and overwhelms the text of Freud; a Freudian reading of Shakespeare reduces Shakespeare, or would if we could bear a reduction that crosses the line into absurdities of loss. *Coriolanus* is a far more powerful reading of Marx's *Eighteenth Brumaire of Louis Napoleon* than any Marxist reading of Coriolanus could hope to be.

Shakespeare's eminence is, I am certain, the rock upon which the School of Resentment must at last founder. How can they have it both ways? If it is arbitrary that Shakespeare centers the Canon, then they need to show why the dominant social class selected him rather than, say, Ben Jonson, for that arbitrary role. Or if history and not the ruling circles exalted Shakespeare, what was it in Shakespeare that so captivated the mighty Demiurge, economic and social history? Clearly this line of inquiry begins to border on the fantastic; how much simpler to admit that there is a *qualitative* difference, a difference in kind, between Shakespeare and every other writer, even Chaucer, even Tolstoy, or whoever. Originality is the great scandal that resentment cannot accommodate, and Shakespeare remains the most original writer we will ever know.

All strong literary originality becomes canonical. Some years ago, on a stormy night in New Haven, I sat down to reread, yet once more, John Milton's *Paradise Lost*. I had to write a lecture on Milton as part of a series I was delivering at Harvard University, but I wanted to start all over again with the poem: to read it as though I had never read it before, indeed as though no one ever had read it before me. To do so meant dismissing a library of Milton criticism from my head, which was virtually impossible. Still, I tried because I wanted the experience of reading *Paradise Lost* as I had first read it forty or so years before. And while I read, until I fell asleep in the middle of the night, the poem's initial familiarity began to dissolve. It went on dissolving in the several days following, as I read on to the end, and I was left curiously shocked, a little alienated, and yet fearfully absorbed. What was I reading?

Although the poem is a biblical epic, in classical form, the peculiar impression it gave me was what I generally ascribe to literary fantasy or science fiction, not to heroic epic. *Weirdness* was its overwhelming effect. I was stunned by two related but different sensations: the author's competitive and triumphant power, marvelously displayed in a struggle, both implicit and explicit, against every other author and text, the Bible included, and also the sometimes terrifying strangeness of what was being presented. Only after I came to the end did I recall (consciously anyway) William Empson's fierce book *Milton's God*, with its critical observation that *Paradise Lost* seemed to Empson as barbarically splendid as certain African primitive sculptures. Empson blamed the Miltonic barbarism upon Christianity, a doctrine he found abhorrent. Although Empson was politically a Marxist, deeply sympathetic to the Chinese Communists, he was by no means a precursor of the School of Resentment. He historicized freestyle with striking aptitude, and he continually showed awareness of the conflict between social classes, but he was not tempted to reduce *Paradise Lost* to an interplay of economic forces. His prime concern remained aesthetic, the proper business of the literary critic, and he fought free of transferring his moral distaste for Christianity (and Milton's God) to an aesthetic judgment against the poem. The barbaric element impressed me as it did Empson; the agonistic triumphalism interested me more.

There are, I suppose, only a few works that seem even more essential to the Western Canon than *Paradise Lost*—Shakespeare's major tragedies, Chaucer's *Canterbury Tales*, Dante's *Divine Comedy*, the Torah, the Gospels, Cervantes' *Don Quixote*, Homer's epics. Except perhaps for Dante's poem, none of these is as embattled as Milton's dark work. Shakespeare undoubtedly received provocation from rival playwrights, while Chaucer charmingly cited fictive authorities and concealed his authentic obligations to Dante and Boccaccio. The Hebrew Bible and the Greek New Testament were revised into their present forms by redactionists who may have shared very little with the original authors whom they were editing. Cervantes, with unsurpassed mirth, parodied unto death his chivalric forerunners, while we do not have the texts of Homer's precursors.

Milton and Dante are the most pugnacious of the greatest Western writers. Scholars somehow manage to evade the ferocity of both poets and even dub them pious. Thus C. S. Lewis was able to discover his own "mere Christianity" in *Paradise Lost*, and John Freccero finds Dante to be a faithful Augustinian, content to emulate the *Confessions* in his "novel of the self." Dante, as I only begin to see, creatively corrected Virgil (among many others) as profoundly as Milton corrected absolutely everyone before him (Dante included) by his own creation. But whether the writer is playful in the struggle, like Chaucer and Cervantes and Shakespeare, or aggressive, like Dante and Milton, the contest is always there. This much of Marxist criticism seems to me valuable: in strong writing there is always conflict, ambivalence, contradiction between subject and structure. Where I part from the Marxists is on the origins of the conflict. From Pindar to the present, the writer battling for canonicity may fight on behalf of a social class, as Pindar did for the aristocrats, but primarily each ambitious writer is out for himself alone and will frequently betray or neglect his class in order to advance his own interests, which center entirely upon *individuation*. Dante and Milton both sacrificed much for what they believed to be a spiritually exuberant and justified political course, but neither of them would have been willing to sacrifice his major poem for any cause whatever. Their way of arranging this was to identify the cause with the poem, rather than the poem with the cause. In doing so, they provided a precedent that is not much followed these days by the academic rabble that seeks to connect the study of literature with the quest for social change. One finds modern American followers of this aspect of Dante and Milton where one would expect to find them, in our strongest poets since Whitman and Dickinson: the socially reactionary Wallace Stevens and Robert Frost.

Those who can do canonical work invariably see their writings as larger forms than any social program, however exemplary. The issue is containment, and great literature will insist upon its self-sufficiency in the face of the worthiest causes: feminism, African-American culturism, and all the other politically correct enterprises of our moment. The thing contained varies; the strong poem, by definition, refuses to be contained, even by Dante's or Milton's God. Dr. Samuel Johnson, shrewdest of all literary critics, concluded rightly that devotional poetry was impossible as compared to poetic devotion: "The good and evil of Eternity are too ponderous for the wings of wit." "Ponderous" is a metaphor for "uncontainable," which is another metaphor. Our contemporary openers-up of the Canon decry overt religion, but they call for devotional verse (and devotional criticism!) even if the object of devotion has been altered to the advancement of women, or of blacks, or of that most unknown of all unknown gods, the class struggle in the United States. It all depends upon your values, but I find it forever odd that Marxists are perceptive in finding competition everywhere else, yet fail to see that it is intrinsic to the high arts. There is a peculiar mix here of simultaneous overidealization and undervaluation of imaginative literature, which has always pursued its own selfish aims.

Paradise Lost became canonical before the secular Canon was established, in the century after Milton's own. The answer to "Who canonized Milton?" is in the first place John Milton himself, but in almost the first place other strong poets, from his friend Andrew Marvell through John Dryden and on to nearly every crucial poet of the eighteenth century and the Romantic period: Pope, Thomson, Cowper, Collins, Blake, Wordsworth, Coleridge, Byron, Shelley, Keats. Certainly the critics, Dr. Johnson and Hazlitt, contributed to the canonization; but Milton, like Chaucer, Spenser, and Shakespeare before him, and like Wordsworth after him, simply overwhelmed the tradition and subsumed it. That is the strongest test for canonicity. Only a very few could overwhelm and subsume the tradition, and perhaps none now can. So the question today is: Can you compel the tradition to make space for you by nudging it from within, as it were, rather than from without, as the multiculturalists wish to do?

The movement from within the tradition cannot be ideological or place itself in the service of any social aims, however morally admirable. One breaks into the canon only by aesthetic strength, which is constituted primarily of an amalgam: mastery of figurative language, originality, cognitive power, knowledge, exuberance of diction. The final injustice of historical injustice is that it does not necessarily endow its victims with anything except a sense of their victimization. Whatever the Western Canon is, it is not a program for social salvation.

The silliest way to defend the Western Canon is to insist that it incarnates all of the seven deadly moral virtues that make up our supposed range of normative values and democratic principles. This is palpably untrue. The *Iliad* teaches the surpassing glory of armed victory, while Dante rejoices in the eternal torments he visits upon his very personal enemies. Tolstoy's private version of Christianity throws aside nearly everything that anyone among us retains, and Dostoevsky preaches anti-Semitism, obscurantism, and the necessity of human bondage. Shakespeare's politics, insofar as we can pin them down, do not appear to be very different from those of his Coriolanus, and Milton's ideas of free speech and free press do not preclude the imposition of all manner of societal restraints. Spenser rejoices in the massacre of Irish rebels, while the egomania of Wordsworth exalts his own poetic mind over any other source of splendor.

The West's greatest writers are subversive of all values, both ours and their own. Scholars who urge us to find the source of our morality and our politics in Plato, or in Isaiah, are out of touch with the social reality in which we live. If we read the Western Canon in order to form our social, political, or personal moral values, I firmly believe we will become monsters of selfishness and exploitation. To read in the service of any ideology is not, in my judgment, to read at all. The reception of aesthetic power enables us to learn how to talk to ourselves and how to endure ourselves. The true use of Shakespeare or of Cervantes, of Homer or of Dante, of Chaucer or of Rabelais, is to augment one's own growing inner self. Reading deeply in the Canon will not make one a better or a worse person, a more useful or more harmful citizen. The mind's dialogue with itself is not primarily a social reality. All that the Western Canon can bring one is the proper use of one's own solitude, that solitude whose final form is one's confrontation with one's own mortality.

We possess the Canon because we are mortal and also rather belated. There is only so much time, and time must have a stop, while there is more to read than there ever was before. From the Yahwist and Homer to Freud, Kafka, and Beckett is a journey of nearly three millennia. Since that voyage goes past harbors as infinite as Dante, Chaucer, Montaigne, Shakespeare, and Tolstoy, all of whom amply compensate a lifetime's rereadings, we are in the pragmatic dilemma of excluding something else each time we read or reread extensively. One ancient test for the canonical remains fiercely valid: unless it demands rereading, the work does not qualify. The inevitable analogue is the erotic one. If you are Don Giovanni and Leporello keeps the list, one brief encounter will suffice.

Contra certain Parisians, the text is there to give not pleasure but the high unpleasure or more difficult pleasure that a lesser text will not provide. I am not prepared to dispute admirers of Alice Walker's *Meridian*, a novel I have compelled myself to read twice, but the second reading was one of my most remarkable literary experiences. It produced an epiphany in which I saw clearly the new principle implicit in the slogans of those who proclaim the opening-up of the Canon. The correct test for the new canonicity is simple, clear, and wonderfully conducive to social change: it must not and cannot be reread, because its contribution to societal progress is its generosity in offering itself up for rapid ingestion and discarding. From Pindar through Hölderlin to Yeats, the self-canonizing greater ode has proclaimed its agonistic immortality. The socially acceptable ode of the future will doubtless spare us such pretensions and instead address itself to the proper humility of shared sisterhood, the new sublimity of quilt making that is now the preferred trope of Feminist criticism.

Yet we must choose: As there is only so much time, do we reread Elizabeth Bishop or Adrienne Rich? Do I again go in search of lost time with Marcel Proust, or am I to attempt yet another rereading of Alice Walker's stirring denunciation of all males, black and white? My former students, many of them now stars of the School of Resentment, proclaim that they teach social selflessness, which begins in learning how to read selflessly. The author has no self, the literary character has no self, and the reader has no self. Shall we gather at the river with these generous ghosts, free of the guilt of past self-assertions, and be baptized in the waters of Lethe? What shall we do to be saved?

The study of literature, however it is conducted, will not save any individual, any more than it will improve any society. Shakespeare will not make us better, and he will not make us worse, but he may teach us how to overhear ourselves when we talk to ourselves. Subsequently, he may teach us how to accept change, in ourselves as in others, and perhaps even the final form of change. Hamlet is death's ambassador to us, perhaps one of the few ambassadors ever sent out by death who does not lie to us about our inevitable relationship with that undiscovered country. The relationship is altogether solitary, despite all of tradition's obscene attempts to socialize it.

My late friend Paul de Man liked to analogize the solitude of each literary text and each human death, an analogy I once protested. I had suggested to him that the more ironic trope would be to analogize each human birth to the coming into being of a poem, an analogy that would connect texts as infants are connected, voicelessness linked to past voices, inability to speak linked to what had been spoken to, as all of us have been spoken to, by the dead. I did not win that critical argument because I could not persuade him of the larger human analogue; he preferred the dialectical authority of the more Heideggerian irony. All that a text, let us say the tragedy of *Hamlet*, shares with death is its solitude. But when it shares with us, does it speak with the authority of death? Whatever the answer, I would like to point out that the authority of death, whether literary or existential, is not primarily a social authority. The Canon, far from being the servant of the dominant social class, is the minister of death. To open it, you must persuade the reader that a new space has been cleared in a larger space crowded by the dead. Let the dead poets consent to stand aside for us, Artaud cried out; but that is exactly what they will not consent to do.

If we were literally immortal, or even if our span were doubled to seven score of years, say, we could give up all argument about canons. But we have an interval only, and then our place knows us no more, and stuffing that interval with bad writing, in the name of whatever social justice, does not seem to me to be the responsibility of the literary critic. Professor Frank Lentricchia, apostle of social change through academic ideology, has managed to read Wallace Stevens's "Anecdote of the Jar" as a political poem, one that voices the program of the dominant social class. The art of placing a jar was, for Stevens, allied to the art of flower arranging, and I don't see why Lentricchia should not publish a modest volume on the politics of flower arranging, under the title *Ariel and the Flowers of Our Climate*. I still remember my shock, thirty-five years or so back, when I was first taken to a soccer match in Jerusalem where the Sephardi spectators were cheering for the visiting Haifa squad, it being of the political right, while the Jerusalem squad was affiliated with the labor party. Why stop with politicizing the study of literature? Let us replace sports writers with political pundits as a first step toward reorganizing baseball, with the Republican League meeting the Democratic League in the World Series. That would give us a form of baseball into which we could not escape for pastoral relief, as we do now. The political responsibilities of the baseball player would be just as appropriate, no more, no less, than the now-trumpeted political responsibilities of the literary critic.

Cultural belatedness, now an all-but-universal world condition, has a particular poignance in the United States of America. We are the final inheritors of Western tradition. Education founded upon the *Iliad*, the Bible, Plato, and Shakespeare remains, in some strained form, our ideal, though the relevance of these cultural monuments to life in our inner cities is inevitably rather remote. Those who resent all canons suffer from an elitist guilt founded upon the accurate enough realization that canons always do indirectly serve the social and political, and indeed the spiritual, concerns and aims of the wealthier classes of each generation of Western society. It seems clear that capital is necessary for the cultivation of aesthetic values. Pindar, the superb last champion of archaic lyric, invested his art in the celebratory exercise of exchanging odes for grand prices, thus praising the wealthy for their generous support of his generous exaltation of their divine lineage. This alliance of sublimity and financial and political power has never ceased, and presumably never can or will.

There are, of course, prophets, from Amos to Blake and beyond to Whitman, who rise up to cry out against this alliance, and doubtless a great figure, equal to a Blake, will some day come again; but Pindar rather than Blake remains the canonical norm. Even such prophets as Dante and Milton compromised themselves as Blake would or could not, insofar as pragmatic cultural aspirations may be said to have tempted the poets of the *Divine Comedy* and *Paradise Lost*. It has taken me a lifetime of immersion in the study of poetry before I could understand why Blake and Whitman were compelled to become the hermetic, indeed esoteric poets that they truly were. If you break the alliance between wealth and culture—a break that marks the difference between Milton and Blake, between Dante and Whitman—then you pay the high, ironic price of those who seek to destroy canonical continuities. You become a belated Gnostic, warring against Homer, Plato, and the Bible by mythologizing your misreading of tradition. Such a war can yield limited victories; a *Four Zoas* or a *Song of Myself* are triumphs I call limited because they drive their inheritors to perfectly desperate distortions of creative desire. The poets who walk Whitman's open road most successfully are those who resemble him profoundly but not at all superficially, poets as severely formal as Wallace Stevens, T. S. Eliot, and Hart Crane. Those who

seek to emulate his apparently open forms all die in the wilderness, inchoate rhap-
sodists and academic impostors sprawling in the wake of their delicately hermetic
father. Nothing is got for nothing, and Whitman will not do your work for you.
A minor Blakean or an apprentice Whitmanian is always a false prophet, making no
way straight for anyone.

I am not at all happy about these truths of poetry's reliance upon worldly power;
I am simply following William Hazlitt, the authentic left-winger among all great critics. Hazlitt, in his wonderful discussion of Coriolanus in *Characters of Shakespeare's
Plays*, begins with the unhappy admission that "the cause of the people is indeed but
little calculated as a subject for poetry: it admits of rhetoric, which goes into
argument and explanation, but it presents no immediate or distinct images to the
mind." Such images, Hazlitt finds, are everywhere present on the side of tyrants and
their instruments.

Hazlitt's clear sense of the troubled interplay between the power of rhetoric and
the rhetoric of power has an enlightening potential in our fashionable darkness.
Shakespeare's own politics may or may not be those of Coriolanus, just as Shakespeare's
anxieties may or may not be those of Hamlet or of Lear. Nor is Shakespeare the tragic
Christopher Marlowe, whose work and life alike seem to have taught Shakespeare the
way not to go. Shakespeare knows implicitly what Hazlitt wryly makes explicit: the
Muse, whether tragic or comic, takes the side of the elite. For every Shelley or Brecht
there are a score of even more powerful poets who gravitate naturally to the party
of the dominant classes in whatever society. The literary imagination is contaminated
by the zeal and excesses of societal competition, for throughout Western history the
creative imagination has conceived of itself as the most competitive of modes, akin to
the solitary runner, who races for his own glory.

The strongest women among the great poets, Sappho and Emily Dickinson, are
even fiercer agonists than the men. Miss Dickinson of Amherst does not set out to
help Mrs. Elizabeth Barrett Browning complete a quilt. Rather, Dickinson leaves
Mrs. Browning far behind in the dust, though the triumph is more subtly conveyed
than Whitman's victory over Tennyson in "When Lilacs Last in the Dooryard
Bloom'd," where the Laureate's "Ode on the Death of the Duke of Wellington" is
overtly echoed so as to compel an alert reader's recognition of how far the Lincoln
elegy surpasses the lament for the Iron Duke. I do not know whether Feminist criticism will succeed in its quest to change human nature, but I rather doubt that any
idealism, however belated, will change the entire basis of the Western psychology of
creativity, male and female, from Hesiod's contest with Homer down to the agon
between Dickinson and Elizabeth Bishop.

As I write these sentences, I glance at the newspaper and note a story on the
anguish of feminists forced to choose between Elizabeth Holtzman and Geraldine
Ferraro for a Senate nomination, a choice not different in kind from a critic pragmatically needing to choose between the late May Swenson, something close to a
strong poet, and the vehement Adrienne Rich. A purported poem may have the most
exemplary sentiments, the most exalted politics, and may also be not much of a
poem. A critic may have political responsibilities, but the first obligation is to raise
again the ancient and quite grim triple question of the agonist: more than, less than,
equal to? We are destroying all intellectual and aesthetic standards in the humanities

and social sciences, in the name of social justice. Our institutions show bad faith in this: no quotas are imposed upon brain surgeons or mathematicians. What has been devalued is learning as such, as though erudition were irrelevant in the realms of judgment and misjudgment.

The Western Canon, despite the limitless idealism of those who would open it up, exists precisely in order to impose limits, to set a standard of measurement that is anything but political or moral. I am aware that there is now a kind of covert alliance between popular culture and what calls itself "culture criticism," and in the name of that alliance cognition itself may doubtless yet acquire the stigma of the incorrect. Cognition cannot proceed without memory, and the Canon is the true art of memory, the authentic foundation for cultural thinking. Most simply, the Canon is Plato and Shakespeare; it is the image of the individual thinking, whether it be Socrates thinking through his own dying, or Hamlet contemplating that undiscovered country. Mortality joins memory in the consciousness of reality-testing that the Canon induces. By its very nature, the Western Canon will never close, but it cannot be forced open by our current cheerleaders. Strength alone can open it up, the strength of a Freud or a Kafka, persistent in their cognitive negations.

Cheerleading is the power of positive thinking transported to the academic realm. The legitimate student of the Western Canon respects the power of the negations inherent in cognition, enjoys the difficult pleasures of aesthetic apprehension, learns the hidden roads that erudition teaches us to walk even as we reject easier pleasures, including the incessant calls of those who assert a political virtue that would transcend all our memories of individual aesthetic experience.

Easy immortalities haunt us now because the current staple of our popular culture has ceased to be the rock concert, which has been replaced by the rock video, the essence of which is an instantaneous immortality, or rather the possibility thereof. The relation between religious and literary concepts of immortality has always been vexed, even among the ancient Greeks and Romans, where poetic and Olympian eternities mixed rather promiscuously. This vexation was tolerable, even benign, in classical literature, but became more ominous in Christian Europe. Catholic distinctions between divine immortality and human fame, firmly founded upon a dogmatic theology, remained fairly precise until the advent of Dante, who regarded himself as a prophet and so implicitly gave his *Divine Comedy* the status of a new Scripture. Dante pragmatically voided the distinction between secular and sacred canon formation, a distinction that has never quite returned, which is yet another reason for our vexed sense of power and authority.

The terms "power" and "authority" have pragmatically opposed meanings in the realms of politics and what we still ought to call "imaginative literature." If we have difficulty in seeing the opposition, it may be because of the intermediate realm that calls itself "spiritual." Spiritual power and spiritual authority notoriously shade over into both politics and poetry. Thus we must distinguish the aesthetic power and authority of the Western Canon from whatever spiritual, political, or even moral consequences it may have fostered. Although reading, writing, and teaching are necessarily social acts, even teaching has its solitary aspect, a solitude only the two could share, in Wallace Stevens's language. Gertrude Stein maintained that one wrote for oneself and for strangers, a superb recognition that I would extend into a parallel apothegm: one reads for oneself and for strangers. The Western Canon does not exist in order to augment preexisting societal elites. It is there to be read by you and by strangers, so that you and those you will never meet can encounter authentic aesthetic power and the authority of what Baudelaire (and Erich Auerbach after him) called "aesthetic dignity." One of the ineluctable stigmata of the canonical is aesthetic dignity, which is not to be hired.

Aesthetic authority, like aesthetic power, is a trope or figuration for energies that are essentially solitary rather than social. Hayden White long ago exposed Foucault's great flaw as being a blindness toward his own metaphors, an ironic weakness in a professed disciple of Nietzsche. For the tropes of the Lovejoyan history of ideas Foucault substituted his own tropes and then did not always remember that his "archives" were ironies, deliberate and undeliberate. So is it with the "social energies" of the New Historicist, who is perpetually prone to forget that "social energy" is no more quantifiable than the Freudian libido. Aesthetic authority and creative power are tropes too, but what they substitute for—call it "the canonical"—has a roughly quantifiable aspect, which is to say that William Shakespeare wrote thirty-eight plays, twenty-four of them masterpieces, but social energy has never written a single scene. The death of the author is a trope, and a rather pernicious one; the life of the author is a quantifiable entity.

All canons, including our currently fashionable counter-canons, are elitist, and as no secular canon is ever closed, what is now acclaimed as "opening up the canon" is a strictly redundant operation. Although canons, like all lists and catalogs, have a tendency to be inclusive rather than exclusive, we have now reached the point at which a lifetime's reading and rereading can scarcely take one through the Western Canon. Indeed, it is now virtually impossible to master the Western Canon. Not only would it mean absorbing well over three thousand books, many, if not most, marked by authentic cognitive and imaginative difficulties, but the relations between these books grow more rather than less vexed as our perspectives lengthen. There are also the vast complexities and contradictions that constitute the essence of the Western Canon, which is anything but a unity or stable structure. No one has the authority to tell us what the Western Canon is, certainly not from about 1800 to the present day. It is not, cannot be, precisely the list I give, or that anyone else might give. If it were, that would make such a list a mere fetish, just another commodity. But I am not prepared to agree with the Marxists that the Western Canon is another instance of what they call "cultural capital." It is not clear to me that a nation as contradictory as the United States of America could ever be the context for "cultural capital," except for those slivers of high culture that contribute to mass culture. We have not had an official high culture in this country since about 1800, a generation after the American Revolution. Cultural unity is a French phenomenon, and to some degree a German matter, but hardly an American reality in either the nineteenth century or the twentieth. In our context and from our perspective, the Western Canon is a kind of survivor's list. The central fact about America, according to the poet Charles Olson, is space, but Olson wrote that as the opening sentence of a book on Melville and thus on the nineteenth century. At the close of the twentieth century, our central fact is time, for the evening land is now in the West's evening time. Would one call the list of survivors of a three-thousand-year-old cosmological war a fetish?

The issue is the mortality or immortality of literary works. Where they have become canonical, they have survived an immense struggle in social relations, but those relations have very little to do with class struggle. Aesthetic value emanates from the struggle between texts: in the reader, in language, in the classroom, in arguments within a society. Very few working-class readers ever matter in determining the survival of texts, and left-wing critics cannot do the working class's reading for it. Aesthetic value rises out of memory, and so (as Nietzsche saw) out of pain, the pain of surrendering easier pleasures in favor of much more difficult ones. Workers have anxieties enough and turn to religion as one mode of relief. Their sure sense that the aesthetic is, for them, only another anxiety helps to teach us that successful literary works are achieved anxieties, not releases from anxieties. Canons, too, are achieved anxieties, not unified props of morality, Western or Eastern. If we could conceive of a universal canon, multicultural and multivalent, its one essential book would not be a scripture, whether Bible, Koran, or Eastern text, but rather Shakespeare, who is acted and read everywhere, in every language and circumstance. Whatever the convictions of our current New Historicists, for whom Shakespeare is only a signifier for the social energies of the English Renaissance, Shakespeare for hundreds of millions who are not white Europeans is a signifier for their own pathos, their own sense of identity with the characters that Shakespeare fleshed out by his language. For them his universality is not historical but fundamental; he puts their lives upon his stage. In his characters they behold and confront their own anguish and their own fantasies, not the manifested social energies of early mercantile London.

The art of memory, with its rhetorical antecedents and its magical burgeonings, is very much an affair of imaginary places, or of real places transmuted into visual images. Since childhood, I have enjoyed an uncanny memory for literature, but that memory is purely verbal, without anything in the way of a visual component. Only recently, past the age of sixty, have I come to understand that my literary memory has relied upon the Canon as a memory system. If I am a special case, it is only in the sense that my experience is a more extreme version of what I believe to be the principal pragmatic function of the Canon: the remembering and ordering of a lifetime's reading. The greatest authors take over the role of "places" in the Canon's theater of memory, and their masterworks occupy the position filled by "images" in the art of memory. Shakespeare and *Hamlet*, central author and universal drama, compel us to remember not only what happens in *Hamlet*, but more crucially what happens in literature that makes it memorable and thus prolongs the life of the author.

The death of the author, proclaimed by Foucault, Barthes, and many clones after them, is another anticanonical myth, similar to the battle cry of resentment that would dismiss "all of the dead, white European males"—that is to say, for a baker's dozen, Homer, Virgil, Dante, Chaucer, Shakespeare, Cervantes, Montaigne, Milton, Goethe, Tolstoy, Ibsen, Kafka, and Proust. Livelier than you are, whoever you are, these authors were indubitably male, and I suppose "white." But they are not dead, compared to any living author whomsoever. Among us now are García Márquez, Pynchon, Ashberry, and others who are likely to become as canonical as Borges and Beckett among the recently deceased, but Cervantes and Shakespeare are of another order of vitality. The Canon is indeed a gauge of vitality, a measurement that attempts to map the incommensurate. The ancient metaphor of the writer's immortality is relevant here and renews the power of the Canon for us. Curtius has an excursus on "Poetry as Perpetuation" where he cites Burckhardt's reverie on "Fame in Literature" as equating fame and immortality. But Burckhardt and Curtius lived and died before the Age of Warhol, when so many are famous for fifteen minutes each. Immortality for a quarter of an hour is now freely conferred and can be regarded as one of the more hilarious consequences of "opening up the Canon."

The defense of the Western Canon is in no way a defense of the West or a nationalist enterprise. If multiculturalism meant Cervantes, who could quarrel with it? The greatest enemies of aesthetic and cognitive standards are purported defenders who blather to us about moral and political values in literature. We do not live by the ethics of the *Iliad*, or by the politics of Plato. Those who teach interpretation have more in common with the Sophists than with Socrates. What can we expect Shakespeare to do for our semiruined society, since the function of Shakespearean drama has so little to do with civic virtue or social justice? Our current New Historicists, with their odd blend of Foucault and Marx, are only a very minor episode in the endless history of Platonism. Plato hoped that by banishing the poet, he would also banish the tyrant. Banishing Shakespeare, or rather reducing him to his contexts, will not rid us of our tyrants. In any case, we cannot rid ourselves of Shakespeare, or of the Canon that he centers. Shakespeare, as we like to forget, largely invented us; if you add the rest of the Canon, then Shakespeare and the Canon wholly invented us. Emerson, in *Representative Men*, got this exactly right: "Shakespeare is as much out of the category of eminent authors, as he is out of the crowd. He is inconceivably wise; the others, conceivably. A good reader can, in a sort, nestle into Plato's brain, and think from thence; but not into Shakespeare's. We are still out of doors. For executive faculty, for creation, Shakespeare is unique."

Nothing that we could say about Shakespeare now is nearly as important as Emerson's realization. Without Shakespeare, no canon, because without Shakespeare, no recognizable selves in us, whoever we are. We owe to Shakespeare not only our representation of cognition but much of our capacity for cognition. The difference between Shakespeare and his nearest rivals is one of both kind and degree, and that double difference defines the reality and necessity of the Canon. Without the Canon, we cease to think. You may idealize endlessly about replacing aesthetic standards with ethnocentric and gender considerations, and your social aims may indeed be admirable. Yet only strength can join itself to strength, as Nietzsche perpetually testified.

Chapter 38

Jacques Derrida (1930–) from "To Whom To Give To," *The Gift of Death* (1992, trans. 1995)

Abraham doesn't speak in figures, fables, parables, metaphors, ellipses, or enigmas. His irony is meta-rhetorical. If he knew what was going to happen, if for example God had charged him with the mission of leading Isaac onto the mountain so that He could strike him with lightning, then he would have been right to have recourse to enigmatic language. But the problem is precisely that he doesn't know. Not that that makes him hesitate, however. His nonknowledge doesn't in any way suspend his own decision, which remains resolute. The knight of faith must not hesitate. He accepts his responsibility by heading off towards the absolute request of the other, beyond knowledge. He decides, but his absolute decision is neither guided nor controlled by knowledge. Such, in fact, is the paradoxical condition of every decision: it cannot be deduced from a form of knowledge of which it would simply be the effect, conclusion, or explicitation. It structurally breaches knowledge and is thus destined to nonmanifestation; a decision is, in the end, always secret. It remains secret in the very instant of its performance, and how can the concept of decision be dissociated from this figure of the instant? From the stigma of its punctuality?

Abraham's decision is absolutely responsible because it answers for itself before the absolute other. Paradoxically it is also irresponsible because it is guided neither by reason nor by an ethics justifiable before men or before the law of some universal tribunal. Everything points to the fact that one is unable to be responsible at the same time before the other and before others, before the others of the other. If God is completely other, the figure or name of the wholly other, then every other (one) is every (bit) other. *Tout autre est tout autre.* This formula disturbs Kierkegaard's discourse on one level while at the same time reinforcing its most extreme ramifications. It implies that God, as the wholly other, is to be found everywhere there is something of the

Although this selection from *The Gift of Death* (originally published in Paris in 1992, and translated into English in 1995), is not directly addressing the question of the canon, Derrida's discussion of a "nonhistory of absolute beginnings which are repeated" could apply to the transmission of meaning across literary history as well.

wholly other. And since each of us, everyone else, each other is infinitely other in its absolute singularity, inaccessible, solitary, transcendent, nonmanifest, originarily nonpresent to my *ego* (as Husserl would say of the *alter ego* that can never be originarily present to my consciousness and that I can apprehend only through what he calls *appresentation* and analogy), then what can be said about Abraham's relation to God can be said about my relation without relation to *every other (one) as every (bit) other* [*tout autre comme tout autre*], in particular my relation to my neighbor or my loved ones who are as inaccessible to me, as secret and transcendent as Jahweh. Every other (in the sense of each other) is every bit other (absolutely other). From this point of view what *Fear and Trembling* says about the sacrifice of Isaac is the truth. Translated into this extraordinary story, the truth is shown to possess the very structure of what occurs every day. Through its paradox it speaks of the responsibility required at every moment for every man and every woman. At the same time, there is no longer any ethical generality that does not fall prey to the paradox of Abraham.[1] At the instant of every decision and through the relation to *every other (one) as every (bit) other*, every one else asks us at every moment to behave like knights of faith. Perhaps that displaces a certain emphasis of Kierkegaard's discourse: the absolute uniqueness of Jahweh doesn't tolerate analogy; we are not all Abrahams, Isaacs, or Sarahs either. We are not Jahweh. But what seems thus to universalize or disseminate the exception or the extraordinary by imposing a supplementary complication upon ethical generality, that very thing ensures that Kierkegaard's text gains added force. It speaks to us of the paradoxical truth of our responsibility and of our relation to the *gift of death* of each instant. Furthermore, it explains to us its own status, namely its ability to be read by all at the very moment when it is speaking to us of secrets in secret, of illegibility and absolute undecipherability. It stands for Jews, Christians, Muslims, but also for everyone else, for every other in its relation to the wholly other. We no longer know who is called Abraham, and he can no longer even tell us.

Whereas the tragic hero is great, admired, and legendary from generation to generation, Abraham, in remaining faithful to his singular love for the wholly other, is never considered a hero. He doesn't make us shed tears and doesn't inspire admiration: rather stupefied horror, a terror that is also secret. For it is a terror that brings us close to the absolute secret, a secret that we share without sharing it, a secret between someone else, Abraham as the other, and another, God as the other, as wholly other. Abraham himself is in secret, cut off both from man and from God.

But that is perhaps what we share with him. But what does it mean to share a secret? It isn't a matter of knowing what the other knows, for Abraham doesn't know anything. It isn't a matter of sharing his faith, for the latter must remain an initiative of absolute singularity. And moreover, we don't think or speak of Abraham from the point of view of a faith that is sure of itself, any more than did Kierkegaard. Kierkegaard keeps coming back to this, recalling that he doesn't understand Abraham, that he wouldn't be capable of doing what he did. Such an attitude in fact seems the only possible one; and even if it is the most widely shared idea in the world, it seems to be required by this monstrosity of such prodigious proportions. Our faith is not assured, because faith can never be, it must never be a certainty. We share with Abraham what cannot be shared, a secret we know nothing about, neither him nor us. To share a secret is not to know or reveal the secret, it is to share we know not what: nothing that can be determined. What is a secret that is a secret about nothing and a sharing that doesn't share anything?

Such is the secret truth of faith as absolute responsibility and as absolute passion, the "highest passion" as Kierkegaard will say; it is a passion that, sworn to secrecy, cannot be transmitted from generation to generation. In this sense it has no history. This untransmissibility of the highest passion, the normal condition of a faith which is thus bound to secrecy, nevertheless dictates to us the following: we must always start over. A secret can be transmitted, but in transmitting a secret as a secret that remains secret, has one transmitted at all? Does it amount to history, to a story? Yes and no. The epilogue of *Fear and Trembling* repeats, in sentence after sentence, that this highest passion that is faith must be started over by each generation. Each generation must begin again to involve itself in it without counting on the generation before. It thus describes the nonhistory of absolute beginnings which are repeated and the very historicity that presupposes a tradition to be reinvented each step of the way, in this incessant repetition of the absolute beginning.

With *Fear and Trembling*, we hesitate between two generations in the lineage of the so-called religions of the Book: we hesitate at the heart of the Old Testament and of the Jewish religion, but also the heart of a founding event or a key sacrifice for Islam. As for the sacrifice of the son by his father, the son sacrificed by men and finally saved by a God that seemed to have abandoned him or put him to the test, how can we not recognize there the foreshadowing or the analogy of another passion? As a Christian thinker, Kierkegaard ends by reinscribing the secret of Abraham within a space that seems, in its literality at least, to be evangelical. That doesn't necessarily exclude a Judaic or Islamic reading, but it is a certain evangelical text that seems to orient or dominate Kierkegaard's interpretation. That text isn't cited; rather, like the "kings and counselors" of "Bartleby the Scrivener," it is simply suggested, but this time without the quotation marks, thus being clearly brought to the attention of those who know their texts and have been brought up on the reading of the Gospels:

> But there was no one who could understand Abraham. And yet what did he achieve? He remained true to his love. But anyone who loves God needs no tears, no admiration; he forgets the suffering in the love. Indeed, so completely has he forgotten it that there would not be the slightest trace of his suffering left if God himself did not remember it, *for he sees in secret* and recognizes distress and counts the tears and forgets nothing.

Thus, either there is a paradox, that the single individual stands in an absolute relation to the absolute, or Abraham is lost.

Note

1. This is the logic of an objection made by Levinas to Kierkegaard: "For Kierkegaard, ethics signifies the general. For him, the singularity of the self would be lost under a rule valid for all; the generality can neither contain nor express the secret of the self. However, it is not at all certain that the ethical is to be found where he looks for it. Ethics as the conscience of a responsibility towards the other . . . does not lose one in the generality, far from it, it

singularizes, it posits one as a unique individual, as the Self. . . . In evoking Abraham he describes the meeting with God as occurring where subjectivity is raised to the level of the religious, that is to say above ethics. But one can posit the contrary: the attention Abraham pays to the voice that brings him back to the ethical order by forbidding him to carry out the human sacrifice, is the most intense moment of the drama. . . . It is there, in the ethical, that there is an appeal to the uniqueness of the subject and sense is given to life in defiance of death" (Emmanuel Levinas, *Noms propres* (Montpellier: Fata Morgana, 1976), p. 113; my translation, DW). Levinas's criticism doesn't prevent him from admiring in Kierkegaard "something absolutely new" in "European philosophy," "a new modality of the True," "the idea of a persecuted truth" pp. 114–15.

Chapter 39

Marjorie Garber (1944–)
from "Greatness," *Symptoms of Culture* (1998)

"Greatness" as a term is today both an inflated and a deflated currency, shading over into categories of notoriety, transcendence, and some version of the postmodern fifteen minutes of fame. The modern cultural fantasy about heroes and greatness is a symptom of desire and loss: a desire for identifiable and objective standards, and a nostalgia for hierarchy, whether of rank or merit.

Sometimes today "greatness"—so often linked, in our national rhetoric, with "America"—functions rhetorically as pure boiler-plate (the politician's statutory "this is a great country") while at other times it seems to be its own, tautologous, ground of self-evident truth. To give one trivial but telling example: the announcement of the U.S. Post Office's plan to issue an Elvis Presley commemorative stamp—thus officially declaring Elvis dead, as well as transcendent—was greeted with pleasure by a 72-year-old Vermont woman who had written the Postmaster General almost every week since the King's death, pushing for an Elvis commemorative. "I can't imagine anybody more deserving to be put on a stamp than my Elvis," she told the *New York Times*. "I'm not one of those who believes he's not dead. He's dead, unfortunately. He was a great man, a great American. I knew that the first time I laid eyes on him in that black leather suit."[1]

Bear in mind that "great" in English once meant "fat." Or thick, or coarse, or bulky—take your pick. It was an aspect of physical size, not of moral weight. The "Great Bed of Ware" in Elizabethan England was 10 feet 11 inches (3.33 metres) square. It was not the bed of a "great man," but rather a convenient lodging for several itinerant travelers. Nor, when applied to persons, did "greatness" necessarily imply quality or merit. Shakespeare has more than one joke on this: Shakespearean characters called "Pompey" tend in fact to present themselves as targets for comic undercutting because of their pretensions to greatness. An amateur actor in the bumbling "Pageant of the Nine Worthies" presented before the court in *Love's*

In this selection from *Symptoms of Culture*, Garber, William R. Kenan Professor of English at Harvard University, tackles a word central to the canon debate: greatness.

Labour's Lost announces, "I Pompey am, Pompey, surnam'd the Big—" and is quickly corrected by a condescending lord: "The Great." Later he acknowledges, with a modesty that would better become his noble audience, "I made a little fault in 'Great.' "[2] A pimp named Pompey in *Measure for Measure* is surnamed "Bum," and his judicial interrogator quips disgustedly that "your bum is the greatest thing about you; so that, in the beastliest sense, you are Pompey the Great."[3]

But "great" also meant "powerful." A "great man" was a mover-and-shaker, a political force, nobly born and to be reckoned with. "Madness in great ones must not unwatched go," declares the politic Claudius about the dangerously unpredictable Hamlet.[4] Hamlet himself, in a phrase that has attracted scholarly attention for its knotted syntax, seeks to find some common ground between the moral or ethical realm and the demands of power politics, observing admiringly of his rival Fortinbras that "Rightly to be great/Is not to stir without great argument" (i.e., strong motivation), "But greatly to find quarrel in a straw/When honour's at the stake."[5] By this reasoning, one can be wrongly as well as rightly great. Significantly, the Shakespearean locus classicus of the concept of greatness is put into the mouth of a social climber rather than a "great man." "Some are born great, some achieve greatness, and some have greatness thrust upon 'em."[6] The words are in fact already a bromide when the pompous Malvolio finds and reads them: he picks up a letter counterfeited in the handwriting of his noble employer, the Lady Olivia, and imagines that they have direct and unambiguous pertinence to him.

The sense in which "greatness" here means high birth rather than merit is underscored by the counterfeit letter's preceding line, "In my stars I am above thee, but be not afraid of greatness." Yet only the second half of the line is commonly remembered in modern citations of this famous phrase, so that, as with so many other Shakespearean phrases taken out of context, the "some are born great" passage is frequently used by 20th-century pundits to mean pretty much the opposite of what the original context implies. As yet another Shakespearean clown will remark of an impostor pretending to be a courtier, "A great man, I'll warrant; I know by the picking on's teeth."[7]

Tooth-picking, warfare, and "marrying up" may be three infallible marks of greatness, not only for the Renaissance but for our own day. But the cultural role of "greatness" has shifted a little in these democratic days. Jane Austen's Mr. Darcy in *Pride and Prejudice* is regarded as "so great a man"[8] not because he is brilliant or accomplished but because he has inherited a large estate. "Perhaps he may be a little whimsical in his civilities," worries Elizabeth Bennet's city uncle, who is doubtful about whether to trust Darcy's invitation to fish on his estate. "Your great men often are."[9] In the same novel the unlikeable but high-born Lady Catherine de Bourgh and her sickly daughter are regarded with awe by the new knight, Sir William Lucas, who "was stationed in the doorway, in earnest contemplation of the greatness before him, and constantly bowing" whenever they deigned to look his way.[10]

With the separation from the old world of rank and status, when inherited titles conferred "greatness," came a new ideology of the natural aristocrat, the aristocrat of the mind. A sermon preached before the King of England in 1698 raised the question of *Great Men's Advantages and Obligations to Religion*, where "great men" refers to social rank. But *Great Men are God's Gift*—the title of a memorial discourse on the death of Daniel Webster in 1852—offers a different notion of "greatness." The phrase was much on America's mind. When Ralph Waldo Emerson wrote that "It is natural to believe in great men"[11] he meant men like Plato, Goethe, Napoleon, and, indeed, Shakespeare. Their greatness consists, as we will see, in the greatness of their books, or in their presumed exemplary status as models of decorum and achievement.

Emerson's own example in cataloguing "great men" has been followed in the twentieth century with varying success. Today one can consult volumes on *Great Men of Science, Great Men of American Popular Song, Great Men of Derbyshire, Great Men of Michigan, Great Men Who Have Added to the Enlightenment of Mankind Through Endowed Professorships at the University of Chicago, Short Sayings of Great Men*, and, my favorite, an instructive fictionalization for children, *Great Men's Sons: Who They Were, What They Did, and How They Turned Out*. There are also of course, in our enlightened century, lists of great women: *Great Women of the Bible, Of Antiquity, Of Faith, Of Medicine, Of India, and Of the Press*, as well as *Great Women Mystery Writers, Great Women Athletes, and Great Women Superheroes*. On library bookshelves F.R. Leavis's *Great Books and a Liberal Education* jostles for space with *Great Books as Life Teachers: Studies of Character Real and Ideal*, and the *Great Book of Couscous*.

In short, by the latter half of the 20th century "great" as a term has become an empty colloquial affirmation cognate with other debased terms like "fantastic," "terrific," and "awesome," which have likewise lost their original specificity in fantasy, terror, and awe. "Baby, you're the greatest," declares Jackie Gleason's character Ralph Kramden, enfolding his long-suffering wife in his arms at the end of practically every episode of "The Honeymooners." Alice's "greatness" consists in tolerating her husband's foibles. But "great" has also become a category of popular celebrity, a headline and a cultural diagnosis. "I am the greatest," announced pugilist and poet Muhammad Ali after a boxing match, crowning himself for our age as definitively as did Napoleon for his. [. . .]

Bart Giamatti was the founder of Yale's great books course on the Western tradition from Homer to Brecht and the author of a study of the earthly paradise in the Renaissance epic. He was a premier and eloquent defender of the concept of "humanism" in literary studies, and an explicit champion both of the traditional literary canon and—as these quotations will have demonstrated—the capacity of "great literature" to inform and shape "human life."

The ideology of "greatness"—an ideology that claims, precisely, to transcend ideological concerns and to locate the timeless and enduring, the fit candidates, though few, for a Hall of Fame, whether in sports or in arts and letters—is, in fact, frequently secured with reference to a philology of origins. Yet a specific examination of the relationship of philology to the politics of mimesis yields, as well, some interesting complications.

Consider the case of Erich Auerbach's landmark study, *Mimesis: The Representation of Reality in Western Literature*, a study that takes as its starting point a sustained meditation on the concept of Homer and "home." "Readers of the *Odyssey*," the book begins, without preamble, "will remember the . . . touching scene in book 19, when Odysseus has at last come home." But where is "home" for Erich Auerbach?

A distinguished professor of romance philology who concluded his career as Sterling Professor at Yale, Auerbach was a Jewish refugee from Nazi persecution who was born in Berlin. Discharged from his position at Marburg University by the Nazi government, he emigrated to Turkey, where he taught at the Turkish State University, until his move to the United States in 1947. His celebrated book, *Mimesis*, was written in Istanbul in the period between May 1942 and April 1945. It was published in Berne, Switzerland, in 1946, and translated into English for the Bolligen Series, published by Princeton University Press, in 1953. The politics of *Mimesis* were thus, at least in part, a politics of exile—and a politics of *nostos* and nostalgia. "Home" was the Western tradition, and the *translatio studii*.

In his Epilogue to *Mimesis*, Auerbach is at pains to point out that "the book was written during the war and at Istanbul, where the libraries are not well equipped for European studies." Thus, he explains, his book necessarily lacks footnotes, and may also assert something that "modern research has disproved or modified."[12] Yet, he remarks, "it is quite possible that the book owes its existence to just this lack of a rich and specialized library. If it had been possible for me to acquaint myself with all the work that has been done on so many subjects, I might never have reached the point of writing."

This last sentiment—that reading criticism and scholarship may sometimes impede the creative process—will doubtless be familiar to all graduate students embarking on the writing of a Ph.D. thesis. Yet, as we will see in a moment, it is also strikingly similar to a certain tactical enhancement of "great literature" and "great-ness" in general through the evacuation of historical context. I want to suggest that the absence of a critical apparatus in a book on the evolution of the great tradition in Western letters is something more, or less, than an accident of historical contingency. Auerbach's research opportunities were limited by his circumstances; his choice of topic was not. The scholar who would later write that "our philological home is the earth; it can no longer be the nation,"[13] sustained his argument through a selection of texts that he alleges were "chosen at random, on the basis of accidental acquain-tance and personal preference."[14] Out of this came a book which claimed, and has been taken, to set forth "the representation of reality in Western literature."

Edward Said has noted that Auerbach's alienation and "displacement" in Istanbul offers a good example of the way in which *not* being "at home" or "in place" with respect to a culture and its policing authority can enable, as well as impede, literary and cultural analysis.[15] But what for Erich Auerbach was a wartime necessity became, for a group of U.S.-based scholars in the same period, a democratic principle of pedagogy.

Let us now move, profiting from Giamatti's and Auerbach's speculations on home and Homer, to a consideration of the specific kind of "greatness" embodied in the concept of the Great Books, the cultural heroes of our time for pundits from Allan Bloom to Harold Bloom. To study "Greats" at Oxford and Cambridge is to read the ancient classics; for this generation of Americans, however, the greats have been updated—slightly.

In search of some wisdom on this topic—of what makes the great books great— I decided to consult the experts: specifically, the editors of the Encyclopedia Britannica Great Books Series, more accurately described as the *Great Books of the Western World*, first collected and published in 1952 in a Founders' Edition under the editorship of Robert Maynard Hutchins and Mortimer J. Adler.

Hutchins's prefatory volume, entitled *The Great Conversation*, makes it clear that, at least in 1952, "There [was] not much doubt about which [were] the most important voices in the Great Conversation."[16] "The discussions of the Board revealed few differences of opinion about the overwhelming majority of the books in the list," which went from Homer to Freud. "The set" wrote Hutchins, "is almost self-selected, in the sense that one book leads to another, amplifying, modifying, or contradicting it."[17] *The Great Conversation*, as Adler and his board conceived it, at the time of the election of President Eisenhower, was, it is not surprising to note, exclusively considered as taking place between European and American men, men who were no long living at the time they were enshrined in the hard covers of "greatness." The explicit politics of the edition was, nonetheless, aggressively democratic: no "scholarly apparatus" was included in the set, since the editors believed that "Great books contain their own aids to reading; that is one reason why they are great. Since we hold"—writes Hutchins—"that these works are intelligible to the ordinary man, we see no reason to interpose ourselves or anybody else between the author and the reader."

The assumption here was one of enlightened "objectivity": given a handsomely produced, uniformly bound set of volumes vetted for "greatness," the reader—unreflectively gendered male, an inevitable commonplace of the times—would be able, the editors thought, with the help of a curious kind of two-volume outline called the *Syntopticon*, "which began as an index and then turned into a means of helping the reader find paths through the books," to "find what great men have had to say about the greatest issues and what is being said about these issues today." A chief obstacle to this process, apparently, was what Hutchins called, in a phrase later to be echoed by the likes of Bill Bennett and Lynne Cheney, "the vicious specialization of scholarship." With the help of this completely objective and apolitical edition "the ordinary reader," we are assured, will be able to break through the obfuscating barrier of "philology, metaphysics, and history," the "cult of scholarship" that forms a barrier between him and the great authors. For example, despite the huge "apparatus" of commentary surrounding *The Divine Comedy* (an apparatus the "ordinary reader" has "heard of" but "never used"), the purchaser and reader of the Great Books will be "surprised to find that he understands Dante without it."

The end-papers of the *Great Books of the Western World*, uniform throughout the 54 volumes, are themselves a treasure trove of information. The first pair of end-papers, in the front of each volume, lists the product being sold, and bought: "The Great Books of the Western World" and the three introductory volumes that frame them, *The Great Conversation, The Great Ideas I*, and *The Great Ideas II*. But what *are* the Great Ideas? In case we are in any doubt, the editors conveniently list them for us in the second set of end-papers, the ones that close the book. Remember that this is an objective, non-political list, assembled by editors who "believe that the reduction of the citizen to an object of propaganda, private and public, is one of the greatest dangers to democracy,"[18] and that "until lately" (again, 1952) "there never was very much doubt in anybody's mind about which the masterpieces were. They were the books that had endured and that the common voice of mankind called the finest creations, in writing, of the Western mind."

The Great Ideas, the preoccupations of the great authors who wrote the Great Books and participate in the ongoing Great Conversation in which the ordinary citizen is encouraged to think he should also take part—these Great Ideas are listed in the second set of end-papers in alphabetical order, from Angel to World. I will restrict myself to two comments about them, one of which will be quite self-evident, the other, perhaps, less so.

You will notice that in the course of this list, which includes ideas like Citizen, Constitution, Courage, Democracy, and Education, there appear, occasionally, words with a more disquieting ring: "evil," "pain," "contingency," "other," and the great cornerstone or individualism, and therefore of humanist hero-making, "death."

But all of these words are tamed and contained—and here we should indeed think of Cold War containment theory—by being presented as part of a dyad. Angel, Animal, and Aristocracy stand alone; but Good and Evil, Life and Death, Necessity and Contingency, One and Many, Pleasure and Pain, Same and Other, Virtue and Vice, Universal and Particular are tethered together like the horses of the charioteer. It is perhaps too much to say that cutting free each of the dark twins in this dyad would produce an entirely different profile of "great ideas" and great books; but it is *not* too much to say that the last forty-odd years of literary and cultural theory have explored, precisely, the dangerous complacencies of these binarisms, the politics of their masquerade as opposites rather than figures for one another, the master–slave relation that informs them.

My second observation about "The Great Ideas" is one that addresses the question of packaging. On one page of this list the ideas run alphabetically from Angel to Mathematics, and on the other they run from Matter to World. In each case the list fills up the entire page, with one decorative squiggle at the beginning, and one at the end. Angel to Mathematics, Matter to World. It is of some small interest, however, that the two series volumes that contain the Great Ideas, the *Syntopticon* Volumes I and II, choose slightly different moments to begin and end. Volume I ends not with Mathematics but with Love; Volume II thus starts with Man.

Volume I: Angel to Love; Volume II: Man to World. You'll have to admit this gives a somewhat different spin to the alphabetical iconography of greatness. Matter and Mathematics are worthy enough categories in themselves, but seem somehow so material, lacking the humanist grandeur of Love and Man. Nor is this an accident of division based upon the length of the individual articles. Angel to Love, Chapters 1 to 50, the contents of volume one, covers 750 pages; Chapters 51 to 102, Man to World, in the second volume, covers 809 pages. It seems reasonable to think that an editorial decision has been taken—and a perfectly appropriate one, given the presumptions of the Great Books project. The titles of the prefatory volumes will be an icon of the whole.

The very trope usually ascribed to deconstructors, and to a deconstructive playfulness, the trope of chiasmus, is here quietly employed to anchor the ideology of the series; the relationship of "Man" to "Love" (*not* the relationship of "Matter" to "Mathematics") will serve as a fulcrum, a micro-relation mediating the macro-relation of "Angel" to "World." Readers of Tillyard's *Elizabethan World Picture* and Lovejoy's *Great Chain of Being* will here recognize a familiar structure. But what I find so scandalous about this whole enterprise is its blithe claim that the absence of a scholarly apparatus is *preferable* because, apparently, *non-ideological*.

I quote again from Hutchins's Preface: "We believe that the reduction of the citizen to an object of propaganda, private and public, is one of the greatest dangers to democracy. . . . The reiteration of slogans, the distortion of the news, the great storm of propaganda that beats upon the citizen twenty-four hours a day all his life long mean either that democracy must fall a prey to the loudest and most persistent propagandists or that the people must save themselves by strengthening their minds so that they can appraise the issues for themselves."[19] And again, "The Advisory Board recommended that no scholarly apparatus should be included in the set. No 'introductions' giving the Editors' views of the authors should appear. The books should speak for themselves, and the reader should decide for himself."[20] Angel to Love; Man to World. [. . .]

For me the question is really not one of elegiac loss, but of the political uses of nostalgia. Are great books most in need of being called "great" when their link with the culture is most tenuous? Has political life as we commonly understand it—from Wilbur to Nixon to Reagan and Bush and Clinton—become an arena in which what is *imitated* is mimesis? (Bush pretending that he buys his socks at JC Penneys in an attempt to stimulate the economy? Reagan "remembering" wartime events that he saw, or acted, in Hollywood B-pictures? Bill Clinton gaining political momentum from a photograph of him as a young boy shaking the hand of JFK, as if the ghost of the slain president were literally "electing" or choosing his successor?) Is "greatness" largely or entirely an effect—and if so, what kind of effect? A stage effect, a psychoanalytic effect—or an effect of nostalgia? It's not something extra, but something missing.

What is at issue is overcompensation, and an anxious fantasy of wholeness. As with Oz the Great and Terrible; as with Genet's Chief of Police and his fantasy of the giant phallus. Mortimer Adler, updating his list of "Great Books, Past and Present" in 1988 lists 36 new white male authors who published between 1900 and 1945, and an additional 18 authors—also all male and all white—for the period 1945 to the present. But he is worried about his capacity to see clearly: "Could it be that my nineteenth-century mentality . . . blind[s] me to the merit of work that represents the artistic and intellectual culture of the last forty or so years?"[21] Adler's concern is that he may fail to identify some of the great works; but he is entirely convinced, not only that they are there to be found, but that "greatness" can be pinpointed, however tautologous the test. "If we say that a good book is a book that is worth reading carefully once, and that a better book than that—a great book—is one that is worth reading carefully a second or third time, then the greatest books are those worth reading over and over again—endlessly." And, he implies, we can make a list.

Wilbur, Oz, the Great Books, the Great Tradition. Greatness is an effect of decon-textualization, of the decontextualizing of the sign—and of a fantasy of control, a fantasy of the *sujet supposé savoir*, of a powerful agency, divine or other. "If you build it, he will come." "A miracle has happened and a sign has occurred here on earth, right on our farm, and we have no ordinary pig." Someone knows; someone—someone *else* is in control. The political logic of this is as disturbing as its psychology. It's a lesson that has not been lost on contemporary political "spinmeisters" from Reagan's Peggy Noonan to Bush's Lee Atwater to Clinton's technocratic masters of the Web.

"Good" books, like "competent" politicians, are in our inflated culture somehow not good enough. From the canon debate to the political arena, "greatness" has become an increasingly problematic standard. If we have greatness thrust upon us in either sphere, we should recognize it as an ideological category, a redundancy effect, a "recognition factor," as the pundits say. It seems clear that anxieties about greatness in literature are closely tied to anxieties about national, political, and cultural greatness, and that the more anxious the government, the more pressure is placed upon the humanities to textualize and naturalize the category of the "great." This is no reason to discard such a category entirely, even if it were possible to do so. But it is a good reason to be wary, and to pay some attention to that man behind the curtain—or, if anyone tries to sell you one, to be cautious about lionizing "some pig"—however terrific, radiant, and humble—in a poke.

Notes

1. B. Drummond Ayres, Jr. "Millions of Elvis Sightings Certain in '93." *The New York Times* (January 11,1992) p. 6.
2. Shakespeare, *Love's Labour's Lost*, 5.2.545–546, 55. *The Arden Shakespeare*, ed. Richard David (London: Methuen, 1968).
3. Shakespeare, *Measure for Measure* 3.1.214–216. *The Arden Shakespeare*, ed. J.W. Lever (London and New York: Methuen, 1987).
4. Shakespeare, *Hamlet* 3.1.190. *The Arden Shakespeare*, ed. Harold Jenkins (London and New York: Routledge, 1982).
5. *Hamlet* 4.4.52–56.
6. Shakespeare, *Twelfth Night* 2.5.145–146. *The Arden Shakespeare*, ed. J.M. Lothian and T.W. Craik (London: Methuen, 1975).
7. Shakespeare, *The Winter's Tale* 4.4.740–742 (The Clown is here describing the rogue Autolycus dressed in Prince Florizel's clothes.)
8. Jane Austen, *Pride and Prejudice* (London and New York: Penguin Books, 1985; orig. pub. 1813), p. 96. The phrase is Elizabeth Bennet's; she is wondering, early in the novel, "how to suppose that she could be an object of admiration to so great a man." p. 96.
9. Ibid., p. 278.
10. Ibid., p. 194.
11. Ralph Waldo Emerson, "Uses of Great Men," in *Representative Men* (Boston: Houghton Mifflin, 1903).
12. Erich Auerbach, *Mimesis: The Representation of Reality in Western Literature*, trans. Willard Trask, (Garden City, NY: Doubleday Anchor, 1957), p. 492.
13. Erich Auerbach, "Philology and Weltliteratur," trans. M. and E.W. Said, *Centennial Review*, 13 (Winter 1969), p. 17. "Culture often has to do with an aggressive sense of nation, home, community, and belonging." Edward Said, *The World, the Text, and the Critic* (Cambridge, MA: Harvard University Press, 1983), p. 12.
14. Auerbach, *Mimesis*, p. 491.
15. Said, *The World, the Text, and the Critic*, p. 8.
16. Robert Maynard Hutchins, *The Great Conversation*, in *Great Books of the Western World*, ed. Robert Maynard Hutchins and Mortimer J. Adler, 54 volumes (Chicago: Encyclopedia Britannica), p. xvii.
17. Ibid., p. xvi.
18. Ibid., p. xiii.
19. Ibid.
20. Ibid., p. xxv.
21. Mortimer J. Adler, "Reforming Education: The opening of the American Mind," ed. Geraldine Van Doren (New York: Collier Books, 1988), p. 350.

Chapter 40

Richard Rorty (1931–)
from "On The Inspirational Value of Great Works of Literature," *Achieving Our Country: Leftist Thought in Twentieth-Century America* (1998)

The self-protective project described in this familiar Horatian tag is exemplified by one strain of thought in Fredric Jameson's influential *Postmodernism, or The Cultural Logic of Late Capitalism*. In one of the most depressing passages of that profoundly antiromantic book, Jameson says that "the end of the bourgeois ego, or monad, . . . means . . . the end . . . of style, in the sense of the unique and the personal, the end of the distinctive individual brush stroke."[1] Later he says that

> if the poststructuralist motif of the "death of the subject" means anything socially, it signals the end of the entrepreneurial and inner-directed individualism with its "charisma" and its accompanying categorial panoply of quaint romantic values such as that of the "genius" . . . Our social order is richer in information and more literate . . . This new order no longer needs prophets and seers of the high modernist and charismatic type, whether among its cultural products or its politicians. Such figures no longer hold any charm or magic for the subjects of a corporate, collectivized, post-individualistic age; in that case, goodbye to them without regret, as Brecht might have put it: woe to the country that needs geniuses, prophets, Great Writers, or demiurges![2]

Adoption of this line of thought produces what I shall call "knowingness." Knowingness is a state of soul which prevents shudders of awe. It makes one immune to romantic enthusiasm.

This state of soul is found in the teachers of literature in American colleges and universities who belong to what Harold Bloom calls the "School of Resentment."

In this essay, Richard Rorty, Professor of Comparative Literature at Stanford University, worried that the rise of cultural studies in literature departments might be equivalent to the earlier rise of logical positivism in philosophy departments, argues for the "inspirational value of literature," and its ability to "produce hope."

These people have learned from Jameson and others that they can no longer enjoy "the luxury of the old-fashioned ideological critique, the indignant moral denunciation of the other."[3] They have also learned that hero-worship is a sign of weakness, and a temptation to elitism. So they substitute Stoic endurance for both righteous anger and social hope. They substitute knowing theorization for awe, and resentment over the failures of the past for visions of a better future.

Although I prefer "knowingness" to Bloom's word "resentment," my view of these substitutions is pretty much the same as his. Bloom thinks that many rising young teachers of literature can ridicule anything but can hope for nothing, can explain everything but can idolize nothing. Bloom sees them as converting the study of literature into what he calls "one more dismal social science"—and thereby turning departments of literature into isolated academic backwaters. American sociology departments, which started out as movements for social reform, ended up training students to clothe statistics in jargon. If literature departments turn into departments of cultural studies, Bloom fears, they will start off hoping to do some badly needed political work, but will end up training their students to clothe resentment in jargon.

I think it is important to distinguish know-nothing criticisms of the contemporary American academy—the sort of thing you get from columnists like George Will and Jonathan Yardley, and politicians like William Bennett and Lynne Cheney—from the criticisms currently being offered by such insiders as Bloom and Christopher Ricks. The first set of critics believe everything they read in scandal-mongering books by Dinesh D'Souza, David Lehman, and others. They do not read philosophy, but simply search out titles and sentences to which they can react with indignation. Much of their work belongs to the current conservative attempt to discredit the universities—which itself is part of a larger attempt to discredit all critics of the cynical oligarchy that has bought up the Republican Party. The insiders' criticism, on the other hand, has nothing to do with national politics. It comes from people who are careful readers, and whose loathing for the oligarchy is as great as Jameson's own.

I myself am neither a conservative nor an insider. Because my own disciplinary matrix is philosophy, I cannot entirely trust my sense of what is going on in literature departments. So I am never entirely sure whether Bloom's gloomy predictions are merely peevish, or whether he is more far-sighted than those who dismiss him as a petulant eccentric. But in the course of hanging around literature departments over the past decade or so, I have acquired some suspicions that parallel his.

The main reason I am prey to such suspicions is that I have watched, in the course of my lifetime, similarly gloomy predictions come true in my own discipline. Philosophers of my generation learned that an academic discipline can become almost unrecognizably different in a half-century—different, above all, in the sort of talents that get you tenure. A discipline can quite quickly start attracting a new sort of person, while becoming inhospitable to the kind of person it used to welcome.

Bloom is to Jameson as A. N. Whitehead was to A. J. Ayer in the 1930s. Whitehead stood for charisma, genius, romance, and Wordsworth. Like Bloom, he agreed with Goethe that the ability to shudder with awe is the best feature of human beings. Ayer, by contrast, stood for logic, debunking, and knowingness. He wanted philosophy to be a matter of scientific teamwork, rather than of imaginative breakthroughs by heroic figures. He saw theology, metaphysics, and literature as devoid of what he called "cognitive significance," and Whitehead as a good logician who had been ruined by poetry. Ayer regarded shudders of awe as neurotic symptoms. He helped create the philosophical tone which Iris Murdoch criticized in her celebrated essay "Against Dryness."

In the space of two generations, Ayer and dryness won out over Whitehead and romance. Philosophy in the English-speaking world became "analytic," antimetaphysical, unromantic, and highly professional. Analytic philosophy still attracts first-rate minds, but most of these minds are busy solving problems which no nonphilosopher recognizes as problems: problems which hook up with nothing outside the discipline.[4] So what goes on in anglophone philosophy departments has become largely invisible to the rest of the academy, and thus to the culture as a whole. This may be the fate that awaits literature departments.

Analytic philosophy is not exactly one more dismal social science, but its desire to be dryly scientific, and thereby to differentiate itself from the sloppy thinking it believes to be prevalent in literature departments, has made it stiff, awkward, and isolated. Those who admire this kind of philosophy often claim that philosophy professors are not only a lot drier but also a lot smarter nowadays than in the past. I do not think this is so. I think they are only a little meaner. Philosophy is now more adversarial and argumentative than it used to be, but I do not think that it is pursued at a higher intellectual level.

As philosophy became analytic, the reading habits of aspiring graduate students changed in a way that parallels recent changes in the habits of graduate students of literature. Fewer old books were read, and more recent articles. As early as the 1950s, philosophy students like myself who had, as undergraduates, been attracted to philosophy as a result of falling in love with Plato or Hegel or Whitehead, were dutifully writing Ph.D. dissertations on such Ayer-like topics as the proper analysis of subjunctive conditional sentences. This was, to be sure, an interesting problem. But it was clear to me that if I did not write on some such respectably analytic problem I would not get a very good job. Like the rest of my generation of philosophy Ph.D.'s, I was not exactly cynical, but I did know on which my side my bread was likely to be buttered. I am told, though I cannot vouch for the fact, that similar motives are often at work when today's graduate students of literature choose dissertation topics.

Nowadays, when analytic philosophers are asked to explain their cultural role and the value of their discipline, they typically fall back on the claim that the study of philosophy helps one see through pretentious, fuzzy thinking. So it does. The intellectual moves which the study of analytic philosophy trained me to make have proved very useful. Whenever, for example, I hear such words as "problematize" and "theorize," I reach for my analytic philosophy.

Still, prior to the rise of analytic philosophy, ridiculing pretentious fuzziness was only *one* of the things that philosophy professors did. Only some philosophers made this their specialty: Hobbes, Hume, and Bentham, for example, but not Spinoza, Hegel, T. H. Green, or Dewey. In the old days, there was another kind of philosopher— the romantic kind. This is the kind we do not get any more, at least in the English-speaking world. Undergraduates who want to grow up to be the next Hegel, Nietzsche, or Whitehead are not encouraged to go on for graduate work in anglophone philosophy departments. This is why my discipline has undergone both a paradigm shift and a personality change. Romance, genius, charisma, individual brush strokes, prophets, and demiurges have been out of style in anglophone philosophy for several generations. I doubt that they will ever come back into fashion, just as I doubt that

American sociology departments will ever again be the centers of social activism they were in the early decades of the century.

So much for my analogy between the rise of cultural studies within English departments and of logical positivism within philosophy departments. I have no doubt that cultural studies will be as old hat thirty years from now as was logical positivism thirty years after its triumph. But the victory of logical positivism had irreversible effects on my discipline—it deprived it of romance and inspiration, and left only professional competence and intellectual sophistication. Familiarity with these effects makes me fear that Bloom may be right when he predicts that the victory of cultural studies would have irreversibly bad effects upon the study of literature.

To make clearer the bad effects I have in mind, let me explain what I mean by the term "inspirational value." I can do so most easily by citing an essay by the novelist Dorothy Allison: "Believing in Literature." There she describes what she calls her "atheist's religion"—a religion shaped, she says, by "literature" and by "her own dream of writing." Toward the close of this essay, she writes:

> There is a place where we are always alone with our own mortality, where we must simply have something greater than ourselves to hold onto—God or history or politics or literature or a belief in the healing power of love, or even righteous anger. Sometimes I think they are all the same. A reason to believe, a way to take the world by the throat and insist that there is more to this life than we have ever imagined.[5]

When I attribute inspirational value to works of literature, I mean that these works make people think there is more to this life than they ever imagined. This sort of effect is more often produced by Hegel or Marx than by Locke or Hume, Whitehead than Ayer, Wordsworth than Housman, Rilke than Brecht, Derrida than de Man, Bloom than Jameson.

Inspirational value is typically *not* produced by the operations of a method, a science, a discipline, or a profession. It is produced by the individual brush strokes of unprofessional prophets and demiurges. You cannot, for example, find inspirational value in a text at the same time that you are viewing it as the product of a mechanism of cultural production. To view a work in this way gives understanding but not hope, knowledge but not self-transformation. For knowledge is a matter of putting a work in a familiar context—relating it to things already known.

If it is to have inspirational value, a work must be allowed to recontextualize much of what you previously thought you knew; it cannot, at least at first, be itself recontextualized by what you already believe. Just as you cannot be swept off your feet by another human being at the same time that you recognize him or her as a good specimen of a certain type, so you cannot simultaneously be inspired by a work and be knowing about it. Later on—when first love has been replaced by marriage—you may acquire the ability to be both at once. But the really good marriages, the inspired marriages, are those which began in wild, unreflective infatuation.

A humanistic discipline is in good shape only when it produces both inspiring works and works which contextualize, and thereby deromanticize and debunk, those inspiring works. So I think philosophy, as an academic discipline, was in better shape when it had room for admirers of Whitehead as well as admirers of Ayer. I think that literature departments were in better shape when people of Bloom's and Allison's sort had a better chance than, I am told, they now have of being allowed to spend their teaching lives reiterating their idiosyncratic enthusiasms for their favorite prophets and demiurges. People of that sort are the ones Jameson thinks outdated, because they are still preoccupied with what he calls the "bourgeois ego." They are people whose motto is Wordsworth's "What we have loved/Others will love, and we will teach them how." This kind of teaching is different from the kind that produces knowingness, or technique, or professionalism.

Of course, if such connoisseurs of charisma were the *only* sort of teacher available, students would be short-changed. But they will also be short-changed if the only sort of teacher available is the knowing, debunking, *nil admirari* kind. We shall always need people in every discipline whose talents suit them for understanding rather than for hope, for placing a text in a context rather than celebrating its originality, and for detecting nonsense rather than producing it. But the natural tendency of professionalization and academicization is to favor a talent for analysis and problem-solving over imagination, to replace enthusiasm with dry, sardonic knowingness. The dismalness of a lot of social science, and of a lot of analytic philosophy, is evidence of what happens when this replacement is complete.

Within the academy, the humanities have been a refuge for enthusiasts. If there is no longer a place for them within either philosophy or literature departments, it is not clear where they will find shelter in the future. People like Bloom and Allison—people who began devouring books as soon as they learned to read, whose lives were saved by books—may get frozen out of those departments. If they are, the study of the humanities will continue to produce knowledge, but it may no longer produce hope. Humanistic education may become what it was in Oxbridge before the reforms of the 1870s: merely a turnstile for admission to the overclass.

I hope that I have made clear what I mean by "inspirational value." Now I should like to say something about the term "great works of literature." This term is often thought to be obsolete, because Platonism is obsolete. By "Platonism" I mean the idea that great works of literature all, in the end, say the same thing—and are great precisely because they do so. They inculcate the same eternal "humanistic" values. They remind us of the same immutable features of human experience. Platonism, in this sense, conflates inspiration and knowledge by saying that only the eternal inspires—that the source of greatness has always been out there, just behind the veil of appearances, and has been described many times before. The best a prophet or a demiurge can hope for is to say once again what has often been said, but to say it in a different way, to suit a different audience.

I agree that these Platonist assumptions are best discarded. But doing so should not lead us to discard the hope shared by Allison, Bloom, and Matthew Arnold—the hope for a religion of literature, in which works of the secular imagination replace Scripture as the principal source of inspiration and hope for each new generation. We should cheerfully admit that canons are temporary, and touchstones replaceable. But this should not lead us to discard the idea of greatness. We should see great works of literature as great because they have inspired many readers, not as having inspired many readers because they are great.

This difference may seem a quibble, but it is the whole difference between pragmatist functionalism and Platonist essentialism. For a functionalist, it is no surprise that some putatively great works leave some readers cold; functionalists do not expect the same key to open every heart. For functionalists like Bloom, the main reason for drawing up a literary canon, "ordering a lifetime's reading," is to be able to offer suggestions to the young about where they might find excitement and hope. Whereas essentialists take canonical status as indicating the presence of a link to eternal truth, and lack of interest in a canonical work as a moral flaw, functionalists take canonical status to be as changeable as the historical and personal situations of readers. Essentialist critics like de Man think that philosophy tells them how to read nonphilosophy. Functionalist critics like M. H. Abrams and Bloom read philosophical treatises in the same way they read poems—in search of excitement and hope. [. . .]

Unfortunately, in contemporary American academic culture, it is commonly assumed that once you have seen through Plato, essentialism, and eternal truth you will naturally turn to Marx. The attempt to take the world by the throat is still, in the minds of Jameson and his admirers, associated with Marxism. This association seems to me merely quaint, as does Jameson's use of the term "late capitalism"—a term which equivocates nicely between economic history and millenarian hope. The main thing contemporary academic Marxists inherit from Marx and Engels is the conviction that the quest for the cooperative commonwealth should be scientific rather than utopian, knowing rather than romantic.

This conviction seems to me entirely mistaken. I take Foucault's refusal to indulge in utopian thinking not as sagacity but as a result of his unfortunate inability to believe in the possibility of human happiness, and his consequent inability to think of beauty as the promise of happiness. Attempts to imitate Foucault make it hard for his followers to take poets like Blake or Whitman seriously. So it is hard for these followers to take seriously people inspired by such poets—people like Jean Jaurès, Eugene Debs, Vaclav Havel, and Bill Bradley. The Foucauldian academic Left in contemporary America is exactly the sort of Left that the oligarchy dreams of: a Left whose members are so busy unmasking the present that they have no time to discuss what laws need to be passed in order to create a better future.

Emerson famously distinguished between the party of memory and the party of hope. Bloom has remarked that this distinction is now, in its application to American academic politics, out of date: the party of memory, he says, is the party of hope. His point is that, among students of literature, it is only those who agree with Hölderlin that "what abides was founded by poets" who are still capable of social hope. I suspect he is right at least to this extent: it is only those who still read for inspiration who are likely to be of much use in building a cooperative commonwealth. So I do not see the disagreement between Jamesonians and Bloomians as a disagreement between those who take politics seriously and those who do not. Instead, I see it as between people taking refuge in self-protective knowingness about the present and romantic utopians trying to imagine a better future.

Notes

1. Fredric Jameson, *Postmodernism, or The Cultural Logic of Late Capitalism* (Durham, NC: Duke University Press, 1991), p. 15.

2. Ibid., p. 306.
3. Ibid., p. 46.
4. The best of these minds, however, are more inclined to dissolve problems than to solve them. They challenge the presuppositions of the problems with which the profession is currently occupied. This is what Ludwig Wittgenstein did in his *Philosophical Investigations*, and similar challenges are found in the work of the contemporary analytic philosophers I most admire—for example, Annette Baier, Donald Davidson, and Daniel Dennett. Such innovators are always viewed with some suspicion: those brought up on the old problems would like to think that their clever solutions to those problems are permanent contributions to human knowledge. Forty-odd years after its publication, *Philosophical Investigations* still makes many philosophers nervous. They view Wittgenstein as a spoilsport.
5. Dorothy Allison, "Believing in Literature," in Allison, *Skin: Talking about Sex, Class, and Literature* (Ithaca, NY: Firebrand Books, 1994), p. 181.

Chapter 41

Robert Scholes (1929–)
from "A Flock of Cultures: A Trivial Proposal," *The Rise and Fall of English* (1998)

> *The pigs were ranged on one side, the dogs on another, and then from a third a flock of cultures crept up from time to time.*
>
> —From the French painters first attempt to set William Carlos
> Williams's *The Great American Novel*

Our problem as I see it—that is, the problem of college instruction in general and any humanistic core for such studies in particular—can be put in the form of two questions. It is my hope that those concerned about education, whether they are on the "right" or the "left," might agree that it is reasonable to see our problems in this manner. One question is how we can put students in touch with a usable cultural past. The other is how we can help students attain an active relationship with their cultural present. These two questions are intimately related, of course. We cannot answer one without taking a position on the other. Therefore, I shall try to consider them both, though my proposal is concerned mainly with the second. To approach the matter of a usable cultural past, I shall have to begin with questions of canonicity. This may at first seem like just another assault on Western Civ and the Great Books, but I ask for your patience. This is a different kind of critique, I believe, and it will have a different outcome than is usual. To begin with, however, we will need to have a clear understanding of the cultural role of canons. Let us begin at the beginning.

In ancient Greek we find the two words from which the modern English word *canon* (in its two spellings, *canon* and *cannon*) has descended: *kanna*: reed; and *kanōn*: straight rod, bar, ruler, reed (of a wind organ), rule, standard, model, severe critic,

Robert Scholes is Research Professor in the Department of Modern Culture and Media at Brown University. In this essay, Scholes argues that Western Civilization and the Great Books lack the coherency needed for a sound educational program; Scholes proposes a "new trivium" in response.

metrical scheme, astrological table, limit, boundary, assessment for taxation (Liddell and Scott). Like *canon*, our word *cane* is also clearly a descendant of the ancient *kanna*, but its history has been simpler and more straightforward than that of its cognate. The second of the two Greek words, on the other hand, has from ancient times been the repository of a complex set of meanings, mainly acquired by metaphorical extensions of the properties of canes, which are a set of hollow or tubular grasses, some of which are regularly jointed (like bamboo), and some of which have flat outside coverings. The tubular channel characteristic of reeds or canes leads to the associations of the word *canon* with functions that involve forcing liquids or gases through a channel or pipe, while their regularity and relative rigidity lead toward those meanings that involve measuring and controlling (ruling—in both senses of that word). And it is likely that the ready applicability of canes as a weapon of punishment (as in our verb *to cane*, or beat with a stick) supported those dimensions of the meaning of *kanōn* that connote severity and the imposition of power.

In Latin we find the same sort of meanings for the word *canon* as were attached to the Greek *kanōn*, with two significant additions, both coming in later Latin. These two additions are due to historical developments that generated a need for new terms. On the one hand, the rise of the Roman Catholic Church as an institution required a Latin term that could distinguish the accepted or sacred writings from all others, so that "works admitted by the rule or canon" came themselves to be called canonical or, in short, the Canon. In this connection we also find a new verb, *canonizare*, to canonize. On the other hand, with the importation of gunpowder and the development of artillery, the tubular signification of the word led to its becoming the name (in late Latin) for large guns (Lewis and Short). A common theme in these extensions is power.

For our purposes, what is significant in this is the way that *canon* in Latin also combined the meanings of rule or law with the designation of a body of received texts. In its Christian signification, however, *canon* came to mean not only a body of received texts, essentially fixed by institutional fiat, it also came to mean a body of individuals raised to heaven by the perfection of their lives. In this latter signification, the canon of saints was not closed but open, with new saints always admissible by approved institutional procedures. This distinction is important because in current literary disputes over the canon, both models are sometimes invoked, one on behalf of a relatively fixed canon and the other on behalf of a relatively open one. In any case, our current thinking about canonicity cannot afford to ignore the grounding of the modern term in a history explicitly influenced by Christian institutions. This influence is apparent, for instance, in the way Thomas Carlyle described the heroic figures of literature in 1840:

> Nay here in these ages, such as they are, have we not two mere Poets, if not deified, yet we may say beatified? Shakespeare and Dante are Saints of Poetry: really, if we think of it, *canonized*, so that it is impiety to meddle with them. The unguided instinct of the world, working across all these perverse impediments, has arrived at such result. Dante and Shakespeare are a peculiar Two. They dwell apart, in a kind of royal solitude; none equal, none second to them: in the general feeling of the world, a certain transcendentalism, a glory as of complete perfection, invests these two. They *are* canonized, though no Pope or Cardinals took a hand in doing it! (Carlyle, 107, emphasis in original)

At this point we must backtrack a bit to note that *canon* also has a more purely secular pedigree going back to Alexandrian Greek, in which the word kanōn was used by rhetoricians to refer to a body of superior texts: *hoi kanōnes* "were the works which the Alexandrian critics considered as the most perfect models of style and composition, equivalent to our modern term 'The Classics' " (Donnegan). Exactly how the interplay between the rhetorical and the religious uses of the notion of *canon* functioned two millennia ago is a matter well beyond the scope of the present inquiry. What we most need to learn from the ancient significations of *canon*, however, is that they ranged in meaning all the way from "a text possessing stylistic virtues that make it a proper model" to "a text that is a repository of the Law and the Truth, being the word of God." We should remember also that the word, as a transitive verb, referred to "a process of inclusion among the saints."

In the vernacular languages, the meanings of *canon* found in late Latin were simply extended. In French, for instance, we can find the following in a modern dictionary: *canon* (1) Gun, barrel of a gun, cannon; cylinder, pipe, tube; leg (of trousers); and *canon* (2) Canon. *Canon des Écritures*, the sacred canon; *école de droit canon*, school of canon law (Baker). The French is especially useful in reminding us that the word for gun and the word for the law and the sacred texts are simply branches of a single root rather than two totally different words. The fact that in English we regularized separate spellings for the guns and the laws in the later eighteenth century has tended to obscure the common heritage of both of these spellings in the ancient extensions of a word for reed or cane. In English the most relevant meanings of the word *canon* for our purposes are these: A rule, law, or decree of the Church; a general rule, a fundamental principle; the collection or list of the books of the Bible accepted by the Christian Church as genuine and inspired; hence, any set of sacred books; a list of saints acknowledged and canonized by the Church (*OED*).

Guns and ruling are associated in more ways than one. The English, of course, seem particularly responsible for institutionalizing the cane as an instrument for beating docility into subject peoples and Greek into schoolboys. The *OED* illustrates the use of cane as a verb with a quotation from a Victorian newspaper: "I had a little Greek caned into me." Many a native in India had Shakespeare as well as other canonical texts caned into him by the curricular arm of the British Raj. The Empire was based on its cannon, canon, and canes—to a startling degree.

The use of *canon* to mean a body of sacred texts comes to us from Latin rather than Greek, and specifically from the Latin of the Roman Church, where it is an extension of the notion of a canon as rule or law. The most common extension of this sense of the word in literary studies has until very recently been in reference to the works written by any single author. We speak of the Shakespeare canon or the Defoe canon, meaning no more than the works really written by these authors as opposed to those that might be erroneously attributed to them. Inevitably, however, some of the religious connotations of canonicity flow into this secular use. Where there is a canon, there is both power and sanctity. Above all, however, there is discipline. A textual canon is always a disciplinary function. A canon is in every sense a phallocratic object.

First the law, then the sacred texts. As religious practices and beliefs are institutionalized in a church, the canonical texts are separated from the apocrypha, or the angelic from the satanic verses, as matters are put in the Islamic canon. Canonical texts are held to be fully authorized, ultimately attributable to God. They are, therefore, not only sacred but authoritative, truthful. What is excluded from the religious canon turns into mere literature—a principle that we should note, for it says much about literature as a field of study that is not yet a discipline. Perhaps I should at this point make my own position clearer. I have no case to make against either canons or disciplines—in fact the whole thrust of my argument in this book is in favor of making English studies more rather than less disciplined. Disciplines, after all, are the essentials of academic life, and I am an academician. I only want to emphasize that canons and disciplines need one another. They go together. And *discipline*, like *canon*, is a word that scarcely conceals its potential for abuses of power. We need disciplines in order to think productively. We also need to challenge them in order to think creatively. [. . .]

The formal study of literature as a branch of the arts emerges only after the rise of science has demonstrated the way a discipline can coalesce around certain carefully defined objects and methods of study. The study of literature as a discipline (as opposed to the study of Greek and Latin grammar and a mixed bag of classical texts) begins with English works like Lord Kames's *Elements of Criticism* but is really consolidated by the German Romantics in texts like Schiller's *Letters on the Aesthetic Education of Man*, Schelling's *Philosophy of Art* (especially the last section on "The Verbal Arts"), and the section on poetry that closes Hegel's *Aesthetics: Lectures on Fine Art*. In these texts, and in their less systematic English counterparts by Coleridge, Shelley, and others, the notion of literature as a branch of the fine arts, characterized by *Imagination*—the absolutely crucial word—became sufficiently clear and stable to support a field of study. [. . .]

In drawing out the connections between canons and forms of institutionalized power, I may have seemed to be headed toward some quasi-Foucaultian critique of power itself, along with a plea for the elimination of all canons. Nothing could be further from my intent, however, for I am persuaded that the connection between institutions and canons is inevitable. Furthermore, our awareness of the existence of canons and our understanding of the processes by which they are maintained and altered makes it possible for us to influence canons through the institutions that support them and to change the institutions through their canons. What I am opposed to is the pretense that there may be some cosmic canon that transcends all institutions because it is based on an unexaminable and unchallengeable Absolute. This, I contend, is the case with notions like Great Books and Western Civ, in which a flock of cultures marches under the banner of a canonical eagle. I also want to suggest that some shifting between canons of texts and canons of methods has been a regular part of cultural history, so that we should regard it as a normal feature of our lives. I believe that we are at a point in cultural and textual studies where a realignment between these two types of canonicity is essential to the health of English studies. At this moment, however, my main point is that there has never been a canon of Great Books.

There is no canon of Great Books, in my view, because there is no intellectual core to the notion of Great Books in the first place. Literary study, though far from being a quantifying science, obtained a degree of coherence by organizing itself around

Romantic concepts of Art, Imagination, and Spirit. Other textual studies organize themselves by time, by genre, or by other systems of connection among their objects, just as biology has organized itself around the concepts of life, the cell, and so on. But notions like those of Great Books and Western Civ have no disciplinary focus and hence no academic core. There is, just to consider the most basic matters, absolutely no notion of bookish Greatness that has any coherence whatsoever. Allan Bloom would tell us, I suppose, that all the Great Books exhibit something called Greatness of Soul, but the concept of Great Souls is just as vague—in both adjective and noun—as what it is supposed to define. Nor is the notion of Western Civilization much of an improvement, though Gandhi thought it would be a good idea for the West to attempt it. There can be no notion of textual greatness, I am arguing, apart from a set of texts organized by a discipline. Of course, there have been great philosophers—but only since philosophy has been a discipline could we perceive them as such. Nor is their "greatness" of the same kind as that of Mozart, Shakespeare, or Tintoretto. All these are great only in contexts, partly narrative ones, that allow them to be perceived as such.

Western Civ, I maintain, lacks the coherence for pedagogically sound instruction. Such coherence as it might have, I would add, comes from a philosophy which even its adherents no longer claim to accept. One of the things we need to remember when considering concepts like Western Civ is that they originated in the Eurocentric thinking of German philosophers. The greatest of these, of course, was Hegel, who systematized the notion of cultural progress from East to West in ways that still haunt most of our thinking on these subjects. Let us listen to him a moment:

> The History of the World travels from East to West, for Europe is absolutely the end of History, Asia the beginning. . . . Although the earth forms a sphere, History forms no circle around it, but has on the contrary a determinate East, viz., Asia. Here rises the outward physical sun, and in the West it sinks down; here consentaneously rises the sun of self-consciousness, which diffuses a nobler brilliance. The History of the World is the *discipline* of the uncontrolled natural will, bringing it into obedience to a Universal principle and conferring subjective freedom. The East knew and to the present day knows only that One is Free; the Greek and Roman world, that some are free; the German World knows that All are free. (Hegel, 1956, 103–4, emphasis added)

It would be easy to mock the smug Eurocentrism of Hegelian thought, but this would involve ignoring some of the complications and nuances of that thought. In this discussion, however, I intend to pay more attention to some of the Hegelian nuances that are often lost in later adaptations of that Eurocentric perspective. What Hegel meant by the German world, in this instance, was Europe after the fall of Rome, a Europe that had been overrun by Germanic tribes moving from east to west: the Angles, the Saxons, the Franks, the Goths, the Lombards. He also meant a Europe in which ultimately Protestantism would come to elevate the materialism of the Roman Catholic Church to a more spiritual level, finally realizing Christ's message that every human soul is free and worthy of development. He describes this process, in a memorable passage, as subjecting Christianity to "the terrible discipline of culture":

> Secularity appears now [he was writing of the sixteenth century] as gaining a consciousness of its intrinsic worth—becomes aware of its having a value of its own in the

morality, rectitude, probity and activity of man. The consciousness of independent
validity is aroused through the restoration of Christian freedom. The Christian principle has now passed through the terrible discipline of culture, and it first attains truth
and reality through the Reformation. This third period of the German World extends
from the Reformation to our own times. (Hegel, 1956, 344) [. . .]

The point I am trying so laboriously to make is that any presentation of Europe's
cultural past must itself be laboriously thought out and carefully presented. When disconnected texts are collected in surveys of Great Books, one of the first things lost is
history itself. When texts that speak to one another—that address the same problems,
that work in the same medium or genre—are studied, then such courses can make
sense. They will make the greatest sense, however, if they take a narrative structure that
finally connects them to the present. To return to the example I have been working
with, Hegel is important to us because our thought is still shaped by ideas he formulated so powerfully—and because we need to reject some of those ideas (on the actual
history of the "Orient," for example) in order to understand our own situation.

In my view, every discipline should offer courses in its own history, or in some
coherent segment of that history ending with the present time. But there can be no
coherent overview of the historical whole, no single historical core of Great Books
embodying something called Western Civilization. And if any single discipline's history were to be privileged as the best embodiment of the ideal that Western Civ fails
to reach, that would certainly be the History of Art from Egypt to America, including the powerful influence of African and Oceanian art upon Euro-American
modernism—but visual art, like music, is regularly ignored in courses called Western
Civ—as if "Civ" were a purely verbal matter. I would privilege sculpture and painting
because they are so palpable, so representable, so suited to a generation attuned to
visual texts. In the history of art, what my teacher George Kubler called so beautifully
"the shape of time" can be grasped as a structure to which other historical events and
texts can be attached. However—and here my discourse will take its final turn toward
the specific and practical—I also want to argue that historical studies themselves
should be preceded or accompanied by another core, designed to help students situate themselves in their own culture, and, in particular, designed to make the basic
processes of language itself intelligible and fully available for use. Toward the establishment of such a core I now wish to make the "trivial proposal" mentioned in the
title of this chapter.

This proposal will be trivial, perhaps, in sense the that it will make a much smaller
claim than that made by Great Books or Western Civ curricula. It will be trivial,
however, in another sense: it is an attempt to rethink in modern terms the trivium
that was the core of medieval education. This will also be a radical proposal, in that
I propose to go back to the roots of our liberal arts tradition and reinstate grammar,
dialectic, and rhetoric at the core of college education. These three subjects, you
remember, constituted the preliminary studies to the medieval quadrivium of arith-
metic, geometry, astronomy, and music. Our culture is too complicated for education
to be quadrivial now, but not to accommodate a trivial core. To envision such a thing,
we need only rethink what grammar, dialectic, and rhetoric might mean in modern
terms. My own rethinking of these terms has taken the form of seeing all three of the

trivial arts as matters of textuality, with the English language at the center of them, but noting their extension into media that are only partly linguistic. I offer the results here, with a certain humility, as trivial in yet another sense. This is crude, provisional thinking, meant to stimulate refinements and alternatives rather than to lay down any curricular law.

This modern trivium, like its ancestor, would be organized around a canon of concepts, precepts, and practices rather than a canon of texts. In particular, each trivial study would encourage textual production by students in appropriate modes. Since this is a modern trivium, such production would include, where appropriate, not only speaking and writing but work in other media as well. Similarly, texts for reading, interpretation, and criticism would be drawn from a range of media, ancient and modern. I will present my trivial proposal in the form of a set of courses, each of which would be based not on a canon of sacred texts but on certain crucial concepts to be understood not simply in a theoretical way but in their application to the analysis of specific cultural or textual objects. This means that the specific texts selected could have considerable variety from course to course and place to place, though it may well be that certain texts should prove so useful that they would be widely adopted for use in textual curricula. In some cases, even, "classic" texts from philosophy and literature will present themselves as the most useful things available—which may tell us something about why they have become classics in the first place. At any rate, the specific titles given in the following descriptions are meant to be illustrative rather than prescriptive.

My first trivial topic is grammar, traditionally the driest and narrowest of academic subjects. I propose to change all that by means of a course or study that follows the implications of the grammar of the pronouns all the way to the subject and object positions of discourse. I see grammar, conceived in this generous manner, as an alternative to traditional composition courses, taking perhaps two semesters of work, the first of which might be called Language and Human Subjectivity. The basis of this course would be the way that their mother tongue presents human beings with a set of words and grammatical rules in which they may attain subjectivity at the cost of being subjected. The very heart of such a course would be the grammar of the pronouns, beginning with *I* and *you*, as opposed to *he, she*, and *it*. But this grammar must be connected to the philosophical questions of subject and object and the ethical relationship of *I* and *thou*. The virtual loss of *thou* in English, except in certain religious contexts, would make one point of discussion. In designing such a course I would be careful to use a mixture of theoretical texts and illustrative embodiments of the problems of subjectivity. The necessary theory is conveniently embodied, for instance, in such discussions as those of the linguist Emile Benveniste on "The Nature of Pronouns" and "Subjectivity in Language" (*Problems in General Linguistics*, 1971); in Hegel's dialectic of Master and Servant in the *Phenomenology of Spirit*, in Freud's *Das Ich und das Es*, which is usually translated as *The Ego and the Id* but which is just as properly translated as *The I and the It*; and in other works by Piaget, Vygotsky, and Lacan, for example. [. . .]

The second trivial topic in the core curriculum I am proposing would be dialectic. In its modern dress, and because the word *dialectic* has drifted far from its earlier usage, a course in this trivial topic might be called System and Dialectic. Such a course would have as its object of study discourses that work at a high level of abstraction and systematization, in which texts are constructed not so much by representing objects as by abstracting from them their essential qualities or their principles of composition. This is preeminently the domain of philosophy itself, and especially of the tradition of Continental philosophy from the pre-Socratics to Derrida. It may well be that literature departments would need help from their friends in philosophy to mount courses that approach this topic effectively, but several decades of literary theory ought to have made them readier to undertake such a project themselves than they were some years ago.

The intent of such study would be, in part, to make available to students the tradition of clear and systematic thinking that has been so crucial to the history of what Richard Rorty has called "the rich North Atlantic nations"—so that such students may learn to employ the resources of logic and dialectic in their own thinking and writing. A further intent, however, would be to introduce students to those countertrends, arising mainly within philosophy itself, that seek to criticize or even undo that very tradition. Put more specifically, such absolutely essential philosophers as Plato, Aristotle, Kant, and Hegel might be read and discussed in speech and writing, along with such antithetical writers as Nietzsche, Wittgenstein, Heidegger, Derrida, Rorty, and Davidson. Such a course might have a particular theme, such as philosophies of science, which would bring Aristotle, Bacon, Locke, Kuhn, and Feyerabend into prominence, or government, which would make Plato, Machiavelli, Hobbes, Montesquieu, and others important—or education, or language, or justice, or freedom. The point would be for students to learn both how to use and how to criticize discourse that takes reason, system, and logical coherence as its principles of articulation.

The last of the trivial topics I am proposing might well be taught first in any sequence of core courses, because it deals with more familiar matters and perhaps even with more immediately accessible material. I am not offering a rigid order or sequence of courses here, in any case, but trying to suggest how one might go about revitalizing the old trivium, the third division of which, you will remember, was rhetoric. I would be inclined to call a modern course in rhetoric something like Persuasion and Mediation. Such a course would obviously include the traditional arts of manipulation of audiences but would also point toward the capacities and limits of the newer media, especially those that mix verbal and visual textuality to generate effects of unprecedented power. Such a course would embrace the traditional topics of rhetoric but would extend them in certain specific directions. One might well wish to begin with Aristotle's *Rhetoric*, but in this kind of course the *Poetics* would also have a place as a discussion of both another type of manipulation and a specific medium (tragic drama) that mediates human experience in a particular way, incorporating the hegemonic codes of a particular cultural situation. From here one might go on to Nietzsche's *Birth of Tragedy* and Brecht on "Epic Theatre." In this connection it would be especially effective to move from the rhetoric of theater to the rhetoric of film and visual spectatorship in general, in which the gendering of subjects and objects of viewing could be considered (as in Laura Mulvey, Teresa de Lauretis, and John Berger, for instance), along with other ideological analyses of the rhetoric of the mass media in both direct (overt) and indirect (covert) manipulation of viewers. Plays, films, and television texts would be the objects of rhetorical analysis in such a course, along with such more overtly persuasive texts as political speeches and advertisements.

In such a curricular core of study, students might well encounter as many "classic" texts as in more traditional core curricula, but these texts would not be studied simply "because they are there" but rather as the means to an end of greater mastery of cultural processes by the students themselves. By putting language and textuality at the center of education, we would not be making some gesture of piety toward the medieval roots of education, but we would certainly be acknowledging the cultural past of our institutions. More important, however, we would be responding to the "linguistic turn" of so much of modern thought and to the media saturation that is the condition of our students' lives as well as of our own. Already, in such a trivium, the cultural past will have begun to be presented as a body of texts that can help students to understand their current cultural situation—just as they help their teachers (who also, of course, continue to be students). This trivium should serve, as well, to whet the appetite for other courses that attend more specifically to the historical narratives of one or another mode of cultural activity. If the pigs and the dogs learn to communicate and negotiate with one another, perhaps they can turn this flock of cultures into a nest of singing birds, and make such music as will stir the corrupting carcass of Western Civilization itself. That, at least, is my hope.

Chapter 42

Azar Nafisi (1950–)
from Sections 16, 17, 18, and 19 of Part II,
"Gatsby," *Reading Lolita in Tehran:*
A Memoir in Books (2002)

It was late; I had been at the library. I was spending a great deal of time there now, as it was becoming more and more difficult to find "imperialist" novels in bookstores. I was emerging from the library with a few books under my arm when I noticed him standing by the door. His two hands were joined in front of him in an expression of reverence for me, his teacher, but in his strained grimace I could feel his sense of power. I remember Mr. Nyazi always with a white shirt, buttoned up to the neck—he never tucked it in. He was stocky and had blue eyes, very closely cropped light brown hair and a thick, pinkish neck. It seemed as if his neck were made of soft clay; it literally sat on his shirt collar. He was always very polite.

"Ma'am, may I talk to you for a second?" Although we were in the middle of the semester, I had not as yet been assigned an office, so we stood in the hall and I listened. His complaint was about Gatsby. He said he was telling me this for my own good. For my own good? What an odd expression to use. He said surely I must know how much he respected me, otherwise he would not be there talking to me. He had a complaint. Against whom, and why me? It was against Gatsby. I asked him jokingly if he had filed any official complaints against Mr. Gatsby. And I reminded him that any such action would in any case be useless as the gentleman was already dead.

But he was serious. No, Professor, not against Mr. Gatsby himself but against the novel. The novel was immoral. It taught the youth the wrong stuff; it poisoned their minds—surely I could see? I could not. I reminded him that Gatsby was a work of fiction and not a how-to manual. Surely I could see, he insisted, that these novels and their characters became our models in real life? Maybe Mr. Gatsby was all right for

Born in Iran, and educated there and in England, Azar Nafisi returned in 1979 to Tehran, where she taught literature at the University of Tehran, until she was fired. Currently Professorial Lecturer in the School of Advanced International Studies at Johns Hopkins University, Nafisi describes in this excerpt from *Reading Lolita in Tehran* the experience of reading and teaching American literature during the Iranian Revolution.

the Americans, but not for our revolutionary youth. For some reason the idea that this man could be tempted to become Gatsby-like was very appealing to me.

There was, for Mr. Nyazi, no difference between the fiction of Fitzgerald and the facts of his own life. *The Great Gatsby* was representative of things American, and America was poison for us; it certainly was. We should teach Iranian students to fight against American immorality, he said. He looked earnest; he had come to me in all goodwill.

Suddenly a mischievous notion got hold of me. I suggested, in these days of public prosecutions, that we put *Gatsby* on trial: Mr. Nyazi would be the prosecutor, and he should also write a paper offering his evidence. I told him that when Fitzgerald's books were published in the States, there were many who felt just as he did. They may have expressed themselves differently, but they were saying more or less the same thing. So he need not feel lonely in expressing his views.

The next day I presented this plan to the class. We could not have a proper trial, of course, but we could have a prosecutor, a lawyer for the defense and a defendant; the rest of the class would be the jury. Mr. Nyazi would be the prosecutor. We needed a judge, a defendant and a defense attorney.

After a great deal of argument, because no one volunteered for any of the posts, we finally persuaded one of the leftist students to be the judge. But then Mr. Nyazi and his friends objected: this student was biased against the prosecution. After further deliberation, we agreed upon Mr. Farzan, a meek and studious fellow, rather pompous and, fortunately, shy. No one wanted to be the defense. It was emphasized that since I had chosen the book, I should defend it. I argued that in that case, I should be not the defense but the defendant and promised to cooperate closely with my lawyer and to talk in my own defense. Finally, Zarrin, who was holding her own conference in whispers with Vida, after a few persuasive nudges, volunteered. Zarrin wanted to know if I was Fitzgerald or the book itself. We decided that I would be the book: Fitzgerald may have had or lacked qualities that we could detect in the book. It was agreed that in this trial the rest of the class could at any point interrupt the defense or the prosecution with their own comments and questions.

I felt it was wrong for me to be the defendant, that this put the prosecutor in an awkward position. At any rate, it would have been more interesting if one of the students had chosen to participate. But no one wanted to speak for *Gatsby*. There was something so obstinately arrogant about Mr. Nyazi, so inflexible, that in the end I persuaded myself I should have no fear of intimidating him.

A few days later, Mr. Bahri came to see me. We had not met for what seemed like a long time. He was a little outraged. I enjoyed the fact that for the first time, he seemed agitated and had forgotten to talk in his precise and leisurely manner. Was it necessary to put this book on trial? I was somewhat taken aback. Did he want me to throw the book aside without so much as a word in its defense? Anyway, this is a good time for trials, I said, is it not?

17

All through the week before the trial, whatever I did, whether talking to friends and family or preparing for classes, part of my mind was constantly occupied on shaping my arguments for the trial. This after all was not merely a defense of *Gatsby* but of a whole way of looking at and appraising literature—and reality, for that matter. Bijan, who seemed quite amused by all of this, told me one day that I was studying *Gatsby* with the same intensity as a lawyer scrutinizing a textbook on law. I turned to him and said, You don't take this seriously, do you? He said, Of course I take it seriously. You have put yourself in a vulnerable position in relation to your students. You have allowed them—no, not just that; you have forced them into questioning your judgment as a teacher. So you have to win this case. This is very important for a junior member of the faculty in her first semester of teaching. But if you are asking for sympathy, you won't get it from me. You're loving it, admit it—you love this sort of drama and anxiety. Next thing you know, you'll be trying to convince me that the whole revolution depends on this.

But it does—don't you see? I implored. He shrugged and said, Don't tell me. I suggest you put your ideas to Ayatollah Khomeini.

On the day of the trial, I left for school early and roamed the leafy avenues before heading to class. As I entered the Faculty of Persian and Foreign Languages and Literature, I saw Mahtab standing by the door with another girl. She wore a peculiar grin that day, like a lazy kid who has just gotten an A. She said, Professor, I wondered if you would mind if Nassrin sits in on the class today. I looked from her to her young companion; she couldn't have been more than thirteen or fourteen years old. She was very pretty, despite her own best efforts to hide it. Her looks clashed with her solemn expression, which was neutral and adamantly impenetrable. Only her body seemed to express something: she kept leaning on one leg and then the other as her right hand gripped and released the thick strap of her heavy shoulder bag.

Mahtab, with more animation than usual, told me that Nassrin's English was better than most college kids', and when she'd told her about *Gatsby*'s trial, she was so curious that she'd read the whole book. I turned to Nassrin and asked, What did you think of *Gatsby*? She paused and then said quietly, I can't tell. I said, Do you mean you don't know or you can't tell me? She said, I don't know, but maybe I just can't tell you.

That was the beginning of it all. After the trial, Nassrin asked permission to continue attending my classes whenever she could. Mahtab told me that Nassrin was her neighbor. She belonged to a Muslim organization but was a very interesting kid, and Mahtab was working on her—an expression the leftists used to describe someone they were trying to recruit.

I told Nassrin she could come to my class on one condition: at the end of term, she would have to write a fifteen-page paper on *Gatsby*. She paused as she always did, as if she didn't quite have sufficient words at her command. Her responses were always reluctant and forced; one felt almost guilty for making her talk. Nassrin demurred at first, and then she said: I'm not that good. You don't need to be good, I said. And I'm sure you are—after all, you're spending your free time here. I don't want a scholarly paper; I want you to write your own impressions. Tell me in your own words what *Gatsby* means to you. She was looking at the tip of her shoes, and she muttered that she would try.

From then on, every time I came to class I would look for Nassrin, who usually followed Mahtab and sat beside her. She would be busy taking notes all through the session, and she even came a few times when Mahtab did not show up. Then suddenly she stopped coming, until the last class, when I saw her sitting in a corner, busying herself with the notes she scribbled.

Once I had agreed to accept my young intruder, I left them both and continued. I needed to stop by the department office before class to pick up a book Dr. A had left for me. When I entered the classroom that afternoon, I felt a charged silence follow me in. The room was full; only one or two students were absent—and Mr. Bahri, whose activities, or disapproval, had kept him away. Zarrin was laughing and swapping notes with Vida, and Mr. Nyazi stood in a corner talking to two other Muslim students, who repaired to their seats when they caught sight of me. Mahtab was sitting beside her new recruit, whispering to her conspiratorially.

I spoke briefly about the next week's assignment and proceeded to set the trial in motion. First I called forth Mr. Farzan, the judge, and asked him to take his seat in my usual chair, behind the desk. He sauntered up to the front of the class with an ill-disguised air of self-satisfaction. A chair was placed near the judge for the witnesses. I sat beside Zarrin on the left side of the room, by the large window, and Mr. Nyazi sat with some of his friends on the other side, by the wall. The judge called the session to order. And so began the case of the Islamic Republic of Iran versus *The Great Gatsby*.

Mr. Nyazi was called to state his case against the defendant. Instead of standing, he moved his chair to the center of the room and started to read in a monotonous voice from his paper. The judge sat uncomfortably behind my desk and appeared to be mesmerized by Mr. Nyazi. Every once in a while he blinked rather violently.

A few months ago, I was finally cleaning up my old files and I came across Mr. Nyazi's paper, written in immaculate handwriting. It began with "In the Name of God," words that later became mandatory on all official letterheads and in all public talks. Mr. Nyazi picked up the pages of his paper one by one, gripping rather than holding them, as if afraid that they might try to escape his hold. "Islam is the only religion in the world that has assigned a special sacred role to literature in guiding man to a godly life," he intoned. "This becomes clear when we consider that the Koran, God's own word, is the Prophet's miracle. Through the Word you can heal or you can destroy. You can guide or you can corrupt. That is why the Word can belong to Satan or to God.

"Imam Khomeini has relegated a great task to our poets and writers," he droned on triumphantly, laying down one page and picking up another. "He has given them a sacred mission, *much* more exalted than that of the materialistic writers in the West. If our Imam is the shepherd who guides the flock to its pasture, then the writers are the faithful watchdogs who must lead according to the shepherd's dictates."

A giggle could be heard from the back of the class. I glanced around behind me and caught Zarrin and Vida whispering. Nassrin was staring intently at Mr. Nyazi and absentmindedly chewing her pencil. Mr. Farzan seemed to be preoccupied with an invisible fly, and blinked exaggeratedly at intervals. When I turned my attention back to Mr. Nyazi, he was saying, "Ask yourself which you would prefer: the guardianship of a sacred and holy task or the materialistic reward of money and position that has corrupted—" and here he paused, without taking his eyes off his paper, seeming to drag the sapless words to the surface—"that has *corrupted*," he repeated, "the Western writers and deprived their work of spirituality and purpose. *That* is why our Imam says that the pen is mightier than the sword."

The whispers and titters in the back rows had become more audible. Mr. Farzan was too inept a judge to pay attention, but one of Mr. Nyazi's friends cried out: "Your Honor, could you please instruct the gentlemen and ladies in the back to respect the court and the prosecutor?"

"So be it," said Mr. Farzan, irrelevantly.

"Our poets and writers in this battle against the Great Satan," Nyazi continued, "play the same role as our faithful soldiers, and they will be accorded the same reward in heaven. We students, as the future guardians of culture, have a heavy task ahead of us. Today we have planted Islam's flag of victory inside the nest of spies on our own soil. Our task, as our Imam has stated, is to purge the country of the decadent Western culture and . . . "

At this point Zarrin stood up. "Objection, Your Honor!" she cried out.

Mr. Farzan looked at her in some surprise. "What do you object to?"

"This is supposed to be about *The Great Gatsby*," said Zarrin. "The prosecutor has taken up fifteen precious minutes of our time without saying a single word about the defendant. Where is this all going?"

For a few seconds both Mr. Farzan and Mr. Nyazi looked at her in wonder. Then Mr. Nyazi said, without looking at Zarrin, "This is an Islamic court, not *Perry Mason*. I can present my case the way I want to, and I am setting the context. I want to say that as a Muslim I cannot accept *Gatsby*."

Mr. Farzan, attempting to rise up to his role, said, "Well, please move on then."

Zarrin's interruptions had upset Mr. Nyazi, who after a short pause lifted his head from his paper and said with some excitement, "You are right, it is not worth it . . . "

We were left to wonder what was not worth it for a few seconds, until he continued. "I don't have to read from a paper, and I don't need to talk about Islam. I have enough evidence—every page, *every* single page," he cried out, "of this book is its own condemnation." He turned to Zarrin and one look at her indifferent expression was enough to transform him. "All through this revolution we have talked about the fact that the West is our enemy, it is the Great Satan, not because of its military might, not because of its economic power, but because of, because of"—another pause—"because of its sinister assault on the very roots of our culture. What our Imam calls cultural aggression. This I would call a rape of our culture," Mr. Nyazi stated, using a term that later became the hallmark of the Islamic Republic's critique of the West. "And if you want to see cultural rape, you need go no further than this very book." He picked his *Gatsby* up from beneath the pile of papers and started waving it in our direction.

Zarrin rose again to her feet. "Your Honor," she said with barely disguised contempt, "these are all baseless allegations, falsehoods . . . "

Mr. Nyazi did not allow his honor to respond. He half rose from his seat and cried out: "Will you let me finish? You will get your turn! I will tell you why, I will tell you why . . . " And then he turned to me and in a softer voice said, "Ma'am, no offense meant to you."

I, who had by now begun to enjoy the game, said, "Go ahead, please, and remember I am here in the role of the book. I will have my say in the end."

"Maybe during the reign of the corrupt Pahlavi regime," Nyazi continued, "adultery was the accepted norm."

Zarrin was not one to let go. "I object!" she cried out. "There is no factual basis to this statement."

"Okay," he conceded, "but the values were such that adultery went unpunished. This book preaches illicit relations between a man and woman. First we have Tom and his mistress, the scene in her apartment—even the narrator, Nick, is implicated. He doesn't like their lies, but he has no objection to their fornicating and sitting on each other's laps, and, and, those parties at Gatsby's . . . remember, ladies and gentlemen, this Gatsby is the hero of the book—and who is he? He is a charlatan, he is an adulterer, he is a liar . . . this is the man Nick celebrates and feels sorry for, this man, this destroyer of homes!" Mr. Nyazi was clearly agitated as he conjured the fornicators, liars and adulterers roaming freely in Fitzgerald's luminous world, immune from his wrath and from prosecution. "The only sympathetic person here is the cuckolded husband, Mr. Wilson," Mr. Nyazi boomed. "When he kills Gatsby, it is the hand of God. He is the only victim. He is the genuine symbol of the oppressed, in the land of, of, of the Great Satan!"

The trouble with Mr. Nyazi was that even when he became excited and did not read from his paper, his delivery was monotonous. Now he mainly shouted and cried out from his semi-stationary position.

"The one good thing about this book," he said, waving the culprit in one hand, "is that it exposes the immorality and decadence of American society, but we have fought to rid ourselves of this trash and it is high time that such books be banned." He kept calling Gatsby "this Mr. Gatsby" and could not bring himself to name Daisy, whom he referred to as "that woman." According to Nyazi, there was not a single virtuous woman in the whole novel. "What kind of model are we setting for our innocent and modest sisters," he asked his captive audience, "by giving them such a book to read?"

As he continued, he became increasingly animated, yet he refused throughout to budge from his chair. "Gatsby is dishonest," he cried out, his voice now shrill. "He earns his money by illegal means and tries to buy the love of a married woman. This book is supposed to be about the American dream, but what sort of a dream is this? Does the author mean to suggest that we should all be adulterers and bandits? Americans are decadent and in decline because this is their dream. They are going down! This is the last hiccup of a dead culture!" he concluded triumphantly, proving that Zarrin was not the only one to have watched *Perry Mason*.

"Perhaps our honorable prosecutor should not be so harsh," Vida said once it was clear that Nyazi had at last exhausted his argument. "Gatsby dies, after all, so one could say that he gets his just deserts."

But Mr. Nyazi was not convinced. "Is it just Gatsby who deserves to die?" he said with evident scorn. "*No!* The whole of American society deserves the same fate. What kind of a dream is it to steal a man's wife, to preach sex, to cheat and swindle and to . . . and then that guy, the narrator, Nick, he claims to be moral!"

Mr. Nyazi proceeded in this vein at some length, until he came to a sudden halt, as if he had choked on his own words. Even then he did not budge. Somehow it did not occur to any of us to suggest that he return to his original seat as the trial proceeded.

18

Zarrin was summoned next to defend her case. She stood up to face the class, elegant and professional in her navy blue pleated skirt and woolen jacket with gold buttons, white cuffs peering out from under its sleeves. Her hair was tied back with a ribbon in a low ponytail and the only ornament she wore was a pair of gold earrings. She circled slowly around Mr. Nyazi, every once in a while making a small sudden turn to emphasize a point. She had few notes and rarely looked at them as she addressed the class.

As she spoke, she kept pacing the room, her ponytail, in harmony with her movements, shifting from side to side, gently caressing the back of her neck, and each time she turned she was confronted with Mr. Nyazi, sitting hard as rock on that chair. She began with a passage I had read from one of Fitzgerald's short stories. "Our dear prosecutor has committed the fallacy of getting too close to the amusement park," she said. "He can no longer distinguish fiction from reality."

She smiled, turning sweetly towards "our prosecutor," trapped in his chair. "He leaves no space, no breathing room, between the two worlds. He has demonstrated his own weakness: an inability to read a novel on its own terms. All he knows is judgment, crude and simplistic exaltation of right and wrong." Mr. Nyazi raised his head at these words, turning a deep red, but he said nothing. "But is a novel good," continued Zarrin, addressing the class, "because the heroine is virtuous? Is it bad if its character strays from the moral Mr. Nyazi insists on imposing not only on us but on all fiction?"

Mr. Farzan suddenly leapt up from his chair. "Ma'am," he said, addressing me. "My being a judge, does it mean I cannot say anything?"

"Of course not," I said, after which he proceeded to deliver a long and garbled tirade about the valley of the ashes and the decadence of Gatsby's parties. He concluded that Fitzgerald's main failure was his inability to surpass his own greed: he wrote cheap stories for money, and he ran after the rich. "You know," he said at last, by this point exhausted by his own efforts, "Fitzgerald said that the rich are different."

Mr. Nyazi nodded his head in fervent agreement. "Yes," he broke in, with smug self-importance, clearly pleased with the impact of his own performance. "And our revolution is opposed to the materialism preached by Mr. Fitzgerald. We do not need Western materialisms, or American goods." He paused to take a breath, but he wasn't finished. "If anything, we could use their technical know-how, but we must reject their morals."

Zarrin looked on, composed and indifferent. She waited a few seconds after Mr. Nyazi's outburst before saying calmly, "I seem to be confronting two prosecutors. Now, if you please, may I resume?" She threw a dismissive glance towards Mr. Farzan's corner. "I would like to remind the prosecutor and the jury of the quotation we were given at our first discussion of this book from Diderot's *Jacques le Fataliste*: 'To me the freedom of [the author's] style is almost the guarantee of the purity of his morals.' We also discussed that a novel is not moral in the usual sense of the word. It can be called moral when it shakes us out of our stupor and makes us confront the absolutes we believe in. If that is true, then *Gatsby* has succeeded brilliantly. This is the first time in class that a book has created such controversy."

"*Gatsby* is being put on trial because it disturbs us—at least some of us," she added, triggering a few giggles. "This is not the first time a novel—a non-political novel—has been put on trial by a state." She turned, her ponytail turning with her. "Remember the famous trials of *Madame Bovary, Ulysses, Lady Chatterley's Lover* and *Lolita*? In each case the novel won. But let me focus on a point that seems to trouble his honor the judge as well as the prosecutor: the lure of money and its role in the novel.

"It is true that Gatsby recognizes that money is one of Daisy's attractions. He is in fact the one who draws Nick's attention to the fact that in the charm of her voice is the jingle of money. But this novel is not about a poor young charlatan's love of money." She paused here for emphasis. "Whoever claims this has not done his home-work." She turned, almost imperceptibly, to the stationary prosecutor to her left, then walked to her desk and picked up her copy of *Gatsby*. Holding it up, she addressed Mr. Farzan, turning her back on Nyazi, and said, "No, Your Honor, this novel is not about 'the rich are different from you and me,' although they are: so are the poor, and so are you, in fact, different from me. It is about wealth but not about the vulgar materialism that you and Mr. Nyazi keep focusing on."

"You tell them!" a voice said from the back row. I turned around. There were giggles and murmurs. Zarrin paused, smiling. The judge, rather startled, cried out, "Silence! Who said that?" Not even he expected an answer.

"Mr. Nyazi, our esteemed prosecutor," Zarrin said mockingly, "seems to be in need of no witnesses. He apparently is both witness and prosecution, but let us bring our witnesses from the book itself. Let us call some of the characters to the stand. I will now call to the stand our most important witness.

"Mr. Nyazi has offered himself to us as a judge of Fitzgerald's characters, but Fitzgerald had another plan. He gave us his own judge. So perhaps we should listen to him. Which character deserves to be our judge?" Zarrin said, turning towards the class. "Nick, of course, and you remember how he describes himself: 'Everyone sus-pects himself of at least one of the cardinal virtues, and this is mine: I am one of the few honest people that I have ever known.' If there is a judge in this novel, it is Nick. In a sense he is the least colorful character, because he acts as a mirror.

"The other characters are ultimately judged in term of their honesty. And the rep-resentatives of wealth turn out to be the most dishonest. Exhibit A: Jordan Baker, with whom Nick is romantically involved. There is a scandal about Jordan that Nick cannot at first remember. She had lied about a match, just as she would lie about a car she had borrowed and then left out in the rain with the top down. 'She was incur-ably dishonest,' Nick tells us. 'She wasn't able to endure being at a disadvantage, and given this unwillingness I suppose she had begun dealing in subterfuges when she was very young in order to keep that cool insolent smile turned to the world and yet satisfy the demands of her hard jaunty body.'

"Exhibit B is Tom Buchanan. His dishonesty is more obvious: he cheats on his wife, he covers up her crime and he feels no guilt. Daisy's case is more complicated because, like everything else about her, her insincerity creates a certain enchantment: she makes others feel they are complicit in her lies, because they are seduced by them. And then, of course, there is Meyer Wolfshiem, Gatsby's shady business partner. He fixes the World Cup. 'It never occurred to me that one man could start to play with the faith of fifty million people—with the single-mindedness of a burglar blowing a safe.' So the question of honesty and dishonesty, the way people are and the way they present themselves to the world, is a sub-theme that colors all the main events in the novel. And who are the most dishonest people in this novel?" she asked, again focusing on the jury. "The rich, of course," she said, making a sudden turn towards Mr. Nyazi. "The very people our prosecutor claims Fitzgerald approves of.

"But that's not all. We are not done with the rich." Zarrin picked up her book and opened it to a marked page. "With Mr. Carraway's permission," she said, "I should like to quote him on the subject of the rich." Then she began to read: "They were careless people, Tom and Daisy—they smashed up things and creatures and then retreated back into their money or their vast carelessness, or whatever it was that kept them together, and let other people clean up the mess they had made . . .

"So you see," said Zarrin, turning again to Mr. Farzan, "this is the judgment the most reliable character in the novel makes about the rich. The rich in this book, represented primarily by Tom and Daisy and to a lesser extent Jordan Baker, are careless people. After all, it is Daisy who runs over Myrtle and lets Gatsby take the blame for it, without even sending a flower to his funeral." Zarrin paused, making a detour around the chair, seemingly oblivious to the judge, the prosecutor and the jury.

"The word *careless* is the key here," she said. "Remember when Nick reproaches Jordan for her careless driving and she responds lightly that even if she is careless, she counts on other people being careful? *Careless* is the first adjective that comes to mind when describing the rich in this novel. The dream they embody is an alloyed dream that destroys whoever tries to get close to it. So you see, Mr. Nyazi, this book is no less a condemnation of your wealthy upper classes than any of the revolutionary books we have read." She suddenly turned to me and said with a smile, "I am not sure how one should address a book. Would you agree that your aim is not a defense of the wealthy classes?"

I was startled by Zarrin's sudden question but appreciated this opportunity to focus on a point that had been central to my own discussions about fiction in general. "If a critique of carelessness is a fault," I said, somewhat self-consciously, "then at least I'm in good company. This carelessness, a lack of empathy, appears in Jane Austen's negative characters: in Lady Catherine, in Mrs. Norris, in Mr. Collins or the Crawfords. The theme recurs in Henry James's stories and in Nabokov's monster heroes: Humbert, Kinbote, Van and Ada Veen. Imagination in these works is equated with empathy; we can't experience all that others have gone through, but we can understand even the most monstrous individuals in works of fiction. A good novel is one that shows the complexity of individuals, and creates enough space for all these characters to have a voice; in this way a novel is called democratic—not that it advocates democracy but that by nature it is so. Empathy lies at the heart of *Gatsby*, like so many other great novels—the biggest sin is to be blind to others' problems and pains. Not seeing them means denying their existence." I said all this in one breath, rather astonished at my own fervor.

"Yes," said Zarrin, interrupting me now. "Could one not say in fact that this blindness or carelessness towards others is a reminder of another brand of careless people?" She threw a momentary glance at Nyazi as she added, "Those who see the world in black and white, drunk on the righteousness of their own fictions."

"And if," she continued with some warmth, "Mr. Farzan, in real life Fitzgerald was obsessed with the rich and with wealth; in his fiction he brings out the corrupt and decaying power of wealth on basically decent people, like Gatsby, or creative and lively people, like Dick Diver in *Tender Is the Night*. In his failure to understand this, Mr. Nyazi misses the whole point of the novel."

Nyazi, who for some time now had been insistently scrutinizing the floor, suddenly jumped up and said, "I object!"

"To what, exactly, do you object?" said Zarrin with mock politeness.

"Carelessness is not enough!" he shot back. "It doesn't make the novel more moral. I ask you about the sin of adultery, about lies and cheating, and you talk about carelessness?"

Zarrin paused and then turned to me again. "I would now like to call the defendant to the stand." She then turned to Mr. Nyazi and, with a mischievous gleam in her eyes, said, "Would you like to examine the defendant?" Nyazi murmured a defiant no. "Fine. Ma'am, could you please take the stand?" I got up, rather startled, and looked around me. There was no chair. Mr. Farzan, for once alert, jumped up and offered me his. "You heard the prosecutor's remarks," Zarrin said, addressing me. "Do you have anything to say in your defense?"

I felt uncomfortable, even shy, and reluctant to talk. Zarrin had been doing a great job, and it seemed to me there was no need for my pontifications. But the class was waiting, and there was no way I could back down now.

I sat awkwardly on the chair offered me by Mr. Farzan. During the course of my preparations for the trial, I had found that no matter how hard I tried, I could not articulate in words the thoughts and emotions that made me so excited about *Gatsby*. I kept going back to Fitzgerald's own explanation of the novel: "That's the whole burden of this novel," he had said, "the loss of those illusions that give such color to the world so that you don't care whether things are true or false as long as they partake of the magical glory." I wanted to tell them that this book is not about adultery but about the loss of dreams. For me it had become of vital importance that my students accept *Gatsby* on its own terms, celebrate and love it because of its amazing and anguished beauty, but what I had to say in this class had to be more concrete and practical.

"You don't read *Gatsby*," I said, "to learn whether adultery is good or bad but to learn about how complicated issues such as adultery and fidelity and marriage are. A great novel heightens your senses and sensitivity to the complexities of life and of individuals, and prevents you from the self-righteousness that sees morality in fixed formulas about good and evil . . ."

"But, ma'am," Mr. Nyazi interrupted me. "There is nothing complicated about having an affair with another man's wife. Why doesn't Mr. Gatsby get his own wife?" he added sulkily.

"Why don't you write your own novel?" a muffled voice cracked from some indefinable place in the middle row. Mr. Nyazi looked even more startled. From this point on, I hardly managed to get a word in. It seemed as if all of a sudden everyone had discovered that they needed to get in on the discussion.

At my suggestion, Mr. Farzan called for a ten-minute recess. I left the room and went outside, along with a few students who felt the need for fresh air. In the hall I found Mahtab and Nassrin deep in conversation. I joined them and asked them what they thought of the trial.

Nassrin was furious that Nyazi seemed to think he had a monopoly on morality. She said she didn't say she'd approve of Gatsby, but at least he was prepared to die for his love. The three of us began walking down the hallway. Most of the students had gathered around Zarrin and Nyazi, who were in the midst of a heated argument. Zarrin was accusing Nyazi of calling her a prostitute. He was almost blue in the face with anger and indignation, and was accusing her in turn of being a liar and a fool.

"What am I to think of your slogans claiming that women who don't wear the veil are prostitutes and agents of Satan? You call this morality?" she shouted. "What about Christian women who don't believe in wearing veils? Are they all—every single one of them decadent floozies?"

"But this is an Islamic country," Nyazi shouted vehemently. "And this is the law, and whoever . . ."

"The law?" Vida interrupted him. "You guys came in and changed the laws. Is it the law? So was wearing the yellow star in Nazi Germany. Should all the Jews have worn the star because it was the blasted law?"

"Oh," Zarrin said mockingly, "don't even try to talk to him about that. He would call them all Zionists who deserved what they got." Mr. Nyazi seemed ready to jump up and slap her across the face.

"I think it's about time I used my authority," I whispered to Nassrin, who was standing by, transfixed. I asked them all to calm down and return to their seats. When the shouts had died down and the accusations and counteraccusations had more or less subsided, I suggested that we open the floor to discussion. We wouldn't vote on the outcome of the trial, but we should hear from the jury. They could give us their verdict in the form of their opinions.

A few of the leftist activists defended the novel. I felt they did so partly because the Muslim activists were so dead set against it. In essence, their defense was not so different from Nyazi's condemnation. They said that we needed to read fiction like *The Great Gatsby* because we needed to know about the immorality of American culture. They felt we should read more revolutionary material, but that we should read books like this as well, to understand the enemy.

One of them mentioned a famous statement by Comrade Lenin about how listening to "Moonlight Sonata" made him soft. He said it made him want to pat people on the back when we needed to club them, or some such. At any rate, my radical students' main objection to the novel was that it distracted them from their duties as revolutionaries.

Despite, or perhaps because of, the heated arguments, many of my students were silent, although many gathered around Zarrin and Vida, murmuring words of encouragement and praise. I discovered later that most students had supported Zarrin, but very few were prepared to risk voicing their views, mainly because they lacked enough self-confidence to articulate their points as "eloquently," I was told, as the defense and the prosecutor. Some claimed in private that they personally liked the book. Then why didn't they say so? Everyone else was so certain and emphatic in their position, and they couldn't really say why they liked it—they just did.

Just before the bell rang, Zarrin, who had been silent ever since the recess, suddenly got up. Although she spoke in a low voice, she appeared agitated. She said sometimes she wondered why people bothered to claim to be literature majors. Did it mean anything? she wondered. As for the book, she had nothing more to say in its defense. The novel was its own defense. Perhaps we had a few things to learn from it, from

Mr. Fitzgerald. She had not learned from reading it that adultery was good or that we should all become shysters. Did people all go on strike or head west after reading Steinbeck? Did they go whaling after reading Melville? Are people not a little more complex than that? And are revolutionaries devoid of personal feelings and emotions? Do they never fall in love, or enjoy beauty? This is an amazing book, she said quietly. It teaches you to value your dreams but to be wary of them also, to look for integrity in unusual places. Anyway, she enjoyed reading it, and that counts too, can't you see?

In her "can't you see?" there was a genuine note of concern that went beyond her disdain and hatred of Mr. Nyazi, a desire that even *he* should see, definitely see. She paused a moment and cast a look around the room at her classmates. The class was silent for a while after that. Not even Mr. Nyazi had anything to say.

I felt rather good after class that day. When the bell rang, many had not even noticed it. There had been no formal verdict cast, but the excitement most students now showed was the best verdict as far as I was concerned. They were all arguing as I left them outside the class—and they were arguing not over the hostages or the recent demonstrations or Rajavi and Khomeini, but over Gatsby and his alloyed dream.

19

Our discussions of *Gatsby* for a short while seemed as electric and important as the ideological conflicts raging over the country. In fact, as time went by, different versions of this debate did dominate the political and ideological scene. Fires were set to publishing houses and bookstores for disseminating immoral works of fiction. One woman novelist was jailed for her writings and charged with spreading prostitution. Reporters were jailed, magazines and newspapers closed and some of our best classical poets, like Rumi and Omar Khayyam, were censored or banned.

Like all other ideologues before them, the Islamic revolutionaries seemed to believe that writers were the guardians of morality. This displaced view of writers, ironically, gave them a sacred place, and at the same time it paralyzed them. The price they had to pay for their new preeminence was a kind of aesthetic impotence.

Personally, the *Gatsby* "trial" had opened a window into my own feelings and desires. Never before—not during all my revolutionary activities—did I feel so fervently as I did now about my work and about literature. I wanted to spread this spirit of goodwill, so I made a point the next day of asking Zarrin to stay after class, to let her know how much I had appreciated her defense. I'm afraid it fell on deaf ears, she said somewhat despondently. Don't be so sure, I told her.

A colleague, passing me two days later in the hall, said: I heard shouts coming from the direction of your class the other day. Imagine my surprise when instead of Lenin versus the Imam I heard it was Fitzgerald versus Islam. By the way, you should be thankful to your young protégé. Which one? I asked him with a laugh. Mr. Bahri—he seems to have become your knight in shining armor. I hear he quieted down the voices of outrage and somehow convinced the Islamic association that you had put America on trial.

Permissions

Index